RADIATION HEAT TRANSFER

SERIES IN THERMAL AND FLUIDS ENGINEERING

JAMES P. HARTNETT and THOMAS F. IRVINE, JR., Editors
JACK P. HOLMAN, Senior Consulting Editor

Cebeci and Bradshaw	• Momentum Transfer in Boundary Layers
Chang	• Control of Flow Separation: Energy Conservation, Operational Efficiency, and Safety
Chi	• Heat Pipe Theory and Practice: A Sourcebook
Eckert and Goldstein	• Measurements in Heat Transfer, 2nd edition
Edwards, Denny, and Mills	• Transfer Processes: An Introduction to Diffusion, Convection, and Radiation
Fitch and Surjaatmadja	• Introduction to Fluid Logic
Ginoux	• Two-Phase Flows and Heat Transfer with Application to Nuclear Reactor Design Problems
Hsu and Graham	• Transport Processes in Boiling and Two-Phase Systems, Including Near-Critical Fluids
Hughes	• An Introduction to Viscous Flow
Kreith and Kreider	• Principles of Solar Engineering
Lu	• Introduction to the Mechanics of Viscous Fluids
Moore and Sieverding	• Two-Phase Steam Flow in Turbines and Separators: Theory, Instrumentation, Engineering
Richards	• Measurements of Unsteady Fluid Dynamic Phenomena
Sparrow and Cess	• Radiation Heat Transfer, augmented edition
Tien and Lienhard	• Statistical Thermodynamics
Wirz and Smolderen	• Numerical Methods in Fluid Dynamics

PROCEEDINGS

Keairns	• Fluidization Technology
Spalding and Afgan	• Heat Transfer and Turbulent Buoyant Convection: Studies and Applications for Natural Environment, Buildings, Engineering Systems

Augmented Edition

RADIATION HEAT TRANSFER

E. M. SPARROW

University of Minnesota

R. D. CESS

State University of New York at Stony Brook

HEMISPHERE PUBLISHING CORPORATION

Washington   London

McGRAW-HILL BOOK COMPANY

New York   St. Louis   San Francisco   Auckland
Bogotá   Düsseldorf   Johannesburg   London
Madrid   Mexico   Montreal   New Delhi   Panama
Paris   Sao Paulo   Singapore   Sydney
Tokyo   Toronto

RADIATION HEAT TRANSFER

45678 KGPKGP 898765

Library of Congress Cataloging in Publication Data

Sparrow, Ephraim M
    Radiation heat transfer.

    (Series in thermal fluids engineering)
    Includes bibliographies references and indexes.
    1. Heat—Radiation and absorption.   2. Heat engineer-
ing.   I. Cess, R. D., joint author.   II. Title.
QC331.S67 1977   536′.33   77-24158
ISBN 0–07–059910–6

CONTENTS

v

PART THREE

**RADIANT ENERGY TRANSFER THROUGH
ABSORBING, EMITTING, AND SCATTERING MEDIA**   201

PREFACE
TO THE AUGMENTED EDITION

In the decade that has passed since the original publication of *Radiation Heat Transfer*, significant advances have been made in the treatment of radiative transfer in the presence of participating media. Whereas the gray medium continues to be a useful pedagogical vehicle for introducing certain concepts, it is no longer regarded as a quantitative research tool. Rather, to obtain meaningful quantitative results for radiatively participating media, analytical models which take account of wavelength dependence have to be employed.

The need to update the treatment of radiatively participating media has motivated this augmented edition of *Radiation Heat Transfer*. A new chapter, Chapter 11, has been added which sets forth analytical models to account of wavelength dependence. In keeping with the rest of the book, Chapter 11 has been written to present key ideas, without being exhaustive.

With regard to radiation among surfaces in the absence of participating media, the research of the past decade has, in the main, been focused on specific problems. The presently available analytical models are essentially the same as those that were on hand when the first edition of *Radiation Heat Transfer* was written. Therefore, from the point of view of the classroom, the material on surface-to-surface radiation contained in Part Two of the book continues to be entirely suitable.

A modest structural change has been incorporated into the new edition to facilitate the use of the problems. Formerly, the problems for all of the chapters were contained in Appendix C. In the new edition, the problems are situated at the end of the chapter to which they are relevant.

E. M. Sparrow
R. D. Cess

ix

PREFACE

Under the impact of modern technology, there has been a dramatic growth of research activity in various aspects of radiation heat transfer. The newer applications have called for heat transfer predictions of higher precision and of greater detail than were deemed necessary in the past. For instance, whereas it was formerly sufficient to compute an apparent mean temperature of a radiatively participating gas, it is now frequently desired to know the detailed distribution of temperature throughout the gas. Furthermore, many of the vital new problem areas involve strong interactions between radiation heat transfer and other modes, such as conduction and convection.

The present emphasis on precision and detail has brought forth new approaches for the computation of radiant transfer between surfaces and within participating media. Many of the new analytical methods require, for their execution, radiation properties that take account of wave length, direction, polarization, temperature, surface condition, and so forth. The measurement of radiation properties having the requisite detail is also emerging as an area of research activity.

This book is aimed at providing a contemporary account of radiation heat transfer. It has been written to fulfill two general functions: as a textbook for a college course in radiation heat transfer, and as a reference source for both research workers and applications engineers.

With a view toward providing the greatest possible flexibility, the book is organized into three more or less autonomous parts. Part One sets forth the basic characteristics of thermal radiation and of the radiation properties of surfaces and participating media. The description of the radiation properties endeavors to take account of the most recently appearing information. Analytical methods for the computation of radiant interchange among surfaces are treated in detail in Part Two. Consideration is given both to diffuse surfaces and to surfaces that possess directional characteristics. An entire chapter in Part Two is devoted to applications, including interactions with conduction and convection. Part Three deals with radiatively participating media. A general analytical formulation of the energy transport in such media is developed. This is then successively applied to situations involving purely radiative transport, simultaneous radiation and conduction, and simultaneous radiation and convection.

Although Part One is basic to a thorough understanding of radiation heat transfer, the succeeding parts have been written so that they may be utilized without prior knowledge of Part One. Furthermore, it is intended that Part Three be essentially autonomous of Part Two and vice-versa. Therefore, these parts may be taken in any order that pleases the reader.

In preparing the textual material, each of the authors has concentrated his writing activities in his special area of interest and experience. Correspondingly, one of us is primarily responsible for Chapters 2 to 6, while the other is primarily responsible for Chapter 1 and Chapters 7 to 10. The contents of Chapters 1 and 2 have been influenced by discussions with Professor E. R. G. Eckert and Professor L. S. Wang; Professor Wang made several valuable suggestions in connection with Part Three as well. Their insights are gratefully acknowledged. Also, we would like to thank Professor George Leppert of Stanford University, who made helpful suggestions in the manuscript stage.

It is a particular pleasure to acknowledge the contribution of those persons whose assistance made the actual preparation of the manuscript a painless undertaking. In this connection, heartfelt thanks are extended to Mrs. Lucille R. Laing and Mr. Stone Lin of the Heat Transfer Laboratory of the University of Minnesota, who assisted with the typing of the manuscript and the preparation of the figures. Equally sincere thanks for similar services are extended to Mrs. Sylvia H. Hershey and Messrs. George Hoschel and David Montrose of the Mechanics Department of the State University of New York at Stony Brook.

E. M. Sparrow
R. D. Cess

RADIATION HEAT TRANSFER

PART ONE

THERMAL RADIATION AND RADIATION PROPERTIES

This book is concerned primarily with *energy exchange* by the mechanism of thermal radiation. Admittedly, a thorough understanding of the fundamental physical processes involved in radiative transfer requires a knowledge of both quantum mechanics and electromagnetic theory. However, from the viewpoint of the engineer, it is often sufficient to make use of specific results from these disciplines without pursuing the details of their derivation. This point of view is adopted in the presentation of thermal radiation and of radiation properties as set forth in Part One of this book.

Two chapters are included in Part One. The first is entitled Thermal Radiation, the second Radiation Properties of Surfaces. Inasmuch as in later parts of the book equal attention is extended to surface-to-surface radiation and to radiation in absorbing, emitting, and scattering media, it is appropriate to explain the absence of a separate chapter dealing with the radiation properties of such participating media. It is a fact that at the present time the amount of information available on the properties of participating media is less than that available on the properties of surfaces. Therefore the information on the properties of participating media has been incorporated into Chapter 1 and into several chapters of Part Three. This latter placement of material was prompted by the fact that, in some instances, a more incisive discussion of the properties is possible after the governing equations of radiative transport have been developed.

In essence, Chapter 1 contains a general exposition of the character-

1

istics of thermal radiation, introduces the properties of surfaces, and then treats the properties of participating media in greater depth. Detailed consideration is extended to the radiation properties of surfaces in Chapter 2, including basic definitions and interrelationships, available experimental information, and results of electromagnetic theory.

CHAPTER 1

THERMAL RADIATION

As previously mentioned, this chapter will be concerned with an exposition of the essential characteristics of thermal radiation as related to both radiation exchange between surfaces and to radiation transfer through participating media. Section 1-1 describes the basic laws and definitions associated with thermal radiation and sets forth the properties of the black body. In Section 1-2 attention is directed toward the parameters that characterize surface radiation, and this section will be kept quite brief in cognizance of the more detailed discussion that is contained in Chapter 2. The final section, Section 1-3, embodies a more detailed exposition of the parameters and properties that pertain to absorbing, emitting, and scattering media.

1-1 Basic Laws and Definitions

The physics of radiation. Radiant energy may be alternatively envisioned as being transported either by electromagnetic waves or by photons. Neither point of view completely describes the nature of observed phenomena. Nevertheless, these separate concepts have considerable utility. For example, radiation properties of certain classes of surfaces may be predicted through use of electromagnetic theory, whereas quantum theory is employed in determining the properties of absorbing and emitting media.

Radiation travels at the speed of light. Thus, from the viewpoint of electromagnetic theory, the waves travel at this speed. Alternatively, from a quantum point of view, energy is transported by photons, all of which travel at the speed of light (this differs from molecular transport in that all the photons have the same speed). There is, however, a distribution of energy among the photons. The energy associated with each photon is $h\nu$, where h is Planck's constant and ν is the frequency of the radiation. Each photon also possesses a momentum $h\nu/c$, where c is the velocity of light within the medium through which the radiation travels.

Three parameters may be employed in characterizing radiation. They are the frequency ν, the wavelength λ, and the wave or photon speed c. Of these, only two are independent, since they are related by

$$c = \lambda\nu \tag{1-1}$$

The choice as to whether to employ ν or λ as a characteristic parameter is somewhat arbitrary, although ν has one advantage in that it does not change when radiation travels from one medium to another. The speed of light c within a given medium is related to that in a vacuum, c_0, by

$$c = \frac{c_0}{n} \tag{1-2}$$

where n is the index of refraction.

Consideration is next given to the wavelength classification of thermal radiation. Electromagnetic phenomena encompass numerous types of radiation, from the short wavelength γ rays and X rays to long wavelength radio waves. Each classification corresponds to a specific means by which the radiation is produced. For example, X rays are produced by bombarding a metal with electrons. Thermal radiation is defined as *radiant energy emitted by a medium that is due solely to the temperature of the medium*; that is, it is the temperature of the medium which governs the emission of thermal radiation.

The wavelength range encompassed by thermal radiation is approximately $0.3\ \mu$ to $50\ \mu$, where $\mu = 10^{-4}$ cm. In turn, this wavelength range includes three subranges, the ultraviolet, the visible, and the infrared. These subranges are illustrated schematically in Fig. 1-1.

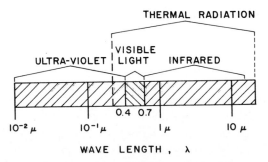

Fig. 1-1   Classifications of radiation.

Planck's law. It can be demonstrated by application of the second law of thermodynamics that there is a maximum amount of radiant energy that can be emitted at a given temperature and a given wavelength. The emitter of such radiation is called a black body. The energy emission per unit of time and per unit area from a black body in a frequency range $d\nu$ will be denoted by $e_{b\nu} \, d\nu$. The quantity $e_{b\nu}$ is termed the spectral (or monochromatic) emissive power of the black-body radiation. In addition to being a function of ν, the black-body emissive power is also a function of the absolute temperature T of the black body. The explicit form of $e_{b\nu}(T)$ is given by Planck's law as

$$e_{b\nu}(T) = \frac{2\pi h\nu^3 n^2}{c_0^2[\exp{(h\nu/kT)} - 1]} \tag{1-3}$$

which is derivable through use of quantum statistics (ref. 1). In this equation h and k are the Planck and Boltzmann constants, respectively. The index of refraction n refers to the medium bounding the black body, and it is important to recognize this dependence of the black-body emissive power on the bounding medium. A listing of various constants relating to black-body radiation is given in Table 1-1.

Table 1-1 Black-Body Radiation Constants

Boltzmann's constant: $k = 1.380 \times 10^{-16}$ erg/°K
Planck's constant: $h = 6.625 \times 10^{-27}$ erg sec
Speed of light: $c_0 = 2.998 \times 10^{10}$ cm/sec
Stefan-Boltzmann constant: $\sigma = 5.668 \times 10^{-5}$ erg/sec cm² °K⁴
$\qquad\qquad\qquad\qquad\qquad\quad = 1.714 \times 10^{-9}$ Btu/hr ft² °R⁴
$c_1 = 3.740 \times 10^{-5}$ erg cm²/sec $= 1.187 \times 10^8$ Btu μ⁴/ft² hr
$c_2 = 1.4387$ cm °K $= 25,896$ μ °R

Often it is convenient to recast equation (1-3) in terms of wavelength. This rephrasing is useful, however, only when the index of refraction of the bounding medium is independent of frequency. This is exactly the case for a vacuum ($n = 1$) and is a very good approximation for gases ($n \simeq 1$). For quartz, the index of refraction varies from 1.676 to 1.516 in the wavelength range from 0.185 to 2.324 μ.

Proceeding on the assumption that n is independent of frequency (or wavelength), there then follows from equations (1-1) and (1-2)

$$\nu = \frac{c_0}{n\lambda} \qquad d\nu = -\frac{c_0 \, d\lambda}{n\lambda^2}$$

In addition, equation (1-3) may be recast in terms of λ through the relation

$$e_{b\lambda} \, d\lambda = -e_{b\nu} \, d\nu$$

where $e_{b\lambda}$ refers to emission for a wavelength interval $d\lambda$.   With these, Planck's law becomes

$$e_{b_0}(T) = \frac{2\pi h c_0{}^2}{n^2 \lambda^5 [\exp(hc_0/n\lambda kT) - 1]} \qquad (1\text{-}4)$$

or, alternately,

$$e_{b\lambda}(T) = \frac{c_1}{n^2 \lambda^5 (e^{c_2/n\lambda T} - 1)} \qquad (1\text{-}4a)$$

where c_1 and c_2 are listed in Table 1-1.   This equation may further be put in the form

$$\varepsilon_r = \frac{e_{b\lambda}}{\sigma n^3 T^5} = \frac{c_1/\sigma}{(n\lambda T)^5 (e^{c_2/n\lambda T} - 1)} \qquad (1\text{-}4b)$$

where σ is the Stefan-Boltzmann constant given in Table 1-1.

Inspection of equation (1-4b) reveals that the quantity $e_{b\lambda}/\sigma n^3 T^5$, appearing on the left side, is a function solely of $n\lambda T$.   The relationship between $e_{b\lambda}/\sigma n^3 T^5$ and $n\lambda T$ is embodied in Table 1-2 and is illustrated graphically in Fig. 1-2.   The maximum value of the emissive power, as shown in Fig. 1-2, corresponds to

$$(n\lambda T)_{\max} = 0.2898 \text{ cm } °\text{K}$$

$$= 5216 \ \mu \ °\text{R}$$

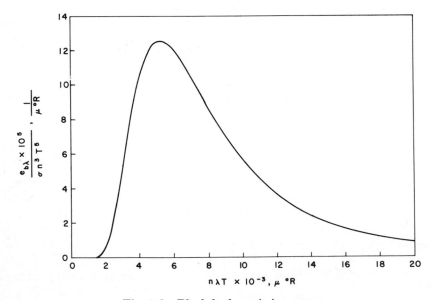

Fig. 1-2   Black-body emissive power.

Table 1-2   Black-Body Radiation Functions*

$n\lambda T$	$\dfrac{e_{b\lambda} \times 10^5}{\sigma n^3 T^5}$	$\dfrac{e_b(0 - n\lambda T)}{n^2\sigma T^4}$	$n\lambda T$	$\dfrac{e_{b\lambda} \times 10^5}{\sigma n^3 T^5}$	$\dfrac{e_b(0 - n\lambda T)}{n^2\sigma T^4}$	$n\lambda T$	$\dfrac{e_{b\lambda} \times 10^5}{\sigma n^3 T^5}$	$\dfrac{e_b(0 - n\lambda T)}{n^2\sigma T^4}$
1000	0.0000394	0	7200	10.089	0.4809	13400	2.714	0.8317
1200	0.001184	0	7400	9.723	0.5007	13600	2.605	0.8370
1400	0.01194	0	7600	9.357	0.5199	13800	2.502	0.8421
1600	0.0618	0.0001	7800	8.997	0.5381	14000	2.416	0.8470
1800	0.2070	0.0003	8000	8.642	0.5558	14200	2.309	0.8517
2000	0.5151	0.0009	8200	8.293	0.5727	14400	2.219	0.8563
2200	1.0384	0.0025	8400	7.954	0.5890	14600	2.134	0.8606
2400	1.791	0.0053	8600	7.624	0.6045	14800	2.052	0.8648
2600	2.753	0.0098	8800	7.304	0.6195	15000	1.972	0.8688
2800	3.872	0.0164	9000	6.995	0.6337	16000	1.633	0.8868
3000	5.081	0.0254	9200	6.697	0.6474	17000	1.360	0.9017
3200	6.312	0.0368	9400	6.411	0.6606	18000	1.140	0.9142
3400	7.506	0.0506	9600	6.136	0.6731	19000	0.962	0.9247
3600	8.613	0.0667	9800	5.872	0.6851	20000	0.817	0.9335
3800	9.601	0.0850	10000	5.619	0.6966	21000	0.702	0.9411
4000	10.450	0.1051	10200	5.378	0.7076	22000	0.599	0.9475
4200	11.151	0.1267	10400	5.146	0.7181	23000	0.516	0.9531
4400	11.704	0.1496	10600	4.925	0.7282	24000	0.448	0.9589
4600	12.114	0.1734	10800	4.714	0.7378	25000	0.390	0.9621
4800	12.392	0.1979	11000	4.512	0.7474	26000	0.341	0.9657
5000	12.556	0.2229	11200	4.320	0.7559	27000	0.300	0.9689
5200	12.607	0.2481	11400	4.137	0.7643	28000	0.265	0.9718
5400	12.571	0.2733	11600	3.962	0.7724	29000	0.234	0.9742
5600	12.458	0.2983	11800	3.795	0.7802	30000	0.208	0.9765
5800	12.282	0.3230	12000	3.637	0.7876	40000	0.0741	0.9881
6000	12.053	0.3474	12200	3.485	0.7947	50000	0.0326	0.9941
6200	11.783	0.3712	12400	3.341	0.8015	60000	0.0165	0.9963
6400	11.480	0.3945	12600	3.203	0.8081	70000	0.0092	0.9981
6600	11.152	0.4171	12800	3.071	0.8144	80000	0.0055	0.9987
6800	10.808	0.4391	13000	2.947	0.8204	90000	0.0035	0.9990
7000	10.451	0.4604	13200	2.827	0.8262	100000	0.0023	0.9992
							0	1.0000

* From Dunkle (ref. 2). See Fig. 1-2 for units.

A quantity that has frequent application in radiation transfer analyses is the black-body emission in the range from $\lambda = 0$ to λ. This is expressed by

$$\int_0^\lambda e_{b\lambda}(T)\,d\lambda$$

The foregoing may be rephrased in dimensionless form as

$$\frac{e_b(0 - n\lambda T)}{n^2\sigma T^4} = \int_0^{n\lambda T} \left(\frac{e_{b\lambda}}{n^3\sigma T^5}\right) d(n\lambda T) \qquad (1\text{-}5)$$

The quantity on the left-hand side of equation (1-5) is the fraction of the total radiant output of the black body that is contained in the range from 0 to $n\lambda T$. A tabulation of this integral is also included in Table 1-2.

Stefan-Boltzmann law. The total black-body emissive power will be denoted by e_b. This represents the energy emitted per unit time and area by a black body over all frequencies (or wavelengths). Thus

$$e_b(T) = \int_0^\infty e_{bv}(T)\, d\nu$$

Upon substituting equation (1-3) into this relation, one has

$$e_b(T) = n^2 \left(\frac{2\pi k^4}{c_0{}^2 h^3}\right) T^4 \int_0^\infty \frac{x^3}{e^x - 1}\, dx \qquad (1\text{-}6)$$

where the assumption has again been made that the index of refraction of the bounding medium is independent of frequency. The definite integral appearing in the preceding equation has the value

$$\int_0^\infty \frac{x^3}{e^x - 1}\, dx = \frac{\pi^4}{15}$$

and with this, equation (1-6) becomes

$$e_b(T) = n^2 \sigma T^4 \qquad (1\text{-}7)$$

where σ is defined by

$$\sigma = \frac{2\pi^5 k^4}{15 c_0{}^2 h^3}$$

The value of σ is given in Table 1-1.

Equation (1-7) is the well-known Stefan-Boltzmann equation for black-body radiation, which is conventionally expressed with $n = 1$. For the sake of brevity, we will, throughout the book, take $n = 1$. However, when dealing with black-body radiation into adjacent media within which $n \neq 1$, proper account must be taken of the index of refraction as embodied in equation (1-7).

Intensity of radiation. To characterize the amount of radiant energy that departs from a surface along a certain path, the concept of a single ray is inadequate. The amount of energy passing in a given direction is described in terms of the *intensity of radiation*, which is denoted by i. With reference to Fig. 1-3, the intensity of radiation is the radiant energy leaving a surface per unit area normal to the pencil of rays, per unit solid angle and per unit time, where the differential solid angle is denoted by $d\omega$.

To illustrate the use of the intensity, let $d\Phi$ represent the radiant

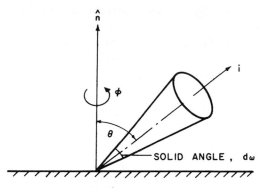

Fig. 1-3   The intensity of radiation.

energy per unit time and unit area leaving a given surface in the direction θ and contained within a solid angle $d\omega$.   Then, clearly

$$i = \frac{d\Phi}{d\omega \cos \theta}$$

The energy flux Φ passing from the surface into the hemispherical space above the surface is then obtained by integration

$$\Phi = \int_{\Omega} i \cos \theta \, d\omega$$

where the symbol Ω denotes integration with respect to solid angle over an entire hemisphere.

As illustrated in Fig. 1-4, the differential solid angle $d\omega$ may be expressed in terms of the angles θ and ϕ of a spherical coordinate system centered on the surface.   Upon recalling that the differential

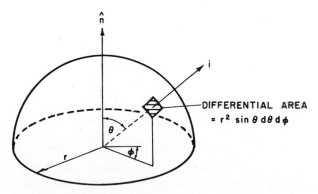

Fig. 1-4   Integration of intensity over solid angle.

solid angle is the surface element on the hemisphere divided by the square of the hemisphere radius, it follows that $d\omega = \sin\theta\, d\theta\, d\phi$. Consequently, integration over the entire hemisphere yields

$$\Phi = \int_0^{2\pi} \int_0^{\pi/2} i \cos\theta \sin\theta\, d\theta\, d\phi \qquad (1\text{-}8)$$

As will be discussed in the following section, as well as in Chapter 2, there are several instances in which the intensity of radiation is independent of direction. For such cases equation (1-8) reduces to

$$\Phi = \pi i \qquad (1\text{-}9)$$

In the foregoing definition of the intensity i, specific reference has been made to radiation *leaving* a surface. When dealing with radiation transfer through absorbing, emitting, and scattering media, one will be concerned with the *net* rate at which energy is locally transferred within the medium. The intensity of radiation will, in this instance, be designated by the symbol I and defined as the local net transfer of radiant energy per unit area normal to the pencil of rays, per unit solid angle, and per unit time.

Although the foregoing discussions of intensity have been concerned with the total energy (over all wavelengths), the formulations apply on a monochromatic basis by appending the subscript λ. Thus the spectral (monochromatic) intensities are i_λ and I_λ, while the expressions

$$i = \int_0^\infty i_\lambda\, d\lambda \qquad I = \int_0^\infty I_\lambda\, d\lambda$$

relate the total and spectral quantities.

1-2   Surface Radiation

Radiation properties of surfaces will be discussed in detail in Chapter 2, and for this reason only preliminary considerations will be presented here. The monochromatic emissive power of a real (non-black) surface will be designated by e_λ. This quantity may be expressed in terms of the black-body emissive power as

$$e_\lambda = \epsilon_\lambda e_{b\lambda} \qquad (1\text{-}10)$$

where ϵ_λ is the monochromatic hemispherical emittance of the surface.

In addition to emitting energy, real surfaces absorb a portion of the energy that is incident on them. Upon letting the monochromatic incident radiation per unit time and area be denoted by H_λ, the amount of energy absorbed is

$$\alpha_\lambda H_\lambda$$

where α_λ is the monochromatic hemispherical absorptance of the surface. Thus α_λ corresponds to the fraction of the incident energy that is absorbed.

An important relation between ϵ_λ and α_λ can be obtained from Kirchhoff's law. Although a number of simple partial proofs of Kirchhoff's law are available in the literature, the only complete derivation applicable for every direction and for each component of polarization appears to be that of Planck (ref. 1). The detailed derivation is quite lengthy and will not be reproduced here.

In essence, Kirchhoff's law states that for a system in thermodynamic equilibrium, the following equality holds for each of the participating surfaces.

$$\alpha_\lambda = \epsilon_\lambda \tag{1-11}$$

Although the above equation was derived under the condition that a given surface is in thermodynamic equilibrium with its surroundings, it is, in fact, a general relationship that applies for nonequilibrium conditions. This is a consequence of α_λ and ϵ_λ being surface properties; that is, they depend solely on the nature of the surface and its temperature. Therefore their magnitude is independent of whether or not equilibrium or nonequilibrium conditions prevail. Although no distinction is usually made with regard to directional effects, it is well to note that equation (1-11) is strictly valid only for each direction and for each component of polarization.

It is obvious from equation (1-10) that $\epsilon_\lambda = 1$ for a black body, and consequently, from equation (1-11), $\alpha_\lambda = 1$ for a black body. Thus a black body is a *perfect absorber of radiant energy*. From a laboratory point of view, this is a much more meaningful description than the definition that a black body is a perfect emitter.

A second conclusion that follows from the general derivation of Kirchhoff's law (ref. 1) is that black-body radiation is isotropic. In other words, if $i_{b\lambda}$ denotes the intensity of black-body radiation, then $i_{b\lambda}$ is independent of direction. Furthermore, since a black body is a perfect absorber, there will be no radiant energy reflected from the surface. Correspondingly, on a monochromatic basis $\Phi_\lambda = e_{b\lambda}$. It then follows from the monochromatic form of equation (1-9) that

$$e_{b\lambda} = \pi i_{b\lambda} \tag{1-12}$$

This provides the relationship between the black-body monochromatic emissive power and black-body monochromatic intensity of radiation. Similarly, on a total basis,

$$e_b = \pi i_b \qquad (1\text{-}13)$$

Although it was previously stated that ϵ_λ and α_λ are surface properties, this is not strictly correct. These quantities will, in addition, depend on the index of refraction of the medium bounding the surface. In most applications this index of refraction may be taken to be unity, corresponding to vacuum and to gases. Thus tabulated emittance data, such as are given in Tables 2-1 (page 38) and 2-2 (page 44), correspond to $n = 1$. If, however, the bounding medium is a substance such as glass ($n \simeq 1.5$), the values of ϵ_λ and α_λ may differ significantly from those for $n = 1$.

1-3   Absorbing, Emitting, and Scattering Media

Consideration will now be given to media that transmit radiation and within which radiant energy is absorbed and scattered as well as emitted. While gases are commonly classified among such media, other substances, such as window glass, absorb and emit radiation throughout their volume.

Physics of absorption, emission, and scattering.   A brief qualitative discussion of the physics of absorption, emission, and scattering will now be presented. This will be restricted specifically to the case in which the medium is a gas. The absorption or emission of thermal radiation is associated with transitions between energy levels of the atoms or molecules that constitute the gas. These transitions are classified as bound-bound, bound-free, and free-free.

A *bound-bound transition* occurs when a photon is absorbed or emitted by a gas molecule, such that the resulting change of energy level of the molecule is associated with electronic, or vibrational, or rotational energy states. For monatomic gases, the transitions involve only electronic states, whereas for molecular gases, all three energy states may be involved. Furthermore, bound-bound transitions are associated with discrete spectral lines, with the photon energy $h\nu$ being distributed over a narrow line width $\Delta\nu$. There is a natural line width that is the result of Heisenberg's uncertainty principle, while line broadening can result from several effects, for example, Doppler broadening and collision broadening.

Electronic transitions involve high frequencies and thus give rise to absorption-emission lines in the ultraviolet, visible, and near infrared.

Vibrational and rotational transitions are associated with the infrared and the far infrared, respectively. However, rotational transitions may occur simultaneously with vibrational transitions, resulting in a rotational perturbation on the basic vibrational frequency. As a consequence, a number of rotational lines will be present in the vicinity of the vibrational frequency, and this is referred to as a *vibration-rotation band*. Although such bands are composed entirely of absorption-emission lines, the discrete band structure is commonly ignored by averaging over frequency intervals that include many lines.

When the absorption of a photon produces an electron and an ion, the process is designated as a *bound-free transition*. Since the electron is initially in a bound state—while after ionization it is free to take on any value of kinetic energy—the bound-free absorption is a continuous function of frequency. The reverse process, that is, emission, is also continuous and results when a free electron and an ion recombine to produce a photon. Since bound-free transitions occur only when the gas is ionized, radiation absorption and emission stemming from these processes are generally important only in high-temperature applications.

A second type of transition involving an ionized gas occurs when a photon is absorbed or emitted by a free electron. Since the electron may possess any value of kinetic energy before and after transition, this is a *free-free transition*. As in the bound-free case, the absorption and the emission processes are continuous functions of frequency. In general, the absorption and the emission of radiation by free-free transitions are less important than by bound-free transitions.

In addition to absorption, a medium may also *scatter* photons. Scattering is defined as any change in the direction of propagation of the photons. This process is physically due to local inhomogeneities within the medium, and such inhomogeneities may result from suspended solid particles or liquid droplets within the gas. In addition, scattering can also be produced by the gas molecules. When radiant energy is scattered with no change in frequency, the scattering is referred to as *coherent scattering*.

If the scattering of radiation within a gas is strictly molecular scattering (i.e., there are no foreign particles present), it is designated as *Rayleigh scattering*. The Rayleigh theory predicts that the spectral intensity of the scattered radiation will vary as the fourth power of the frequency; that is, the scattering is predominantly at the shorter wavelengths. This accounts for the fact that the sky appears blue, for the preferential scattering in the atmosphere involves the short wavelength blue light. This is also the reason why sunsets are red, for the long wavelength red light suffers less attenuation in traversing the large atmospheric path length. Although Rayleigh scattering is an important mechanism in atmospheric phenomena, it is usually unimportant in engineering applications due to the short path lengths involved in the latter (ref. 3).

Scattering can, however, play an important role in radiation energy transfer when foreign particles are present. Typical examples include dust clouds, fogs, fluidized beds, and low-density insulations. In these cases scattering may encompass the combined effects of reflection, refraction, and diffraction. A theory that is pertinent to such situations is *Mie scattering*, which is concerned with electromagnetic scattering from spherical particles. For a more detailed account of scattering with application to thermal radiation, the reader may consult refs. 4, 5, and 6.

The foregoing discussion of the physics of absorption, emission, and scattering has, by necessity, been kept very brief. It is intended solely as an introductory description of the physical phenomena involved. Attention will now be turned to the formulation of the processes of absorption, scattering, and emission in terms of defined radiation properties.

Absorption and scattering. Consider a monochromatic beam of radiation with intensity I_λ, as illustrated in Fig. 1-5. As the beam traverses a path length ds it will undergo partial attenuation as the result of local absorption within the medium. The amount of absorption is assumed to be directly proportional to both the thickness ds and the incident intensity I_λ. Thus the amount of monochromatic absorption per unit time, per unit area normal to the pencil of rays, and per unit solid angle may be written as

$$\kappa_\lambda I_\lambda \, ds \qquad\qquad\qquad (1\text{-}14)$$

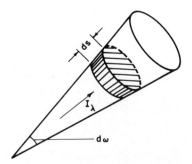

Fig. 1-5   Absorption and scattering of incident radiation.

The constant of proportionality κ_λ is the *monochromatic absorption coefficient** for radiation of wavelength λ. Since the volume per unit surface area of the cross-hatched element in Fig. 1-5 is ds, the monochromatic absorption within the medium per unit time, per unit *volume*, and per unit solid angle is

$$\kappa_\lambda I_\lambda \tag{1-15}$$

By integration of equation (1-15) over all possible values of the solid angle ($\omega = 4\pi$), the local monochromatic absorption per unit time and per unit volume within the medium due to all incident beams is expressed as

$$\kappa_\lambda \int_{4\pi} I_\lambda \, d\omega \tag{1-16}$$

where it has been assumed that the medium is isotropic (i.e., κ_λ is independent of direction).

A parallel development can be carried out for scattering. Again, with reference to Fig. 1-5, the monochromatic beam of intensity I_λ will further be attenuated due to scattering. The monochromatic energy that is scattered per unit time, per unit area normal to the pencil of rays, and per unit solid angle may be characterized as

$$\gamma_\lambda I_\lambda \, ds \tag{1-17}$$

where γ_λ is the *monochromatic scattering coefficient*.† Correspondingly, on a unit volume basis, the expressions

$$\gamma_\lambda I_\lambda \tag{1-18a}$$

$$\gamma_\lambda \int_{4\pi} I_\lambda \, d\omega \tag{1-18b}$$

respectively characterize the scattered energy due to a single incident beam and to all incident beams.

The analogous description of radiation attenuation due both to absorption and to scattering leads to the following representation of the combined effects of both processes

$$\kappa_\lambda I_\lambda \, ds + \gamma_\lambda I_\lambda \, ds = \beta_\lambda I_\lambda \, ds \tag{1-19}$$

where $\beta_\lambda = \kappa_\lambda + \gamma_\lambda$ is the *monochromatic extinction coefficient*. The parallelism between absorption and scattering ceases, however, at this point. Once photons have been absorbed, no further consideration

---

* This is actually a *volumetric* absorption coefficient, as distinct from the often-employed *mass* absorption coefficient.

† The scattering coefficient is often denoted by σ_λ. The present notation was adopted to avoid confusion with the Stefan-Boltzmann constant.

need be given to them.   (The medium emits photons, but this is a
distinctly different process from absorption.)   On the other hand,
scattered photons continue to transport energy throughout the medium,
and these must be taken into account.

To facilitate the description of the radiant energy flux leaving a
volume element as the result of scattering of incident radiation, consider
the schematic representation as shown in Fig. 1-6, where s is a scalar
distance measured along a pencil of rays.   A portion of the incident
beam $I_\lambda(s,\ \theta',\ \phi')$ will be scattered by the cross-hatched volume element
in accord with equation (1-17).   It will be assumed that the scattering
is coherent.

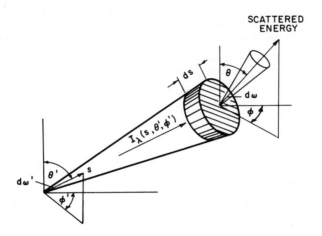

Fig. 1-6   Scattered energy leaving a volume element.

The directional distribution of the scattered energy is characterized
by the *scattering function* $P_\lambda(\theta',\ \phi':\ \theta,\ \phi)$, such that

$$P_\lambda(\theta',\ \phi':\ \theta,\ \phi)\,\frac{d\omega}{4\pi} \qquad\qquad (1\text{-}20)$$

represents the probability that radiation incident in the direction
$(\theta',\ \phi')$ will be scattered in the direction $(\theta,\ \phi)$.   Since the magnitude of
the scattered radiation corresponding to the incident beam $I_\lambda(s,\ \theta',\ \phi')$
is described by equation (1-17), the monochromatic energy per unit
time, per unit solid angle, and per unit area normal to $I_\lambda(s,\ \theta',\ \phi')$ that
is scattered in the direction $(\theta,\ \phi)$ is

$$\gamma_\lambda I_\lambda(s,\ \theta',\ \phi')P_\lambda(\theta',\ \phi':\ \theta,\ \phi)\,\frac{d\omega}{4\pi}\,ds \qquad\qquad (1\text{-}21)$$

The integral of the foregoing quantity over all values of the solid angle must coincide with equation (1-17).   From this, it follows that

$$\frac{1}{4\pi} \int_{4\pi} P_\lambda(\theta', \phi': \theta, \phi)\, d\omega = 1$$

which is consistent with the definition of P_λ as a probability.   An often-used (though not necessarily often-realistic) assumption is that of *isotropic scattering*.   In this case radiation is scattered uniformly in all directions, so that

$$P_\lambda(\theta', \phi': \theta, \phi) = 1$$

It will now be convenient to introduce the concept of a *photon mean free path*.   Consider a beam of energy of intensity I_λ that originates at $s = 0$ (where s is again measured along a pencil of rays), and let $I_{0\lambda}$ be the intensity at $s = 0$ as illustrated in Fig. 1-7.   Furthermore, let dI_λ denote the attenuation of intensity due to absorption and scattering by the cross-hatched element.*   From equation (1-19), one can write

$$dI_\lambda = -\beta_\lambda I_\lambda\, ds$$

or
$$I_\lambda = I_{0\lambda} e^{-\beta_\lambda s}$$

From a photon point of view, the quantity $F(s) = e^{-\beta_\lambda s}$ appearing in the above expression represents the fraction of photons whose free path exceeds a given length s.   In other words, $F(s)$ is the fraction of photons that have neither been absorbed nor scattered in traversing

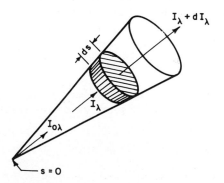

Fig. 1-7   Attenuation due to absorption and scattering.

* In addition, there will be augmentation of I_λ due both to emission by the element and to scattering of all other incident beams from the element.   Although these contributions to dI_λ must be considered in an overall radiation balance (Section 7-3), it is only necessary here to consider attenuation.

the distance s. This is identical with the probability function associated with molecular mean free paths (ref. 7). Proceeding in a manner analogous to the molecular case, a photon free-path distribution function $f(s)$ is defined such that $f(s)\,ds$ is the fraction of photons with free paths in the range s to $s + ds$. The relationship between f and F is readily derived as

$$f(s)\,ds = F(s) - F(s + ds) = -\frac{dF}{ds}\,ds$$

from which it follows that

$$f(s) = \beta_\lambda\, e^{-\beta_\lambda s}$$

In terms of the distribution function f, the *mean* free path λ_p for photons of wavelength λ is expressed by

$$\lambda_p = \int_0^\infty sf(s)\,ds$$

and, consequently,

$$\lambda_p = \frac{1}{\beta_\lambda} \tag{1-22}$$

It should be emphasized that, in contrast to molecular transfer, there are two mechanisms by which a photon free path is terminated; that is, absorption and scattering of the photons. It is evident that, for non-scattering media, equation (1-22) becomes $\lambda_p = 1/\kappa_\lambda$.

Emission. The local monochromatic emission of radiant energy within an absorbing, emitting, and scattering medium will be denoted by J_λ. This quantity is termed the *volumetric emission coefficient*. It represents the local emission per unit time, per unit volume, and per unit solid angle. The emission coefficient may be expressed in terms of the absorption coefficient κ_λ and Planck's Function $e_{b\lambda}(T)$ in the following manner.

For an absorbing, emitting, and scattering medium in thermodynamic equilibrium, Kirchhoff's law yields the result

$$J_\lambda = \frac{\kappa_\lambda}{\pi}\, e_{b\lambda}(T) \tag{1-23}$$

It will now be assumed that the medium is in *local thermodynamic equilibrium*; that is, the populations of the atomic and molecular states that contribute to absorption and emission are given by their equilibrium distributions. This assumption is valid when energy transitions are controlled by molecular collisions rather than by absorption and emission; a further discussion of this point will be given later. It is important to recognize that the assumed equilibrium is solely between particles of the medium and *not* between the medium and the radiation.

Under the assumption of local thermodynamic equilibrium, the monochromatic absorption coefficient κ_λ is a property of the medium. Therefore equation (1-23) constitutes the applicable relation between the absorption and emission coefficients. For an isotropic medium, the emission is independent of direction. Then, upon multiplying equation (1-23) by the total solid angle 4π, there results

$$4\kappa_\lambda e_{b\lambda}(T) \qquad\qquad (1\text{-}24)$$

This represents the local monochromatic emission per unit time and per unit volume.

Induced emission. Although Kirchhoff's law appears to relate emission to the absorption coefficient, there are certain details which it does not describe. These are important in understanding the precise nature of the absorption coefficient κ_λ. In reality, equation (1-14) includes both true absorption and *induced emission*. This latter effect is a consequence of differences in populations of energy states, and it can be evaluated by employing the principle of detailed balancing (ref. 8).

Specifically, if κ'_ν denotes the absorption coefficient characterizing true absorption, then equation (1-14) becomes

$$\kappa'_\nu I_\nu \, ds \; - \; \kappa'_\nu e^{-h\nu/kT} \, I_\nu \, ds \qquad\qquad (1\text{-}25)$$

with the first term characterizing true absorption and the second term representing induced emission. The emission coefficient J_ν is, in turn,

$$J_\nu = \frac{\kappa'_\nu}{\pi} (1 \, - \, e^{-h\nu/kT}) e_{b\nu}(T) \qquad\qquad (1\text{-}26)$$

and this entire quantity represents *spontaneous emission*. Since induced emission is included in the absorption term, it is often termed "negative absorption."

We may note that equations (1-25) and (1-26) reduce directly to the previous expressions (1-14) and (1-23), providing that the absorption coefficient κ_ν is defined as

$$\kappa_\nu = \kappa'_\nu(1 \, - \, e^{-h\nu/kT}) \qquad\qquad (1\text{-}27)$$

That is, the coefficient κ_ν accounts for *both* true absorption and induced emission. In light of this interpretation, no further use will be made of equations (1-25) and (1-26).

Absorption scattering. Under conditions for which the assumption of local thermodynamic equilibrium does not apply, the nonequilibrium process of *absorption scattering* may take place.   For details, the reader is referred to Kulander (ref. 9); only a brief discussion will be given here.

Absorption scattering occurs when radiation-controlled transitions become important relative to collision-controlled transitions (e.g., in low-density applications).   This effect is associated with bound-bound transitions.   The process is one in which the absorption of a photon results in the re-emission of a photon whose frequency lies in the same spectral line as that of the absorbed photon.   Consequently, the process of absorption scattering may be treated in a manner analogous to ordinary scattering.

Total quantities.   The total emission of radiation locally within a medium is obtained through integration of equation (1-24) over all wavelengths, with the result that the total emission per unit time and per unit volume may be expressed as

$$4\kappa_P e_b(T) \tag{1-28}$$

where
$$\kappa_P = \frac{\int_0^\infty \kappa_\lambda e_{b\lambda}(T)\, d\lambda}{e_b(T)} \tag{1-29}$$

κ_P is termed the *Planck mean absorption coefficient*.   This is indeed a misnomer, for κ_P describes emission rather than absorption.   It is clear from equation (1-29) that κ_P is a property of the medium, provided that local thermodynamic equilibrium exists.

Similarly, total absorption and scattering coefficients may be defined from equations (1-16) and (1-18b) as

$$\kappa = \frac{\int_0^\infty \kappa_\lambda \int_{4\pi} I_\lambda\, d\omega\, d\lambda}{\int_{4\pi} I\, d\omega} \tag{1-30}$$

and
$$\gamma = \frac{\int_0^\infty \gamma_\lambda \int_{4\pi} I_\lambda\, d\omega\, d\lambda}{\int_{4\pi} I\, d\omega} \tag{1-31}$$

respectively.   However, these quantities have little utility, for I_λ is generally a function of conditions throughout the entire medium as well as of the surrounding surfaces.   Thus κ and γ are not equilibrium properties.

Absorption coefficient data.   Expressions for the monochromatic absorption coefficient κ_λ can, in principle, be derived through the application of quantum mechanics. However, with the exception of monatomic and diatomic gases, the complexity of the calculations has made this approach generally impractical. As a result, much of the information pertaining to the absorption of thermal radiation by gases (whether in the form of an absorption coefficient or otherwise) is either experimental data or consists of a combination of experiment and analysis.

Consider first radiation absorption by monatomic and diatomic gases. Monatomic gases do not, of course, undergo vibrational or rotational transitions. Furthermore, symmetrical diatomic molecules cannot, as a consequence of the fact that they have no electric dipole moment, significantly absorb and emit by means of vibration-rotation bands. Thus, for monatomic gases and for diatomic gases having symmetric molecules, there will be no absorption or emission in the infrared resulting from bound-bound transitions. It then follows that such gases are transparent to thermal radiation at low and moderate temperature levels.

At high temperatures, however, there can be significant radiation due to electronic bound-bound transitions as well as from bound-free and free-free transitions. Calculated values of the monochromatic absorption coefficient for high temperature air, taken from Meyerott, Sokoloff, and Nicholls (ref. 10), are illustrated in Fig. 1-8. These results actually constitute spectral line absorption due to bound-bound transitions superimposed on the continuous absorption from bound-free and free-free transitions. As previously discussed, however, a continuous distribution of κ_λ is achieved by averaging over frequency intervals that include many lines. It should also be mentioned that absorption due to vibration-rotation bands actually does occur for high-temperature air, since there is dissociation of the symmetrical O_2 and N_2 molecules with partial recombination to form the unsymmetrical NO molecule.

Consider next the diatomic gas CO, which has an unsymmetrical molecule. In the infrared, there are two main vibration-rotation bands. One is a strong (fundamental) band located at 4.7 μ, while the other is a weak (first overtone*) band at 2.3 μ. Experimental values of the monochromatic absorption coefficient as a function of wave number (reciprocal of wavelength) for these two bands are respectively illustrated in Figs. 1-9 and 1-10 for room temperature.

One may note that the ordinate in Figs. 1-9 and 1-10 is the ratio of the absorption coefficient κ_λ to the gas pressure p. For bound-bound

---

* Overtone bands will occur at the frequencies $2\nu_i$, $3\nu_i$, . . ., where ν_i is the fundamental frequency.

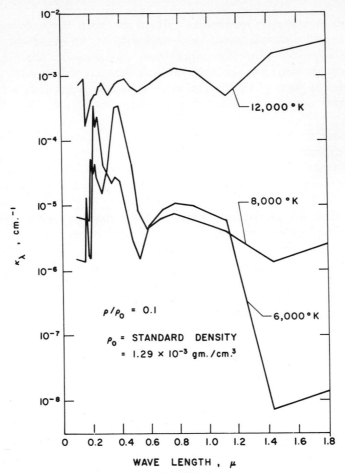

Fig. 1-8 Monochromatic absorption coefficient for air. From
Meyerott, Sokoloff, and Nicholls (ref. 10).

transitions this quantity is independent of pressure (providing the gas
temperature is not high enough to result in dissociation or ionization).
The reason for this behavior is that for a given temperature, absorption
varies directly with gas density ρ, such that the ratio κ_λ/ρ is independent
of density. This ratio is commonly called the mass absorption coeffi-
cient. By applying the perfect gas law, we may alternately state that
the ratio κ_λ/p is independent of pressure.

A quantity that has application with respect to the band approximation (see Section 7-8) is the integrated band absorption S, defined as

$$S = \int_{\Delta\lambda} \frac{\kappa_\lambda}{p}\, d\!\left(\frac{1}{\lambda}\right)$$

(1-32)

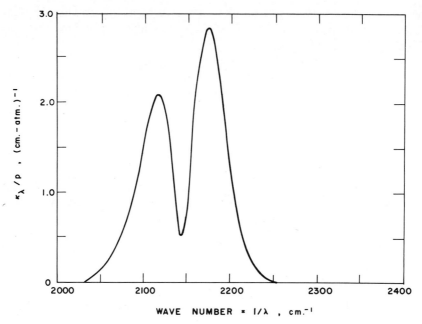

Fig. 1-9 Monochromatic absorption coefficient for CO fundamental band at room temperature.   From Penner (ref. 11).

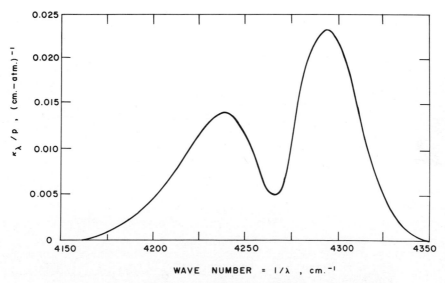

Fig. 1-10 Monochromatic absorption coefficient for CO first
overtone band at room temperature.   From Penner (ref. 11).

where $\Delta\lambda$ is the width of the band.*   By graphical integration of Figs. 1-9 and 1-10, Penner (ref. 11) gives the following S values for the two CO bands at room temperature:

$$S = 230 \text{ cm}^{-2} \text{ atm}^{-1}; \quad 4.7 \text{ } \mu \text{ band}$$

$$S = 1.7 \text{ cm}^{-2} \text{ atm}^{-1}; \quad 2.3 \text{ } \mu \text{ band}$$

The value of S for the fundamental band (4.7 μ) may be extended to other temperatures through the relation (ref. 12)

$$S(T) = \frac{T_0}{T} S(T_0) \tag{1-33}$$

where T_0 denotes a reference temperature (room temperature in this case).

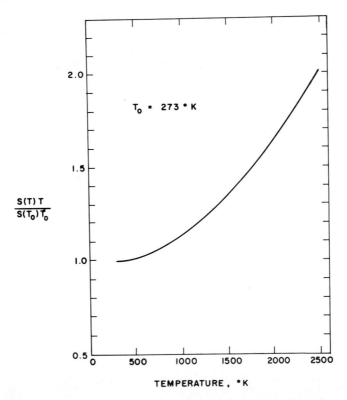

Fig. 1-11   Temperature dependency of $S(T)$ for 2.7 μ NO band. From Breeze et al. (ref. 12).

* Note that the integration is with respect to *wave number* and not wavelength.

The existence of a strong and a weak band in the infrared is characteristic of diatomic gases having unsymmetrical molecules. For NO, the fundamental band is located at 5.3 μ and the first overtone at 2.7 μ. The respective values of the integrated band absorption for NO are (ref. 12)

$$S(T_0) = 76 \text{ cm}^{-2} \text{ atm}^{-1}; \quad 5.3 \text{ μ band}$$

$$S(T_0) = 2.8 \text{ cm}^{-2} \text{ atm}^{-1}; \quad 2.7 \text{ μ band}$$

where $T_0 = 273$ °K. Equation (1-33) is again applicable for the fundamental band, while the temperature dependency for the 2.7 μ band is given in Fig. 1-11.

Two common triatomic gases that absorb significantly in the infrared are CO_2 and H_2O. The CO_2 band structure may be illustrated by considering the monochromatic absorptance α_λ. From the discussion preceding equation (1-22), the expression for α_λ in terms of κ_λ and path length s readily follows as

$$\alpha_\lambda(s) = 1 - e^{-\kappa_\lambda s} \tag{1-34}$$

A sample spectrum, taken from Edwards (ref. 13) and corresponding to $\rho s = 0.5$ lb/ft^2, is shown in Fig. 1-12 and illustrates the four principal bands for CO_2. The 4.3 μ and 15 μ bands actually encompass several additional minor bands.

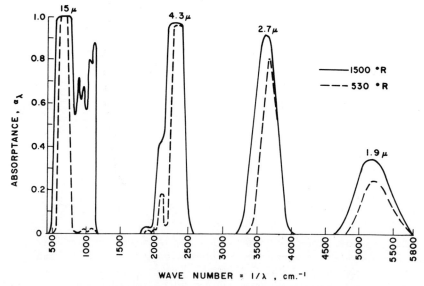

Fig. 1-12   Sample spectrum for CO_2. From Edwards (ref. 13).

A summary of values of integrated band absorption for these four principal CO_2 bands is given in Table 1-3 for $T_0 = 273\,°K$. At other temperatures, equation (1-33) applies for the two fundamental bands, 15 μ and 4.3 μ, while the temperature dependency for the 2.7 μ and 1.9 μ combination bands* is given in Fig. 1-13.

For H_2O, there are a total of five principal vibration-rotation bands in addition to a pure rotation band. The integrated band absorptions for these bands are listed in Table 1-4 for $T_0 = 273\,°K$. Again, equation (1-33) applies for the fundamental bands as well as the pure rotation band, while the temperature dependency for the 1.87 μ and

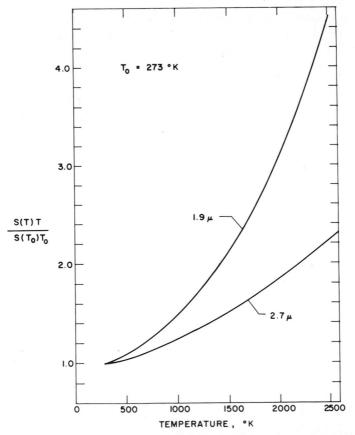

Fig. 1-13   Temperature dependency of $S(T)$ for 2.7 μ and 1.9 μ CO_2 bands.   From Breeze et al. (ref. 12).

* If ν_i and ν_k represent fundamental frequencies, combination bands can occur at the frequencies $\nu_i + \nu_k$, $2\nu_i + \nu_k$, ...

Table 1-3 Integrated Band Absorption for CO_2,
$T_0 = 273\ °K$

Band	$S(T_0)$, $cm^{-2}\ atm^{-1}$
1.9 μ (combination)	2.0*
2.7 μ (combination)	66*
4.3 μ (fundamental)	2970*
15 μ (fundamental)	266†

* Breeze et al. (ref. 12).
† This value was taken from Penner (ref. 11) and corrected to 273 °K.

1.38 μ bands is shown in Fig. 1-14.   (The 1.14 μ band may generally be neglected.)

Goldstein (ref. 16) has presented monochromatic absorption coefficient results for the 1.38 μ, 1.87 μ, 2.7 μ, and 6.3 μ H_2O bands.   An illustration of these results, taken from ref. 17, is shown in Fig. 1-15 for the 2.7 μ band.

With regard to an analytical representation for the monochromatic absorption coefficient, Edwards and Menard (ref. 18) have shown that a satisfactory correlation for total gas absorption can be achieved through use of an exponential band model in which the frequency and temperature dependence of $\kappa_\nu(T)$ are described by

$$\frac{\kappa_\nu(T)}{\rho} = \frac{c_1}{c_2}\left(\frac{180}{T}\right)^n \exp\left[-\frac{|\nu_0 - \nu|}{c_2}\left(\frac{180}{T}\right)^n\right] \qquad (1\text{-}35)$$

where T is in degrees Rankine and the constants c_1, c_2, ν_0, and n refer to a specific gas and to a particular vibration-rotation band of that gas.

Table 1-4 Integrated Band Absorption for H_2O,
$T_0 = 273\ °K$

Band	$S(T_0)$, $cm^{-2}\ atm^{-1}$
1.14 μ (combination)	1.96*
1.38 μ (combination)	21.2*
1.87 μ (combination)	26.0*
2.7 μ (fundamental)	200*
6.3 μ (fundamental)	300†
20 μ (pure rotation)	1840‡

* Ferriso and Ludwig (ref. 14).
† Breeze et al. (ref. 12).
‡ Ludwig et al. (ref. 15).

Values of these constants are given by Edwards for CO_2 (refs. 19 and 20), CH_4 (ref. 19), and H_2O (ref. 21).

In addition to gases, solid substances such as glass and quartz also act as radiation participating media. Measurements of the mono-chromatic absorption coefficient for window glass have been made by Neuroth (ref. 22), and a portion of his results are shown in Fig. 1-16. This figure clearly illustrates the "greenhouse effect"; that is, glass is

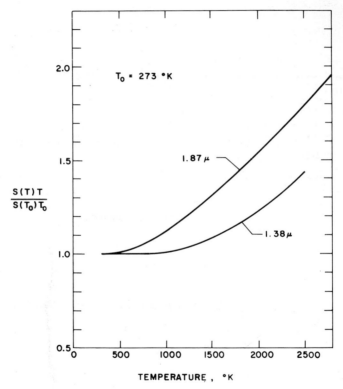

Fig. 1-14   Temperature dependency of $S(T)$ for 1.87 μ and 1.38 μ H_2O bands.   From refs. 12 and 14.

virtually transparent to short wavelength solar radiation, whereas it is nearly opaque to the long wavelength infrared.

While the present discussion has been concerned with results for the monochromatic absorption coefficient and the integrated band absorption, additional property information for participating media is presented in Part Three. Values of the Planck mean absorption coeffi-cient for CO, CO_2, H_2O, and NH_3 are given in Fig. 7-5, while Figs. 8-11 and 8-12 illustrate both monochromatic and Planck mean absorption co-

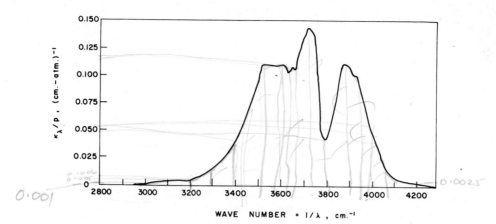

Fig. 1-15 Monochromatic absorption coefficient for 2.7 μ H_2O band at 1000 °K.   From Goldstein (ref. 17).

efficients for high-temperature hydrogen.   Furthermore, various mean absorption coefficients, which have application to certain limiting cases, are discussed in Section 7-8.   As previously mentioned, the reader may consult refs. 4, 5, and 6 for information pertaining to scattering.

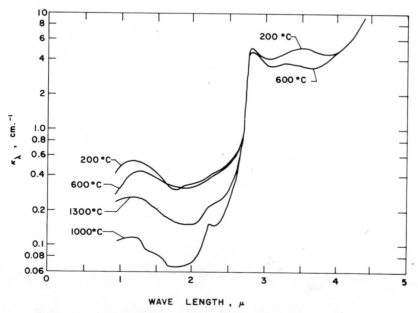

Fig. 1-16 Monochromatic absorption coefficient for window glass. From Neuroth (ref. 22).

REFERENCES

1. M. Planck, *The Theory of Heat Radiation*, Dover Publications, New York, 1959.

2. R. V. Dunkle, Thermal radiation tables and applications. *Trans. ASME*, **76**, 549–552 (1954).

3. M. J. Lighthill, Dynamics of a dissociating gas. Part 2. Quasi-Equilibrium Transfer Theory. *J. Fluid Mechanics*, **8**, 161–182 (1960).

4. T. J. Love, Jr., An investigation of radiant heat transfer in absorbing, emitting, and scattering media. Aeronautical Research Laboratories Report *ARL 63-3*, January 1963.

5. T. J. Love and R. J. Grosh, Radiative heat transfer in absorbing, emitting, and scattering media. *J. Heat Transfer*, **87**, 161–166 (1965).

6. J. H. Chin and S. W. Churchill, Anisotropic, multiply scattered radiation from an arbitrary, cylindrical source in an infinite slab. *J. Heat Transfer*, **87**, 167–172 (1965).

7. R. D. Present, *Kinetic Theory of Gases*, McGraw-Hill, New York, 1958.

8. W. G. Vincenti and C. H. Kruger, Jr., *Introduction to Physical Gas Dynamics*, John Wiley and Sons, New York, 1965.

9. J. L. Kulander, Non-equilibrium radiation. General Electric Co., M.S.D. *TIS R64SD41*, June 1964.

10. R. E. Meyerott, J. Sokoloff, and R. W. Nicholls, Absorption coefficients of air. Air Force Cambridge Research Center, Geophysics Research Paper No. 68, 1960.

11. S. S. Penner, *Quantitative Molecular Spectroscopy and Gas Emissivities*, Addison-Wesley Publishing Co., Reading, Mass., 1959.

12. J. C. Breeze, C. C. Ferriso, C. B. Ludwig, and W. Malkmus, Temperature dependence of the total integrated intensity of vibrational-rotational band systems. *J. Chem. Phys.*, **42**, 402–406 (1965).

13. D. K. Edwards, Radiation interchange in a nongray enclosure containing an isothermal carbon-dioxide-nitrogen gas mixture. *J. Heat Transfer*, **C84**, 1–11 (1962).

14. C. C. Ferriso and C. B. Ludwig, Spectral emissivities and integrated intensities of the 1.87-, 1.38-, and 1.14-μ H_2O bands between 1000° and 2200°K. *J. Chem. Phys.*, **41**, 1668–1674 (1964).

15. C. B. Ludwig, C. C. Ferriso, W. Malkmus, and F. P. Boynton, High-temperature spectra of the pure rotational band of H_2O, *J. Quant. Spectr. Radiative Transfer*, **5**, 697–714 (1965).

16. R. Goldstein, Ph.D. thesis, California Institute of Technology, Pasadena, Calif., June 1964.

17. R. Goldstein, Measurements of infrared absorption by water vapor at temperatures to 1000 °K. *J. Quant. Spectr. Radiative Transfer*, **4**, 343–352 (1964).

18. D. K. Edwards and W. A. Menard, Comparison of models for correlation of total band absorption. *Applied Optics*, **3**, 621–625 (1964).

19. D. K. Edwards and W. A. Menard, Correlations for absorption by methane and carbon dioxide gases. *Applied Optics*, **3**, 847–852 (1964).

20. D. K. Edwards and W. Sun, Correlations for absorption by the 9.4 and 10.4 micron CO_2 bands. *Applied Optics*, **3**, 1501 (1964).

21. D. K. Edwards, B. J. Flornes, L. K. Glassen, and W. Sun, Correlations of absorption by water vapor at temperatures from 300 to 1100 °K. *Applied Optics*, **4**, 715–722 (1965).

22. N. Neuroth, Der Einfluss der Temperatur auf die spektrale Absorption von Glasern in Ultraroter, I (Effect of Temperature on Spectral Absorption of Glasses in the Infrared). *Glastech Ber.*, **25**, 242–249 (1952).

PROBLEMS

1-1.   When the variation of the index of refraction with wavelength is taken into account, show that a multiplicative factor $1 + (\lambda/n)\,(dn/d\lambda)$ appears on the right-hand side of equation (1-4).

1-2.   For each of the following sets of index-of-refraction data for sodium chloride, evaluate the multiplicative factor derived in Problem 1-1.

$\lambda\ (\mu)$	n	$\lambda\ (\mu)$	n
0.20	1.79073	1.4	1.52888
0.22	1.71591	1.6	1.52798
0.24	1.67197	1.8	1.52728
0.26	1.64294	2.0	1.52670
0.28	1.62239		
0.30	1.60714		

1-3.   Show that $e_{b\lambda}$ increases with temperature at any fixed value of λ.

1-4.   Plot $e_{b\lambda}$ as a function of λ for fixed values of temperature equal to 100°F, 1000°F, and 2000°F.   Take $n = 1$.

1-5.   Assume that the surface of the sun has an effective black-body temperature of 10,000°R.   What percentage of the radiant emission of the sun lies in the visible range of the spectrum (0.4 to 0.7 μ)?

1-6.   A laboratory black body operates at 1000°R.   At what wavelength does the maximum value of $e_{b\lambda}$ occur?   What percentages of its output lie in the following wavelength ranges: 0 to 4 μ, 4 to 8 μ, 8 to 12 μ, 12 to 16 μ, 16 μ to ∞?   Take $n = 1$.

1-7. According to equation (1-10), the spectral emissive power of a nonblack body can be found by multiplying $e_{b\lambda}$ times ϵ_λ. Suppose that ϵ_λ is independent of λ (gray-body radiation). Show that under these conditions, the listing of

$$\frac{e_b(0 - n\lambda T)}{n^2 \sigma T^4}$$

given in Table 1-2 also represents the fraction of the total radiant output of the nonblack body in the range from 0 to $n\lambda T$.

1-8. How much larger is the emissive power e_b of a black body when its bounding medium is quartz ($n \approx 1.5$) than when its bounding medium is air?

1-9. Suppose that the intensity of radiation i varies with the angle θ (Fig. 1-4) as

$$i = i_n, \qquad 0 \leqslant \theta \leqslant 60°$$

$$i = i_n - i_n\left(\frac{\theta}{30} - 2\right)^2, \qquad 60 \leqslant \theta \leqslant 90°$$

in which i_n is the intensity in the direction normal to the surface. Derive a relationship between i_n and the radiant flux Φ leaving the surface per unit time and unit area.

1-10. For the conditions of Fig. 1-8, what is the maximum value of the photon mean free path for air at a temperature of 8000°K?

1-11. Show that the Planck mean absorption coefficient for a gas having a single vibration-rotation band can be expressed as

$$\frac{\kappa_P}{p} = \frac{e_{b\lambda_c}(T)}{\sigma T^4} \lambda_c^2 S(T)$$

where λ_c is the wavelength at the band center. Hint: Assume that $e_{b\lambda}$ and λ^2 are constant throughout the narrow wavelength region of the band.

1-12. Neglecting overtone bands, and using the result of Problem 1-11, calculate the temperature for which the Planck mean absorption coefficient of carbon monoxide is a maximum. Compare your result with Fig. 7-5.

1-13. Employing Fig. 1-15, estimate the integrated band absorption for the 2.7 μ fundamental band of H_2O. Compare your result with that given in Table 1-4.

See pg 24

CHAPTER 2

RADIATION PROPERTIES OF SURFACES

In analyzing the interchange of radiant energy among surfaces, it is necessary to know their emission, reflection, and absorption characteristics. If the participating surfaces are black bodies, then, as indicated in the preceding chapter, the radiation properties are well-established. However, commonly encountered engineering surfaces do not generally behave like black bodies. Consequently, means must be found to characterize the radiation properties of nonblack surfaces.

To this end, it is customary to define dimensionless quantities, such as the emittance and the absorptance, which relate the emitting and absorbing capabilities of a given surface to those of a black body. Furthermore, a third dimensionless quantity, the reflectance, is used to relate the reflecting capability of a given surface to that of a perfect reflector. For transmitting substances, it is also necessary to specify the fraction of the incident radiation that passes through the thickness of the material. In this chapter consideration is confined to materials of sufficient thickness to render them essentially opaque.*

In specifying the radiation properties, various degrees of detail are required, depending on the nature of the application. For instance, if it is desired to calculate the radiant energy emitted into the entire hemispherical space above a surface element, it is necessary to know the *hemispherical* emittance. On the other hand, to determine the radiation emitted from an element into a particular direction, the *directional* (angular) emittance is used. Furthermore, depending on whether the calculation is to be performed over all wavelengths or in a specific wavelength range, either the *total* or the *monochromatic* (spectral) emittance is correspondingly required. The distinctions just discussed

---

* For metals, a thickness on the order of a fraction of a micron is sufficient. On the other hand, for electric nonconductors, the required thickness may be as much as a millimeter, depending on the material and the wavelength of the radiation.

among the various types of emittances must also be made for the other radiation properties.

Most engineering calculations continue to be performed using hemispherical and total radiation properties. However, in many of the newer technologies, where results of high accuracy are required and inflated safety factors are inconsistent with design practice, more detailed properties are employed.

The magnitude, the angular distribution, and the wavelength dependence of the radiation properties are all very sensitive to the condition of the surface. Surface condition includes such factors as roughness, oxide layers, and physical and chemical contamination. In the case of dielectric materials, the grain structure of the material may also affect the radiation properties. The qualitative description of a surface by terms such as *smooth, polished, rough*, and *oxidized* is generally insufficient to permit an accurate specification of the radiation properties. For instance, the emittance of *highly polished* aluminum plate at room temperature has been variously reported as 0.04, 0.05, 0.08, and so on. The aforementioned surface descriptions are highly subjective and are therefore open to a broad range of interpretations.

There are extensive tabulations in the literature of measured emittance data and, to a lesser extent, of reflectance and absorptance data. However, because of the very loose specification of the surface condition, it is probable that the tabulated values will not apply with a high degree of precision to a particular surface for which calculations are to be carried out. The situation is especially unsatisfactory with respect to the emittance of metallic surfaces. For electric nonconductors, large errors in the emittance are less likely. It may be concluded that property measurements on the particular surfaces of interest are a necessary prerequisite to the execution of highly precise radiation heat-transfer calculations.

In this chapter a detailed exposition of the radiation properties of surfaces will be presented. Section 2-1 defines the surface properties, shows their interrelation, and sets forth their dependence on wavelength and temperature. An extensive compilation of emittance data is included as an adjunct to this section. Next, in Section 2-2, consideration is given to the angular distribution of the radiation properties. Finally, analytical predictions of the surface properties, based on electromagnetic theory, are presented and discussed in Section 2-3.

2-1 Monochromatic and Total Radiation Properties

Monochromatic radiation properties.   Attention is first directed to the radiant energy that is emitted by a surface element at temperature T_s.   Suppose that the radiation is collected in the entire hemispherical space above the surface and resolved into its wavelength spectrum. The resulting spectral energy density e_λ may be represented by the lower curve in Fig. 2-1.   For purposes of comparison, the black-body

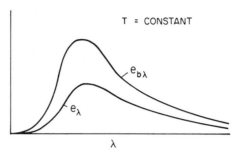

Fig. 2-1   Representative spectral distributions of emitted radiant energy.

energy spectrum $e_{b\lambda}$ corresponding to the same temperature is also shown in the figure.   Inasmuch as the black body is the most effective emitter of thermal radiation, it follows that e_λ cannot exceed $e_{b\lambda}$.

The monochromatic hemispherical emittance ϵ_λ is defined as the ratio of e_λ to $e_{b\lambda}$, where both quantities are taken at the same wavelength and the same temperature.   Thus

$$\epsilon_\lambda = \frac{e_\lambda}{e_{b\lambda}} \tag{2-1}$$

In terms of Fig. 2-1, ϵ_λ may be regarded as the ratio of the ordinates of the two curves at a given wavelength.   In general, the aforementioned ratio of ordinates may vary as a function of the wavelength.   Furthermore, if the e_λ and $e_{b\lambda}$ curves for another temperature level were to be examined, the resulting ratios of e_λ to $e_{b\lambda}$ may well differ from those at the first temperature level.   Therefore, in general, ϵ_λ is a function of both wavelength and temperature.

Next, consider a surface element onto which there impinges radiant energy from the entire hemispherical space above the surface.   The spectral energy density of the radiation incident per unit time and area may be denoted by H_λ.   The monochromatic hemispherical absorptance α_λ of the surface is then defined as

$$\alpha_\lambda = \frac{\text{energy absorbed/time-area-wavelength}}{H_\lambda} \qquad (2\text{-}2)$$

Both of the energy quantities appearing in the foregoing definition are taken at the same wavelength. In general, α_λ depends on both the wavelength and on the surface temperature.

Of the just-discussed incident radiation, a portion may be reflected back into the hemispherical space. This may be characterized by the monochromatic hemispherical reflectance ρ_λ, defined as

$$\rho_\lambda = \frac{\text{energy reflected/time-area-wavelength}}{H_\lambda} \qquad (2\text{-}3)$$

In common with the other properties, ρ_λ may depend both on the surface temperature and on the wavelength. For an opaque (i.e., nontransmitting) material, it is evident that the incident radiation is either absorbed or reflected, that is,

$$\alpha_\lambda + \rho_\lambda = 1 \qquad (2\text{-}4)$$

The monochromatic emittance and absorptance are related by Kirchhoff's law. Strictly speaking, Kirchhoff's law holds for monochromatic radiation in a particular direction and for each component of polarization. These qualifications have generally been passed over in most of the recent statements of Kirchhoff's law. Usually it is stated that

$$\alpha_\lambda = \epsilon_\lambda \qquad (2\text{-}5)$$

for the hemispherical properties. It can be demonstrated that this equality is strictly valid only when the incoming hemispherical radiation is unpolarized and uniformly distributed over all angles. In the subsequent discussion, these restrictions will not be imposed; however, it is well to remember that they exist.

By combining equations (2-4) and (2-5), there results

$$\rho_\lambda = 1 - \epsilon_\lambda \qquad (2\text{-}6)$$

Thus, of the three monochromatic radiation properties, only one is actually independent.

The qualitative variation of the monochromatic properties with wavelength is fairly well-established. For metals, the monochromatic emittance decreases with increasing wavelength. On the other hand, for electric nonconductors, there is a general tendency for the monochromatic emittance to increase with increasing wavelength; but owing to the presence of local emission bands in the infrared, the variation of ϵ_λ with λ is quite irregular.

Representative experimental results that support these trends are shown in Figs. 2-2 and 2-3. The curves actually correspond to radiation emitted in the direction normal to the surface, but the wavelength dependence is the same as for the hemispherical emittance. In the first of these figures, which is taken from the measurements of Dunkle, Gier, and co-workers (ref. 1), results are plotted for aluminum having three different surface finishes. For the bare metal surfaces, the emittance decreases more or less steadily with increasing wavelength throughout the entire range of infrared wavelengths. On the other hand, for the anodized surface, the behavior is altogether different.

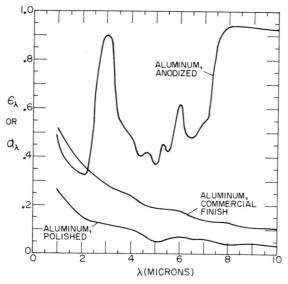

Fig. 2-2   Typical wavelength dependence of ϵ_λ and α_λ for metals.

This is because the anodizing process produces a relatively thick oxide coating on the surface. Since oxides behave like electric nonconductors, anodized aluminum displays some of the traits of a nonmetal, in particular, a high emittance in the far infrared. The thicker the anodic coating, the closer the approach to the behavior of a nonmetal.

It may be observed from Fig. 2-2 that the emittance of metals is quite low, especially in the infrared. Furthermore, by comparing the commercial-finish specimen with the polished one, it is seen that the effect of roughness and/or contaminants is to increase the emittance. Because of the ambiguity of such terms as polished, commercial finish, and anodized, it is unwise to suppose that the emittance values shown in the figure apply with high precision to other similarly described materials.

Figure 2-3 shows emittance results for three electric nonconductors, all of which appear white to the eye. The measurements were made by Sieber (ref. 2). In the visible wavelength range ($\lambda \sim 0.5\,\mu$), these surfaces have a high reflectance (low emittance) and therefore appear to be white. However, in the infrared range, the emittance is generally quite high. For example, the white enamel is very nearly a black body in the infrared. In fact, most white-appearing electric nonconductors are more or less black in the range of infrared wavelengths. As a rule,

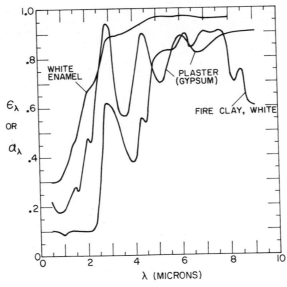

Fig. 2-3   Typical wavelength dependence of ϵ_λ and α_λ for electric nonconductors.

electric nonconductors are characterized by a high infrared emittance, very much higher than that of metals. In addition, the existence of a band structure as displayed in Fig. 2-3 is common for other electric nonconductors.

Further evidence of the foregoing general trends for metals and for electric nonconductors is contained in Table 2-1, which is taken from Eckert and Drake (ref. 3). These data are subject to the same note of caution with regard to surface description as was previously discussed. Additional tabulations of monochromatic radiation properties are included in the book-length compilation of Gubareff and co-workers (ref. 4).

Table 2-1 Monochromatic Emittance Values, ϵ_λ

Surface	\multicolumn{10}{c}{Wavelength (microns)}									
	0.50	0.60	0.95	1.8	2.1	3.6	4.4	5.4	8.8	9.3
Aluminum										
polished			0.26		0.17	0.08		0.05		0.04
oxidized						0.18		0.12		0.11
Duralumin		0.53								
Chromium		0.49	0.43		0.36	0.26		0.17		0.08
Copper										
polished			0.26		0.17	0.18		0.05		0.04
oxidized						0.77		0.83		0.87
Gold										
polished					0.03	0.03		0.02		0.02
Steel										
polished		0.45	0.37		0.23	0.14		0.10		0.07
Iron										
polished		0.45	0.35		0.22	0.13		0.08		0.06
cast, oxidized						0.76		0.66		0.63
galvanized, new	0.66	0.66	0.67	0.42						0.23
galvanized, dirty	0.89	0.89	0.89	0.90						0.28
galvanized,										
whitewashed	0.24	0.21	0.22	0.37						
Magnesium		0.30	0.26		0.23	0.18		0.13		0.07
Monel		0.43	0.29		0.16	0.10				
Molybdenum	0.55		0.43		0.18	0.11		0.08		0.06
Brass										
polished										0.05
oxidized										0.61
Paint										
lampblack		0.97	0.97				0.97		0.96	
red		0.74	0.59				0.70		0.96	
yellow	0.39	0.30					0.59		0.95	
white	0.18	0.14	0.16				0.77		0.95	
aluminum		0.45								
Silver										
polished	0.11		0.04		0.03	0.03		0.02		0.01
Platinum black	0.97		0.97		0.97	0.97	0.96	0.96	0.93	0.93
White paper		0.28	0.25			0.82				0.95
White marble		0.47				0.93				0.95
Graphite	0.78		0.73		0.64	0.54		0.49		0.41

The dependence of the monochromatic properties on temperature may now be discussed. In the case of metals, the monochromatic emittance increases with increasing electrical resistance and, correspondingly, with increasing temperature. In the far infrared, according to electromagnetic theory (to be discussed later) ϵ_λ varies as the square root of the electrical resistivity. For many pure metals, the latter quantity is approximately proportional to the absolute temperature T. Thus, for these materials, ϵ_λ is nearly proportional to $T^{1/2}$. The situation for the case of electric nonconductors is not quite as clear; however, there is evidence that the monochromatic emittance varies only slowly with temperature (refs. 5, 6).

Total radiation properties. The designation *total* is employed to describe radiation quantities that pertain to the entire range of wavelengths. In the case of hemispherically emitted radiation whose spectral energy density is e_λ (e.g., Fig. 2-1), the energy emitted per unit time and area in a wavelength band $d\lambda$ is $e_\lambda d\lambda$. For the entire range of wavelengths, this becomes

$$e = \int_0^\infty e_\lambda \, d\lambda = \int_0^\infty \epsilon_\lambda e_{b\lambda} \, d\lambda \qquad (2\text{-}7)$$

where equation (2-1) has been used to obtain the expression at the right. The quantity e is termed the emissive power; it represents the radiant energy emitted per unit time and area from a surface at temperature T_s into the hemispherical space above the surface. Pictorially speaking, e corresponds to the area under the e_λ curve of Fig. 2-1.

The expression on the right in equation (2-7) provides operational directions for determining e when the monochromatic emittance ϵ_λ of a particular surface is known. At each wavelength λ, one forms the product of ϵ_λ and the black-body spectral energy density $e_{b\lambda}$ corresponding to the surface temperature T_s. The latter is available from Planck's law, equation (1-4a). In this connection it may be observed from Fig. 1-2 or Table 1-2 that $e_{b\lambda}$ is negligibly small for certain values of λ. Thus in performing the integration indicated in equation (2-7) it is only necessary to know ϵ_λ in that range where $e_{b\lambda}$ is of significant magnitude.

For instance, at low or moderate surface temperatures, $e_{b\lambda}$ is of significant magnitude only in the infrared, and, correspondingly, only the infrared distribution of ϵ_λ is needed. For some materials—certain electric nonconductors, for example—ϵ_λ may be regarded as essentially independent of λ in the infrared. In such cases it follows that $e = \epsilon_\lambda \sigma T_s^4$. In other instances it may be permissible to approximate ϵ_λ by a succession of constant values for different wavelength bands.

Next, attention may be turned to the total hemispherical emittance ϵ. This is defined as the ratio of the emissive power e of a given surface to that of a black surface at the same temperature, that is,

$$\epsilon = \frac{e}{e_b} \tag{2-8}$$

or alternatively

$$\epsilon = \frac{\int_0^\infty \epsilon_\lambda e_{b\lambda}\, d\lambda}{\int_0^\infty e_{b\lambda}\, d\lambda} = \frac{\int_0^\infty \epsilon_\lambda e_{b\lambda}\, d\lambda}{\sigma T_s^{\,4}} \tag{2-9}$$

where equations (2-7) and (1-7) have been used. Since all the quantities appearing on the right-hand side of equation (2-9) are, in principle, dependent on the surface temperature, so also may ϵ.

The hemispherical emittance may be determined by experiment directly from its definition, equation (2-8), or by computation from equation (2-9), provided that the monochromatic emittance ϵ_λ is known. In the latter case, the detailed discussion pertaining to the evaluation of equation (2-7) continues to be relevant. When applying the hemispherical emittance to the computation of radiation interchange, it is customary to write equation (2-8) in the form

$$e = \epsilon \sigma T_s^{\,4} \tag{2-10}$$

It is a temptation to regard ϵ as uniquely determined for a specific type of material; for instance, to state that for nickel at temperature T_s there is a unique value of the emittance. However, it is well-established that the radiation properties are also strong functions of the surface condition and, in some cases, of the grain structure. It is for this reason that tabulations of hemispherical emittance values include, in addition to a statement of the type of material, some specification of the surface condition. Unfortunately, as previously noted, such descriptions are generally of a subjective type, thereby diminishing the utility of the tabulated data for future application.

The total hemispherical absorptance α is the fraction of the hemispherically incident radiation that is absorbed by the surface over all wavelengths. If $H_\lambda\, d\lambda$ is the hemispherically incident radiation per unit time and area in a wavelength band $d\lambda$, then the total absorbed and incident energy quantities are, respectively,

$$\int_0^\infty \alpha_\lambda H_\lambda\, d\lambda \quad \text{and} \quad \int_0^\infty H_\lambda\, d\lambda \tag{2-11}$$

Then, in accordance with the aforementioned definition, there follows

$$\alpha = \frac{\displaystyle\int_0^\infty \alpha_\lambda H_\lambda \, d\lambda}{\displaystyle\int_0^\infty H_\lambda \, d\lambda} \tag{2-12}$$

Inspection of equation (2-12) reveals that the absorptance α depends on two factors. First, there is the absorption characteristic of the surface as embodied in α_λ; second, there is the nature of the incident radiation as embodied in H_λ. Therefore, in general, the total absorptance cannot be regarded as being a property of the surface alone. This is in contrast to the fact, evidenced by equation (2-9), that the hemispherical emittance ϵ is a surface property. It would thus appear that ϵ and α are unrelated. Yet in a wide variety of engineering calculations it has long been customary to assume that $\alpha = \epsilon$. The conditions under which it is permissible to deduce α values from ϵ values will now be discussed.

There appear to be four cases where α and ϵ are related to each other in a simple manner. These will be treated successively in the subsequent paragraphs.

(1)   The gray body: ϵ_λ and α_λ are independent of wavelength. In this context the term *gray* is employed to connote the fact that the surface is completely unselective in its spectral characteristics. When ϵ_λ and α_λ are independent of λ, they can be respectively removed from under the integrals in equations (2-9) and (2-12). Consequently,

$$\epsilon = \epsilon_\lambda \qquad \alpha = \alpha_\lambda \tag{2-13}$$

Since $\alpha_\lambda = \epsilon_\lambda$ from Kirchhoff's law, it follows that

$$\alpha = \epsilon \tag{2-14}$$

for a gray body. In the foregoing equation, both α and ϵ correspond to the surface temperature T_s.

There are very few materials for which ϵ_λ and α_λ are constant over the entire range of wavelengths. In spite of this fact, many materials do qualify as gray bodies, at least approximately. To illustrate, suppose that nonnegligible values of $e_{b\lambda}$ and H_λ occur in the same finite range of wavelengths. Then, if ϵ_λ and α_λ are essentially constant in that finite range, the gray-body requirement is fulfilled and $\alpha = \epsilon$. The gray-body condition is likely to be violated when $e_{b\lambda}$ and H_λ lie in different wavelength ranges; for instance, when $e_{b\lambda}$ represents infrared emission and H_λ represents irradiation from a high-temperature source such as the sun.

(2)   Radiation from a black- or gray-body source (temperature T_i) incident on a surface whose temperature is T_s and whose monochromatic emittance ϵ_λ is independent of temperature in the range between T_i

and T_s. H_λ is now evaluated as $ke_{b\lambda}(T_i)$, where $k = 1$ for black-body irradiation and < 1 for gray-body irradiation. Then, returning with this to equation (2-12) and additionally substituting $\alpha_\lambda(T_s) = \epsilon_\lambda(T_s)$, one finds

$$\alpha = \frac{\displaystyle\int_0^\infty \epsilon_\lambda(T_s)e_{b\lambda}(T_i)\,d\lambda}{\displaystyle\int_0^\infty e_{b\lambda}(T_i)\,d\lambda} \qquad (2\text{-}15)$$

Noting that $\epsilon_\lambda(T_s) = \epsilon_\lambda(T_i)$ according to the stipulated conditions and then comparing with equation (2-9), it is evident that

$$\alpha = \epsilon(T_i) \qquad (2\text{-}16)$$

in which the notation $\epsilon(T_i)$ is used to indicate that ϵ is to be evaluated at temperature T_i.

(3)   Radiation from a black- or gray-body source (temperature $T_i = T_s$) incident on a surface whose temperature is also T_s. When this condition is introduced into equation (2-12) along with $\alpha_\lambda = \epsilon_\lambda$, one obtains

$$\alpha = \frac{\displaystyle\int_0^\infty \epsilon_\lambda(T_s)e_{b\lambda}(T_s)\,d\lambda}{\displaystyle\int_0^\infty e_{b\lambda}(T_s)\,d\lambda} \qquad (2\text{-}17)$$

After a comparison with equation (2-9), one sees that

$$\alpha = \epsilon(T_s) \qquad (2\text{-}18)$$

(4)   Radiation from a black- or gray-body source (temperature T_i) incident on a *metallic* surface whose temperature is T_s. Provided that T_i is low enough to exclude appreciable radiation in the visible and in the near infrared wavelength ranges, electromagnetic theory can be applied to evaluate α. This determination, carried out by Eckert (ref. 7), leads to the result

$$\alpha = \epsilon(\tilde{T}) \qquad \tilde{T} = \sqrt{T_i T_s} \qquad (2\text{-}19)$$

If no one of the preceding four conditions is fulfilled, then, strictly speaking, α cannot be determined from tabulated emittance information. Under such circumstances α might still be evaluated directly from equation (2-12), provided that the spectral distributions of H_λ and α_λ were available. When H_λ represents radiation from a black- or gray-body source, then the integration can be carried out for any surface for which the monochromatic absorptance is known. Calculations of this type have been performed by Sieber (ref. 2), the results of which are discussed later. However, in nonelementary configurations of non-black surfaces, H_λ is generally unknown. Consequently, for such systems α cannot be evaluated *a priori*. This is, in fact, the reason why recourse is so often made to the gray-body model as represented by equation (2-14).

The total hemispherical reflectance ρ is the fraction of the total hemispherically incident radiation that is reflected back into the hemispherical space above the surface. In terms of the incident spectral energy density H_λ and the monochromatic reflectance ρ_λ, the total reflected and incident radiant fluxes are readily represented as

$$\int_0^\infty \rho_\lambda H_\lambda \, d\lambda \quad \text{and} \quad \int_0^\infty H_\lambda \, d\lambda \qquad (2\text{-}20)$$

The ratio of these quantities represents the total hemispherical reflectance, that is,

$$\rho = \frac{\displaystyle\int_0^\infty \rho_\lambda H_\lambda \, d\lambda}{\displaystyle\int_0^\infty H_\lambda \, d\lambda} \qquad (2\text{-}21)$$

Although ρ_λ is a property of the surface, H_λ depends on the nature of the incident radiation; hence the total reflectance, like the total absorptance, is not itself a surface property.

For an opaque material, it is apparent that the incident radiation must either be absorbed or reflected. From this, there follows

$$\rho + \alpha = 1 \qquad (2\text{-}22)$$

Thus if α is known, then ρ may be regarded as known, and vice versa. As previously discussed, there are several cases in which α may be rigorously deduced from tabulated values of the emittance ϵ. As a result of the foregoing relationship, equation (2-22), ρ may also be deduced from ϵ for these cases. In particular, for the technically useful gray-body case, the combination of equations (2-14) and (2-22) yields

$$\rho = 1 - \epsilon \qquad (2\text{-}23)$$

In the great majority of engineering applications, ρ is calculated in accordance with equation (2-23).

For situations in which ρ cannot be deduced from ϵ, one may, in principle, return to the defining equation (2-21). When the incident radiation is from a black- or gray-body source, ρ can be evaluated without difficulty provided that the monochromatic reflectance ρ_λ is known. However, as was already discussed in connection with the computation of α, the incident flux H_λ is generally unknown in most systems of practical interest. Consequently, as was the case with the absorptance, the gray-body assumption is very frequently invoked to calculate the reflectance.

Data tabulation for total radiation properties.   Because of their essential role in the analysis of radiant interchange, extensive measurements of the total radiation properties of surfaces have been carried out. The original data are contained in a large number of papers distributed throughout the technical literature.   In recent years much of this information has been compiled and thereby made more accessible. The most extensive tabulation of radiation surface properties was assembled by Gubareff and co-workers (ref. 4).   This is a book-length compilation that cites over 300 literature sources.   Less extensive but important tabulations have been prepared by Hottel (ref. 8) and by Fishenden (ref. 9).

Study of the aforementioned compilations, especially that of Gubareff, provides an interesting insight into the present state of knowledge of the radiation properties.   In many of the tables, a range of values (taken from different investigators) is given for what is purported to be the same material with the same surface condition. This finding reinforces the note of caution that has been sounded relative to the precision that might be expected when applying the tabulated values.

For the convenience of readers of this book, an extensive tabulation of total emittance data has been prepared.   This information is contained in Table 2-2.   The table is divided into two parts, the first of

Table 2-2   Total Emittance Data
I. METALS

Surface	Temperature, °F	ϵ
Aluminum		
polished, 98 percent pure	400–1100	0.04–0.06
commercial sheet	200	0.09
rough plate	100	0.07
heavily oxidized	200–1000	0.20–0.33
Antimony		
polished	100–500	0.28–0.31
Bismuth		
bright	200	0.34
Brass		
highly polished	500	0.03
polished	100	0.07
dull plate	100–500	0.22
oxidized	100–500	0.46–0.56
Chromium		
polished sheet	100–1000	0.08–0.27

Surface	Temperature, °F	ϵ
Cobalt		
unoxidized	500–1000	0.13–0.23
Copper		
highly polished electrolytic	200	0.02
polished	100	0.04
slightly polished	100	0.12
polished, lightly tarnished	100	0.05
dull	100	0.15
black oxidized	100	0.76
Gold		
pure, highly polished	200–1100	0.02–0.035
Inconel		
X, stably oxidized	450–1600	0.55–0.78
B, stably oxidized	450–1750	0.32–0.55
X and B, polished	300–600	0.20
Iron and Steel		
mild steel, polished	300–900	0.14–0.32
steel, polished	100–500	0.07–0.10
sheet steel, ground	1700	0.55
sheet steel, rolled	100	0.66
sheet steel, strong rough oxide	100	0.80
steel, oxidized at 1100 °F	500	0.79
cast iron, with skin	100	0.70–0.80
cast iron, newly turned	100	0.44
cast iron, polished	400	0.21
cast iron, oxidized	100–500	0.57–0.66
iron, red rusted	100	0.61
iron, heavily rusted	100	0.85
wrought iron, smooth	100	0.35
wrought iron, dull oxidized	70–680	0.94
stainless, polished	100	0.07–0.17
stainless, after repeated heating and cooling	450–1650	0.50–0.70
Lead		
polished	100–500	0.05–0.08
gray, oxidized	100	0.28
oxidized at 390 °F	400	0.63
oxidized at 1100 °F	100	0.63
Magnesium		
polished	100–500	0.07–0.13
Manganin		
bright rolled	200	0.05

Table 2-2—*continued*

Surface	Temperature, °F	ϵ
Mercury		
pure, clean	100–200	0.10–0.12
Molybdenum		
polished	100–500	0.06–0.08
polished	1000–2000	0.11–0.18
filament	1000–5000	0.08–0.29
Monel		
after repeated heating and cooling	450–1650	0.45–0.70
oxidized at 1100 °F	400–1100	0.41–0.46
polished	100	0.17
Nickel		
polished	100–500	0.05–0.07
oxidized	100–500	0.35–0.49
wire	500–2000	0.10–0.19
Platinum		
pure, polished plate	400–1100	0.05–0.10
oxidized at 1100 °F	500–1000	0.07–0.11
electrolytic	500–1000	0.06–0.10
strip	1000–2000	0.12–0.14
filament	100–2000	0.04–0.19
wire	400–2500	0.07–0.18
Silver		
polished or deposited	100–1000	0.01–0.03
oxidized	100–1000	0.02–0.04
German silver,* polished	500–1000	0.07–0.09
Tin		
bright tinned iron	100	0.04–0.06
bright	100	0.06
polished sheet	200	0.05
Tungsten		
filament	1000–2000	0.11–0.16
filament	5000	0.39
filament, aged	100–6000	0.03–0.35
polished	100–1000	0.04–0.08
Zinc		
pure polished	100–500	0.02–0.03
oxidized at 750 °F	750	0.11
galvanized, gray	100	0.28
galvanized, fairly bright	100	0.23
dull	100–500	0.21

* German silver is actually an alloy of copper, nickel and zinc.

II. NONMETALS

Surface	Temperature, °F	ϵ
Asbestos		
board	100	0.96
cement	100	0.96
paper	100	0.93–0.95
slate	100	0.97
Brick		
red, rough	100	0.93
silica	1800	0.80–0.85
fireclay	1800	0.75
ordinary refractory	2000	0.59
magnesite refractory	1800	0.38
white refractory	2000	0.29
gray, glazed	2000	0.75
Carbon		
filament	1900–2600	0.53
lampsoot	100	0.95
Clay		
fired	200	0.91
Concrete		
rough	100	0.94
Corundum		
emery rough	200	0.86
Glass		
smooth	100	0.94
quartz glass (2 mm)	500–1000	0.96–0.66
pyrex	500–1000	0.94–0.75
Gypsum	100	0.80–0.90
Ice		
smooth	32	0.97
rough crystals	32	0.99
hoarfrost	0	0.99
Limestone	100–500	0.95–0.83
Marble		
light gray, polished	100	0.93
white	100	0.95
Mica	100	0.75
Paints		
aluminum, various ages and compositions	200	0.27–0.62
black gloss	100	0.90
black lacquer	100	0.80–0.93

Table 2-2—*continued*

Surface	Temperature, °F	ϵ
Paints—*continued*		
white paint	100	0.89–0.97
white lacquer	100	0.80–0.95
various oil paints	100	0.92–0.96
red lead	200	0.93
Paper		
white	100	0.95
writing paper	100	0.98
any color	100	0.92–0.94
roofing	100	0.91
Plaster		
lime, rough	100–500	0.92
Porcelain		
glazed	100	0.93
Quartz	100–1000	0.89–0.58
Rubber		
hard	100	0.94
soft, gray rough	100	0.86
Sandstone	100–500	0.83–0.90
Snow	10–20	0.82
Water		
0.1 mm or more thick	100	0.96
Wood		
oak, planed	100	0.90
walnut, sanded	100	0.83
spruce, sanded	100	0.82
beech	100	0.94
planed	100	0.78
various	100	0.80–0.90
sawdust	100	0.75

which deals with metals and the second with nonmetals. In the tabulation, the surface condition of the various substances is described by standard subjective terms. For those cases in which a temperature range is indicated, the corresponding range of ϵ is given.

Several general features of the tabulated data are worthy of note. Among the metals, it is seen that clean, polished surfaces are characterized by low emittances. Roughness and contamination increase the emittance, while the presence of a heavy oxide layer may produce emittance values that are comparable in magnitude with those of electric nonconductors. From a study of the table as well as from other information, it can be concluded that the total emittance of metallic surfaces increases with increasing temperature. Two factors contribute to this increase. First of all, at higher temperatures, a greater proportion of the emitted radiation is concentrated in the range of short wavelengths, where metals display their largest monochromatic emittance at any temperature. Second, the general level of the monochromatic emittance increases with increasing temperature.

As a group, electric nonconductors are characterized by high values of infrared emittance. The absence of sufficient data over a large temperature range makes it difficult to generalize as to the effect of temperature on the total emittance. On the basis of physical reasoning, it might be expected that materials which appear light to the eye (i.e., low values of ϵ_λ in the visible range) will experience a decrease in total emittance with increasing temperature. On the other hand, surfaces that appear dark to the eye may experience either an increase or a slight decrease in total emittance as the temperature level is raised.

In compilations similar to that of Table 2-2, some authors make a distinction between the total hemispherical emittance and the total normal emittance. The latter quantity pertains to radiation emitted in the direction normal to the surface. As will be demonstrated in Section 2-3, the hemispherical emittance for metals may be from 10 to 30 percent higher than the normal emittance. On the other hand, for electric nonconductors, the hemispherical value may be lower than the normal value by 5 percent or less.

No distinction between the hemispherical and normal emittance has been made in Table 2-2. It is the belief of the authors that the uncertainties that arise from differences in surface condition completely overshadow the distinction between hemispherical and normal. As a matter of interest, it may be noted that most of the data contained in the table are normal emittances.

Total absorptance and reflectance values can be deduced from the tabulated total emittance data provided that the gray-body condition or one of the other previously discussed special cases applies. Otherwise, alternative sources of information must be found. At the time of this writing, it is believed that the Gubareff compilation of spectral and solar absorptances and reflectances is the best available source.

Mention may also be made of total absorptance information that has been computed by Sieber for the case wherein the incident radiation originates at a black- or gray-body source at temperature T_i. The computations were performed by evaluating equation (2-12) with $H_\lambda = e_{b\lambda}(T_i)$ and with α_λ values measured by Sieber himself on various

surfaces at room temperature.   Sieber's results are reproduced in Fig.
2-4, in which the ordinate is the total absorptance corresponding to
normal incidence, while the abscissa is the source temperature T_i.   The
relationship between the normal and hemispherical absorptance is
similar to that previously indicated for the corresponding emittance
quantities.

     For aluminum (curve 8), α increases with increasing values of the
source temperature.   This is the expected behavior for metals.   On the
other hand, those electric nonconductors that appear light to the eye
are characterized by an absorptance which decreases with T_i.   Finally,
a dark nonconducting material such as roofing shingles (curve 7) shows

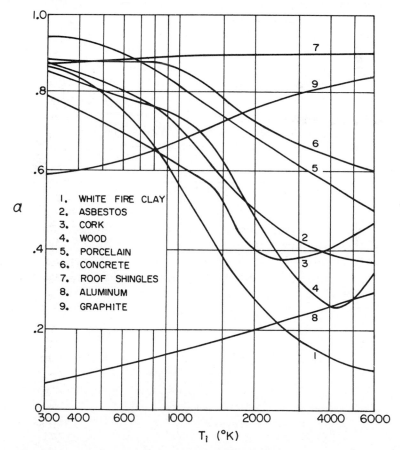

Fig. 2-4   Total absorptance of room-temperature surfaces corresponding to incident energy from a black- or gray-body source at
temperature T_i.

only a small variation of α over the entire temperature range.   These
trends are fully consistent with those that have been discussed for the
total emittance in earlier paragraphs.

2-2   Directional Radiation Properties

Radiant intensity and directional properties.   The preceding section
dealt with properties that describe the radiant energy emitted into, or
absorbed and reflected from, the entire hemispherical space above the
surface.   Consideration is now given to properties that characterize the
radiation traveling in specific angular directions.   Such properties are
alternatively designated as angular or directional.

In describing the directional transport, it is convenient to deal with
the radiant energy contained within a small solid angle $d\omega$.   This
naturally leads to the concept of the intensity.   This quantity has
already been discussed in Chapter 1 but is reintroduced here for convenience.   Let $d\Phi$ represent the radiant energy per unit time and unit
surface area that passes through a solid angle $d\omega$ inclined at an angle θ
with respect to the surface normal.   This situation is represented
pictorially in Fig. 2-5.   Then, with these, the intensity i is defined as

$$i = \frac{d\Phi}{d\omega \cos \theta} \tag{2-24}$$

For monochromatic radiation, a subscript λ is affixed to both $d\Phi$
and i, whereas for total radiation these subscripts are omitted.   In
general, the direction of the intensity is specified by the angles θ and ϕ
of a spherical coordinate system centered on the surface as shown in
Fig. 2-5.   The intensity distributions are related to the corresponding

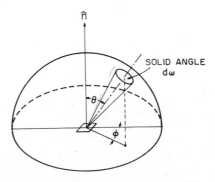

Fig. 2-5   Notation pertinent to the directional properties.

hemispherical radiation fluxes by integration over the hemisphere above the surface. For instance, the emissive power e is connected with the intensity of the emitted radiation as follows:

$$e = \int_\cap i \cos \theta \, d\omega \qquad (2\text{-}25)$$

in which the symbol $\cap$ denotes the hemisphere. Similar relationships apply for the incident and reflected radiant fluxes.

Before proceeding to the definitions of the directional radiation properties, attention may first be turned to the case of *diffuse* radiation, which embodies the most elementary of all directional characteristics. The term diffuse denotes *directional uniformity*. In particular, the intensity of the radiation leaving a diffusely emitting and diffusely reflecting surface is uniform in all angular directions. Similarly, radiation arriving with uniform intensity at a surface is diffusely distributed.

It has already been established in Chapter 1 that the black body is a diffuse emitter of radiant energy. Specifically, if i_b is the radiant intensity of a black body and e_b is its emissive power, then $e_b = \pi i_b$. The same relationship holds for the monochromatic case. Similarly, for any diffuse emitter with uniform intensity i and emissive power e, it follows that

$$e = \pi i \qquad (2\text{-}26)$$

In the case of a diffusely reflecting surface, the intensity of the reflected radiation is uniformly distributed irrespective of the nature of the incident radiation. In other words, regardless of whether the incident radiation arrives as a beam directed along the surface normal, or as a beam grazing the surface, or is uniformly distributed over the hemisphere, the radiation reflected from a diffuse surface is always of uniform intensity.

Thus the role of a diffusely reflecting surface is to obliterate the past history of the incident radiation. If all the participating surfaces in a given physical system are diffuse reflectors, there is no need to trace individual rays as they reflect and re-reflect, since their past history is obliterated at each surface contact. It is this feature that has made the diffusely reflecting surface such an attractive computational model. The hemispherically reflected radiant flux and the corresponding diffuse intensity are connected by equation (2-26).

For reflection, another elementary directional distribution may be singled out for discussion. This is the case of *specular* or mirrorlike reflection. Suppose that the incident beam is contained within a solid angle $d\omega_i$ inclined at an angle ψ with respect to the surface normal.

Then, for a specular surface, the reflected beam is contained within a solid angle $d\omega_r = d\omega_i$ situated on the opposite side of the normal and inclined to it at an angle $\theta = \psi$. This situation is shown pictorially in Fig. 2-6. The intensity of the reflected radiation is zero in all directions except at the specular angle $\theta = \psi$.

Consideration may now be given to the characterization of the directional radiation properties.

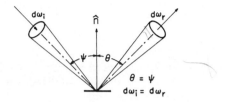

Fig. 2-6   Schematic representation of specular reflection.

The directional emittance and absorptance. The angular emittance is defined as the ratio of the emitted intensity in a particular direction to the intensity of black-body radiation. In general, for arbitrary surface conditions, the distribution of the emitted intensity i depends on the two angles θ and ϕ, Fig. 2-5. On the other hand, the black-body intensity i_b is uniform. Consequently, one can write

$$\epsilon(\theta, \phi) = \frac{i(\theta, \phi)}{i_b} \qquad (2\text{-}27)$$

This definition may be applied either to monochromatic or to total radiation.

For isotropic surfaces, there is no dependence on the angle ϕ and, correspondingly, $\epsilon = \epsilon(\theta)$. In the special case of diffuse emission, $i = \text{constant} = e/\pi$; furthermore, $i_b = e_b/\pi$. With these, equation (2-27) becomes

$$\epsilon(\theta, \phi) = \frac{e}{e_b} = \epsilon \qquad (2\text{-}28)$$

Equation (2-28) states that for a diffuse surface the directional emittance is *independent of angle* and is equal to the hemispherical emittance.

The directional emittance characteristics of real surfaces are revealed by measurements performed on a variety of materials by E. Schmidt and E. Eckert (ref. 10). The measured distributions are reproduced in Figs. 2-7 through 2-9, the first of which pertains to nonmetals and the last two to metals. The figures are in the form of polar diagrams in which the concentric circles are contours of constant ϵ and the radial lines are contours of constant angle θ.   The surfaces employed in the experiments were presumably isotropic and no dependence on the angle ϕ is indicated.   The temperatures of the various test surfaces were on the order of a few hundred degrees Fahrenheit.   The directional emittance results were determined from total radiation measurements.

An inspection of Fig. 2-7 shows that, for nonmetals, the directional emittance remains essentially uniform for inclination angles between 0° and 50–60°; then it drops off quite rapidly to zero.   In the region of near uniformity, the emittance values are in the range from 0.8 to 0.95, which is typical for electric nonconductors.

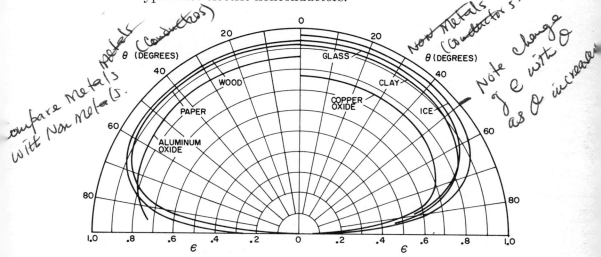

Fig. 2-7   Distribution of the total directional emittance for several electric nonconductors.

The directional emittance characteristics displayed by the metallic surfaces in Figs. 2-8 and 2-9 are somewhat different from those just discussed.   In the range of angles between 0° and 30–40°, ϵ remains quite uniform.   However, at larger angular inclinations ϵ increases quite sharply.   The percentage increase is more pronounced when the level of the emittance is low.   At angles approaching 90° the directional emittance should approach zero.   Such a behavior is displayed by the surfaces represented in Fig. 2-8.   It is expected that the surfaces represented in Fig. 2-9 would have also shown this same trend had measurements been made at larger angles of inclination.

The qualitative characteristics of the directional emittance distributions as presented in Figs. 2-7 through 2-9 are in general accord with the predictions of electromagnetic theory, as will be discussed in Section 2-3.   Interested readers may also wish to consult recent papers

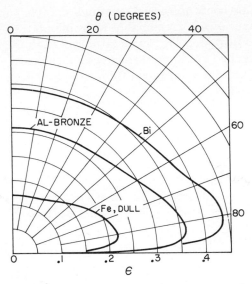

Fig. 2-8 Distribution of the total directional emittance for several metals.

by Edwards and Catton (ref. 11) and by Rolling and co-workers (ref. 12), where the effects of surface roughness and oxidation on the directional emittance of metals are investigated.

The directional absorptance is the fraction of the radiant energy incident on the surface from a specific direction that is absorbed. In accordance with Kirchhoff's law, the directional monochromatic absorptance and the directional monochromatic emittance are equal provided that the incident beam is uniformly polarized—that is,

$$\alpha_\lambda(\theta, \phi) = \epsilon_\lambda(\theta, \phi) \tag{2-29}$$

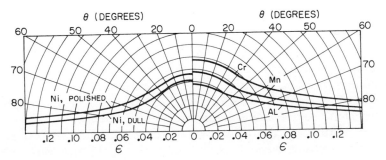

Fig. 2-9 Distribution of the total directional emittance for several metals.

Furthermore, the total directional absorptance $\alpha(\theta, \phi)$ can be deduced from the total directional emittance $\epsilon(\theta, \phi)$ for the same cases as were enumerated in connection with the corresponding hemispherical quantities.

The directional reflectance.   The directional emittance and directional absorptance pertain respectively to emitted and to incident beams of radiation.   On the other hand, the reflectance involves both incoming and outgoing radiation, each of which may have a directional orientation.   Consequently, there are three distinct directional reflectances that may be enumerated according to the manner both in which the surface is illuminated and the reflected radiation is collected.   These reflectances are in addition to the hemispherical reflectance (hemispherical incidence and hemispherical collection) discussed in Section 2-1.

(1)   Angular-hemispherical reflectance, ρ_{ah}. The incident radiation is a beam oriented at a specific angle relative to the surface normal.   The reflected radiation is collected over the entire hemispherical space.

(2)   Hemispherical-angular reflectance, ρ_{ha}. The incident radiation is hemispherically distributed, while the reflected radiation is collected in a specific angular direction.

(3)   Biangular reflectance, ρ_{ba}. The incident radiation is a beam oriented at a specific angle relative to the surface normal, and the reflected radiation is collected in a specific angular direction.

To facilitate the mathematical definitions of these quantities, it is convenient to employ the coordinates illustrated in Fig. 2-10.   The inclination angles of the incident and reflected beams with respect to the surface normal are denoted by ψ and θ, respectively.   The plane of incidence is a plane containing the incoming beam and the surface normal.   The angular orientation of the reflected beam with respect to

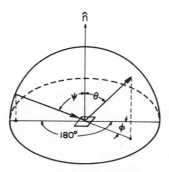

Fig. 2-10   Notation pertinent to the directional reflectance.

the plane of incidence is characterized by ϕ. Furthermore, the subscripts i and r will be affixed to describe quantities associated respectively with the incident and reflected radiant fluxes.

With these, the various reflectances can be represented in mathematical terms. First, for the angular-hemispherical reflectance, one has

$$\rho_{ah}(\psi) = \frac{de_{r,\,h}(\psi)}{de_i} \tag{2-30}$$

in which $de_{r,\,h}$ is the reflected radiant energy that is collected over the entire hemispherical space and de_i is the radiant energy contained in the incident beam. Both quantities are per unit time and unit surface area. Differentials are used in cognizance of the fact that only an infinitesimal amount of energy can be carried in a beam contained within an infinitesimal solid angle. In general, the magnitude of $de_{r,\,h}$ will depend on the angle of inclination ψ of the incoming beam.

Next, the hemispherical-angular reflectance is defined as

$$\rho_{ha}(\theta,\,\phi) = \frac{i_r(\theta,\,\phi)}{e_{i,\,h}/\pi} \tag{2-31}$$

In this expression i_r is the reflected intensity in the direction θ, ϕ and $e_{i,\,h}$ is the hemispherically incident radiant energy per unit time and unit surface area. In the case in which the hemispherically incident radiant energy is diffusely distributed and the surface is isotropic, it can be shown that (e.g., Torrance and Sparrow, ref. 13)

$$\rho_{ah}(\psi) = \rho_{ha}(\theta) \qquad \text{for } \theta = \psi \tag{2-32}$$

The equality of ρ_{ah} and ρ_{ha} is of practical importance, for, owing to the heated cavity reflectometer (Dunkle and co-workers, ref. 14), the latter is much more readily measured than is the former.

In formulating the definition of the biangular reflectance, cognizance is taken of two general classes of surfaces. First, consideration is given to surfaces that tend to be diffusing. Suppose that the incident beam (solid angle $d\omega_i$) carries an energy flux de_i; then, for a diffusing surface, the energy flux carried by a reflected beam confined within a solid angle $d\omega_r = d\omega_i$ will be of smaller order of magnitude, say d^2e_r. With a view toward achieving reasonable magnitudes for the biangular reflectance for such surfaces, there is defined

$$\rho_{ba}(\psi,\,\theta,\,\phi) = \frac{di_r(\psi,\,\theta,\,\phi)}{de_i} \tag{2-33}$$

in which

$$di_r = \frac{d^2e_r}{d\omega_r \cos\theta} \tag{2-33a}$$

A restriction that is sometimes appended to the definition of the bi-angular reflectance is that the solid angles of illumination and of collection, $d\omega_i$ and $d\omega_r$, be equal.

In general, d^2e_r depends on the angle of incidence ψ as well as on the reflection angles θ and ϕ. In the case of a perfectly diffusely reflecting surface, ρ_{ba} is constant over the entire range of reflection angles although it may depend on ψ.

For a surface that reflects strongly in the specular-ray direction, the foregoing definition of the biangular reflectance continues to apply for angular orientations other than the specular. However, for such a surface, the energy reflected in the specular direction is of the same order of magnitude as the energy carried by the incident beam; correspondingly, the term di_r appearing in the numerator of equation (2-33) should be replaced by i_r. It is therefore evident that ρ_{ba} will take on very large values in the specular-ray direction.

A knowledge of the hemispherical distribution of ρ_{ba} provides a complete specification of the directional reflection properties. Un-fortunately the experimental determination of ρ_{ba} is a lengthy and painstaking process. Only a few such investigations have appeared in the engineering literature. The hemispherical-angular reflectance pro-vides less information about the directional characteristics than does the biangular reflectance. However, the measurement of the former is substantially easier to perform (e.g., refs. 11 and 14).

Some insight into the directional reflection characteristics of real surfaces can be gained by inspection of available biangular results. The distributions measured by Eckert (ref. 7) are reproduced in Figs. 2-11 and 2-12. In these experiments, the incident beam impinged normal to the surface, and the reflected radiation was collected at various inclination angles θ in the plane of incidence. In terms of the notation of Fig. 2-10, $\psi = 0°$ and $\phi = 0°$. The radiation source was a black body at a temperature of approximately 550 °F. The reflected energy fluxes were measured on a total basis (i.e., not spectral). The biangular reflectances deduced from the data therefore correspond to total infra-red radiation.

Figures 2-11 and 2-12 are polar diagrams in which contours of constant ρ_{ba} are concentric circles and contours of constant angle θ are radial lines. On such a plot, a perfectly diffuse surface would be characterized by a circle, $\rho_{ba} = $ constant. On the other hand, a specular surface would be characterized by a sharp, high spike in the specular-ray direction, $\theta = 0°$.

It is apparent from an inspection of Fig. 2-11 that all the surfaces represented there display strong specular components. On the other

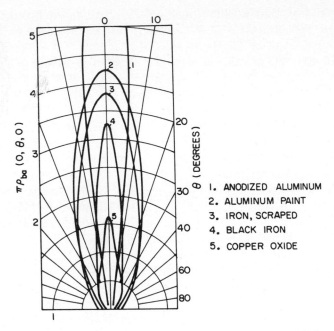

Fig. 2-11   Distribution of the total biangular reflectance in the
plane of incidence for various materials.

hand, the surfaces represented in Fig. 2-12 tend to be somewhat more
diffuse, especially wood.   On the basis of these figures, one might con-
clude that the reflectance distributions of real surfaces lie somewhere
between the diffuse and the specular limits.

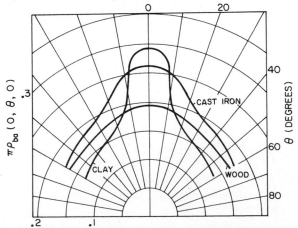

Fig. 2-12   Distribution of the total biangular reflectance in the
plane of incidence for various materials.

More recent experimental investigations, reported in refs. 15 and 16, have been concerned with the effects of wavelength and of surface roughness on the biangular reflectance. Data were collected for reflection angles distributed throughout the entire hemisphere above the surface. Representative results from ref. 16 are reproduced in Fig. 2-13. The figure pertains to a specific test surface with a mechanically measured roughness of 1 μ. The material under investigation was polycrystalline magnesium oxide ceramic, an electric nonconductor.

Figure 2-13 is subdivided into six individual graphs, each of which pertains to a different wavelength λ. The ordinate of each graph is the biangular reflectance $\rho_{ba}(\psi, \theta, \phi)$ normalized by the corresponding biangular reflectance in the specular-ray direction $\theta = \psi$, $\phi = 0°$. Alternatively, the ordinate also represents the reflected intensity in the direction θ, ϕ relative to the reflected intensity at the specular angle, both quantities being for a specific angle of incidence and a specific wavelength. The abscissa is the inclination angle θ. On the portion of the abscissa that runs to the right from 0 to 90°, data points are plotted corresponding to angles $\phi = 0$, 45, and 90°. On the other hand, data for $\phi = 180°$ are plotted on the portion of the abscissa running to the left from 0 to 90°. The open circles correspond to an incidence angle $\psi = 10°$, whereas the darkened points are for an incidence angle $\psi = 45°$.

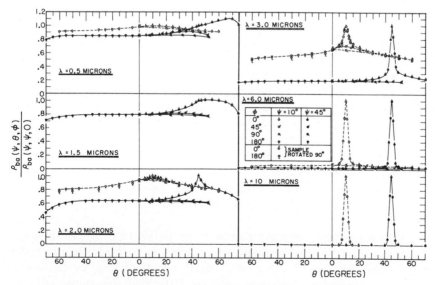

Fig. 2-13   Distribution of the monochromatic biangular reflectance for magnesium oxide ceramic, surface roughness = 1 micron (rms).

In a presentation such as Fig. 2-13, a diffuse surface is represented by a horizontal line with an ordinate value of unity. The other limiting case, specular reflection, is represented by a sharp spike at the specular angle, with a zero ordinate at all other angles of reflection.

By focusing attention on a specific angle of incidence, say $\psi = 10°$, and successively inspecting the results for the various wavelengths, the following trend becomes evident. At short wavelengths, the given surface appears to be very nearly a perfect diffuse reflector. At longer wavelengths, a specular peak emerges which grows increasingly stronger as the wavelength increases. Finally, at the largest wavelength at which data were collected, the surface appears specular although there is a nonspecular component that is too small to be shown on the scale of the figure. The same qualitative trend is in evidence at the larger angle of incidence, but the approach toward specular behavior is more rapid.

Experiments performed employing surfaces of various roughnesses reveal that increased roughness favors the more rapid approach toward diffuse characteristics at short wavelengths. Conversely, decreased roughness favors the more rapid approach toward specular conditions at long wavelengths.

The foregoing experimental data suggest that a given surface may approach either diffuse or specular behavior, depending on wavelength. A similar trend was cited by Birkebak and Eckert (ref. 15) on the basis of biangular reflectance measurements corresponding to near-normal incidence on metallic test surfaces. However, as will be described below, such behavior is restricted to the case of small angles of incidence.

A rather odd feature observable in Fig. 2-13 is that at the shortest wavelength and at the larger angle of incidence, the biangular reflectance displays a peak at an inclination angle θ different from the specular angle. The off-specular peak phenomenon for both metals and nonmetals has been investigated in detail by Torrance and Sparrow (ref. 17). Their results for aluminum surfaces are shown in Fig. 2-14. The figure consists of eight graphs, each of which corresponds to a specific value of the ratio of the mechanically measured surface roughness σ_m to the wavelength λ. The graphs are ordered according to increasing values of σ_m/λ.

The reflectance measurements were performed in the plane of incidence $\phi = 0°$. The ordinate of each graph is the biangular reflectance in the direction $\theta = \theta$, $\phi = 0°$ relative to the biangular reflectance in the specular direction, $\theta = \psi$, $\phi = 0°$. Both of the aforementioned quantities pertain to a specific angle of incidence ψ. All curves therefore attain an ordinate value of unity at the specular angle. The ordinate also represents the ratio of the intensity in the direction $\theta = $

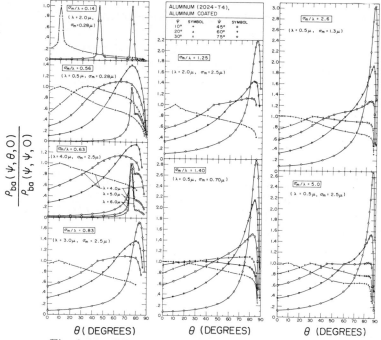

Fig. 2-14   Off-specular peak phenomenon for aluminum.

$\theta, \phi = 0°$ to the intensity in the specular direction, $\theta = \psi, \phi = 0°$. The abscissa is the angle of inclination θ of the reflected ray.   Results are shown for angles of incidence ψ ranging from 10 to 75°.

In this presentation, as in Fig. 2-13, a diffuse surface is represented by a horizontal line at unity ordinate, while a specular surface is represented by a sharp peak at the specular angle.   For small angles of incidence, it appears that the limiting cases of specular and diffuse reflection are respectively approached at small and large values of σ_m/λ. At intermediate and large angles of incidence, the reflectance distribution still approaches the specular limit at small σ_m/λ.   However, as σ_m/λ increases, there appears to be no tendency to approach the diffuse limit.   Instead, the reflected intensity distribution exhibits its maximum value at inclination angles *larger* than the specular angle.   The intensity at this off-specular maximum is, in some cases, three or four times the intensity at the specular angle.

The existence of off-specular peaks for a range of materials, including both metals and a nonmetal, has been demonstrated in ref. 17. Indeed, as cited there, a large number of earlier measurements had revealed the presence of such peaks.   The physical models that have been proposed to explain the off-specular peak phenomenon are reviewed in the reference.   Very recently, by employing geometrical optics,

Torrance (ref. 36) has formulated and evaluated an analytical model that predicts off-specular peaks.

The effect of surface roughness on the monochromatic reflectance distribution has also been investigated analytically by the application of electromagnetic theory. This work will be discussed briefly in Section 2-3.

2-3   Results of Electromagnetic Theory

For optically smooth surfaces that are physically and chemically clean, electromagnetic theory provides a prediction of the mono-chromatic reflectance in the specular direction. From this, with the aid of Kirchhoff's law, the magnitude and the directional distribution of the emittance can be inferred. The radiation properties of the afore-mentioned ideal surfaces are sometimes distinguished by the suffix *ivity* in lieu of the suffix *ance*.

Specular reflectance distribution.   Radiant energy either incident on or leaving from a surface can be resolved into two components of polarization.   One of these components is polarized parallel to the plane of incidence* and the other is polarized perpendicular to the plane of incidence.   Electromagnetic theory predicts the specular reflection characteristics for each component of polarization.   The specular re-flectance ρ^s corresponding to each component is defined as the ratio of the monochromatic intensities of that component in the reflected and incident beams.

The expressions for $\rho_{\perp}^{s}$ (perpendicular-polarized component) and $\rho_{//}^{s}$ (parallel-polarized component) constitute the well-known Fresnel law of reflection; these are derived in standard optics texts.   The resulting equations are

$$\rho_{\perp}^{s}(\theta) = \frac{a^2 + b^2 - 2a \cos \theta + \cos^2 \theta}{a^2 + b^2 + 2a \cos \theta + \cos^2 \theta} \tag{2-34a}$$

$$\rho_{//}^{s}(\theta) = \rho_{\perp}^{s}(\theta) \frac{a^2 + b^2 - 2a \sin \theta \tan \theta + \sin^2 \theta \tan^2 \theta}{a^2 + b^2 + 2a \sin \theta \tan \theta + \sin^2 \theta \tan^2 \theta} \tag{2-34b}$$

The angle θ is the specular-ray direction corresponding to an incident beam whose angle of inclination is $\psi = \theta$.   The quantities a and b are related to the angle of reflection θ and to the refractive index n and

---

* As previously indicated, the plane of incidence includes the incident ray and the surface normal.

the extinction coefficient k of the material on which the radiation is incident.

$$2a^2 = \sqrt{(n^2 - k^2 - \sin^2 \theta)^2 + 4n^2k^2} + (n^2 - k^2 - \sin^2 \theta) \quad (2\text{-}35a)$$

$$2b^2 = \sqrt{(n^2 - k^2 - \sin^2 \theta)^2 + 4n^2k^2} - (n^2 - k^2 - \sin^2 \theta) \quad (2\text{-}35b)$$

The Fresnel relations as embodied by equations (2-34) and (2-35) pertain to the case in which the medium adjacent to the reflecting surface has a refractive index of unity and an extinction coefficient of zero (vacuum and gases). Furthermore, it should be noted that the Fresnel relations are derived for monochromatic radiation. Therefore, strictly speaking, a subscript λ ought to be appended to $\rho_\perp{}^s$ and $\rho_{//}{}^s$. However, this subscript is omitted in order to avoid complicating an already cumbersome notation. All the radiation properties dealt with in this section are to be regarded as monochromatic unless otherwise stated.

When taken together, the optical constants n and k comprise the complex index of refraction N, that is,

$$N = n - ik = n(1 - ik_0) \quad (2\text{-}36)$$

where

$$k_0 = \frac{k}{n} \quad (2\text{-}36a)$$

Since both forms of equation (2-36) are found in the literature, care must be exercised to distinguish between k and k_0. For a dielectric material (electric nonconductor), k is essentially zero;* consequently, N is a real quantity. On the other hand, for both metals and semiconductors, N must be regarded as complex, that is, $k \neq 0$. The terminology optical *constants* is, in itself, somewhat misleading, for both n and k depend on wavelength and temperature. For instance, for metals, these parameters increase sharply with wavelength in the infrared range.

If the incident radiation is unpolarized, then the two components of polarization in the incident beam are of equal intensity. In this case it is readily verified that the monochromatic specular reflectance is the average of $\rho_\perp{}^s$ and $\rho_{//}{}^s$, that is,

$$\rho^s(\theta) = \tfrac{1}{2}[\rho_\perp{}^s(\theta) + \rho_{//}{}^s(\theta)] \quad (2\text{-}37)$$

---

* In spectral regions in the neighborhood of resonance points of bound electrons, atoms, and molecules, crystalline dielectrics possess relatively larger values of the extinction coefficient (ref. 18).

Furthermore, for an opaque material, the incident beam is either absorbed or reflected; consequently, the monochromatic directional absorptance $\alpha(\theta)$ follows as

$$\alpha(\theta) = 1 - \rho^s(\theta) = 1 - \tfrac{1}{2}[\rho_\perp{}^s(\theta) + \rho_{//}{}^s(\theta)] \qquad (2\text{-}38)$$

To illustrate the predicted distribution of the specular reflectance, equations (2-34), (2-35), and (2-37) have been successively evaluated for $0 \leqslant \theta \leqslant 90°$ for the case $N = 4 - ik$, where k ranges from 0 to 20. These results are displayed in Fig. 2-15, in which ρ^s is the ordinate and θ is the abscissa. The curve parameter is the extinction coefficient k.

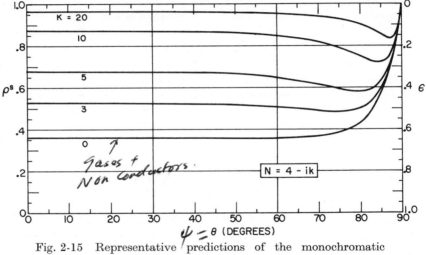

Fig. 2-15   Representative predictions of the monochromatic emittance and the specular reflectance from electromagnetic theory.

$N = $ complex index of refraction

In all cases there is a range of angular inclinations for which the reflectance is essentially independent of angle. In addition, the specular reflectance approaches unity as θ approaches $90°$ (grazing incidence), regardless of the value of k.

From a further inspection of the figure, it is seen that all curves display a minimum value at some moderately large value of θ. However, in the case of the dielectric ($k = 0$), the dip is essentially imperceptible. Metallic surfaces are characterized by high values of the specular reflectance and, for these, the minimum is sharply defined and occurs at large angles of inclination.

The expressions for the specular reflectance take on a particularly

simple form for the case of normal incidence.   By specializing equations
(2-34b) and (2-35) for $\theta = 0°$, it is seen that

$$\rho_{||}{}^s(0) = \rho_\perp{}^s(0) \qquad\qquad (2\text{-}39)$$

and $\qquad\qquad a = n \qquad b = k$

With these, the normal specular reflectance follows from equation
(2-37) as

$$\rho^s(0) \equiv \rho_n{}^s = \frac{(n-1)^2 + k^2}{(n+1)^2 + k^2} \qquad\qquad (2\text{-}40)$$

The subscript n is employed to denote the direction along the normal.
For the special case of dielectric materials, this becomes

$$\rho_n{}^s = \left(\frac{n-1}{n+1}\right)^2 \qquad\qquad (2\text{-}41)$$

If the Fresnel equations are to be used for the prediction of the
specular reflection characteristics of specific surfaces, then the corresponding optical constants n and k of the material must be available.
Information on these parameters may be found in various handbooks
and texts, as well as in research papers; for instance, refs. 19 to 27.
Unfortunately there is a degree of uncertainty in some of the published
values of the optical constants.   For example, in the case of metals, the
presence of oxide layers and surface contaminants may significantly
affect the measured results.   For electric nonconductors, which always
have some degree of transparency and related subsurface reflections,
the grain structure of the material may also influence the data.
    At long wavelengths, the reflectance of *metals* is controlled primarily
by the electrical resistivity of the material.   From electromagnetic
theory, it can be demonstrated that for this condition the optical
constants are given by

$$n = \sqrt{\frac{30\lambda}{r}}, \qquad k_0 = \frac{k}{n} = 1 \qquad\qquad (2\text{-}42)$$

where λ is the wavelength in centimeters (10^4 μ) and r is the electrical
resistivity in ohm-centimeters.   The corresponding normal specular
reflectance follows from equation (2-40) as

$$\rho_n{}^s = \frac{2n^2 - 2n + 1}{2n^2 + 2n + 1} \simeq 1 - 0.365\sqrt{\frac{r}{\lambda}} + 0.0464\frac{r}{\lambda} \qquad\qquad (2\text{-}43)$$

where the approximating polynomial is due to Schmidt and Eckert (ref. 10). When the last term, $0.0464r/\lambda$, is omitted, the remaining two-term representation is frequently identified as the Hagen-Rubens equation (ref. 5).

The theoretical results discussed so far were derived for optically smooth surfaces. Practical engineering surfaces rarely achieve this degree of smoothness. Various authors have pursued the extension of classical electromagnetic theory in an effort to account for surface roughness. Two basic approaches have been employed. The first involves the application of diffraction theory to a statistically rough surface, while the second seeks the solution of Maxwell's equations corresponding to highly complex boundary conditions. These approaches are exemplified respectively by the papers of Davies (ref. 28) and of Rice (ref. 29). Furthermore, recent analytical contributions have been reviewed in a comprehensive text by Beckmann and Spizzichino (ref. 30).

Brief mention will be made here of the results of Davies. The roughness of the surface was characterized in terms of a gaussian distribution with a root-mean-square value σ, and it was assumed that the peak-to-peak spacing is sufficiently large so that no interreflections occur. For the condition where $\sigma/\lambda \ll 1$, the monochromatic specular reflectance is given by

$$\frac{\rho^s(\theta)}{[\rho^s(\theta)]_0} = \exp\left[-\left(\frac{4\pi\sigma\cos\theta}{\lambda}\right)^2\right] \tag{2-44}$$

in which the $[\rho^s(\theta)]_0$ denotes the specular reflectance of an optically smooth surface. The validity of equation (2-44) has been established by a number of experimental investigations (refs. 15, 16, 31, 32). Recently the Davies theory was extended by Porteus (ref. 33) to include roughness distributions other than the gaussian.

Emittance distribution. Owing to the fact that Kirchhoff's law applies directionally and monochromatically for each component of polarization, one can write

$$\epsilon_\perp(\theta) = \alpha_\perp(\theta) = 1 - \rho_\perp{}^s(\theta) \tag{2-45a}$$

$$\epsilon_{//}(\theta) = \alpha_{//}(\theta) = 1 - \rho_{//}{}^s(\theta) \tag{2-45b}$$

Furthermore, the directional emittance of the mixed radiation is the average of $\epsilon_\perp$ and $\epsilon_{//}$, that is,

$$\epsilon(\theta) = \tfrac{1}{2}[\epsilon_\perp(\theta) + \epsilon_{//}(\theta)] \tag{2-46}$$

Although the subscript λ is omitted for notational simplicity, the fore-going equations, as well as those that follow, are to be regarded as applying monochromatically.

The predictions of electromagnetic theory for $\rho_{\perp}{}^{s}$ and $\rho_{//}{}^{s}$ can be employed to provide an expression for the directional emittance. Upon substituting equations (2-34) into equations (2-45) and (2-46), one obtains

$$\epsilon(\theta) = \tfrac{1}{2}\epsilon_{\perp}(\theta)\left[1 + \frac{a^2 + b^2 + \sin^2 \theta}{\cos^2 \theta \, (a^2 + b^2 + 2a \sin \theta \tan \theta + \sin^2 \theta \tan^2 \theta)}\right]$$

$$(2\text{-}47)$$

where

$$\epsilon_{\perp}(\theta) = \frac{4a \cos \theta}{a^2 + b^2 + 2a \cos \theta + \cos^2 \theta} \qquad (2\text{-}48)$$

The quantities a and b are related to the optical constants and to the angle of emission θ by equations (2-35). If n and k are known, equation (2-47) provides a prediction of the directional distribution of the mono-chromatic emittance of an optically smooth surface.

A qualitative picture of the predicted directional distribution can be gleaned from Fig. 2-15, where reference is now made to the right-hand ordinate. In the present context, the abscissa parameter θ is the angle of emission. For an electric nonconductor ($k = 0$), the emittance appears to be essentially uniform over the range of angles between $0°$ and 60–$70°$; then it drops sharply to zero. In the former range of angles, the surface is very nearly a diffuse emitter. These character-istics are in excellent agreement with those of the experimental data shown in Fig. 2-7.

With increasing values of k, the directional distribution displays a maximum at a moderately large value of θ and then returns to zero at $\theta = 90°$. As the level of the emittance decreases, the relative magni-tude of the maximum increases and its location shifts to angular positions close to $90°$. For instance, for the uppermost curve of the figure, the peak emittance value is more than four times the normal emittance. These trends are corroborated by the experimental data for metallic surfaces, Figs. 2-8 and 2-9.

The hemispherical emittance ϵ can be determined from the direc-tional emittance by integrating over the hemisphere. By introducing the definitions of ϵ and $\epsilon(\theta)$ into equation (2-25), there follows

$$\epsilon = \frac{1}{\pi} \int_{\Omega} \epsilon(\theta) \cos \theta \, d\omega \qquad (2\text{-}49)$$

With this, the prediction of electromagnetic theory for $\epsilon(\theta)$, equation (2-47), can be employed to provide a prediction for ϵ. In the case of an electric nonconductor ($k = 0$), the execution of the integration yields (ref. 18)

$$\frac{\epsilon}{\epsilon_n} = \frac{1}{2}\left[\frac{2}{3} + \frac{1}{3n} + \frac{n(n+1)^2(n^2-1)^2}{2(n^2+1)^3}\ln\left(\frac{n+1}{n-1}\right)\right.$$
$$\left. + \frac{n^2(n+1)(n^2+2n-1)}{(n^2+1)^2(n-1)} - \frac{4n^3(n^4+1)}{(n^2+1)^3(n-1)^2}\ln n\right] \qquad (2\text{-}50)$$

in which ϵ_n is the emittance in the direction normal to the surface

$$\epsilon_n = \frac{4n}{(n+1)^2} \qquad (2\text{-}51)$$

In the case of metals ($k \neq 0$), the integration can be carried out in closed form only when n and k_0 are sufficiently large. Under this condition, there results (ref. 18)

$$\epsilon = \frac{1}{2}\left[8n - 8n^2\ln\left(\frac{1+2n+n^2+n^2k_0^2}{n^2+n^2k_0^2}\right)\right.$$
$$+ \frac{8n^2(1-k_0^2)}{k_0}\tan^{-1}\left(\frac{k_0}{1+n+nk_0^2}\right)$$
$$+ \frac{8}{n(1+k_0^2)} - \frac{8\ln(1+2n+n^2+n^2k_0^2)}{n^2(1+k_0^2)^2}$$
$$\left. + \frac{8(1-k_0^2)}{n^2k_0(1+k_0^2)^2}\tan^{-1}\left(\frac{nk_0}{1+n}\right)\right] \qquad (2\text{-}52)$$

When n and k_0 are not large enough to permit the approximations leading to the closed-form solution, then numerical integration of equation (2-49) is required. The corresponding normal emittance is

$$\epsilon_n = \frac{4n}{(n+1)^2 + k^2} \qquad (2\text{-}53)$$

With the foregoing, it is possible to evaluate the ratio of the hemispherical to the normal emittance. This ratio has practical significance inasmuch as it is the normal emittance that is more often determined by experiment, whereas it is the hemispherical emittance that is usually needed in the computation of the radiant interchange. The ϵ/ϵ_n ratio is presented in Fig. 2-16 as a function of the refractive index n for parametric values of the extinction coefficient $k_0 = k/n$. The figure is reproduced from ref. 18. The curve for $k_0 = 0$ corresponds to equation (2-50), while the solid curves for $k_0 > 0$ were determined by numerical integration. Within the scale of the figure, these same

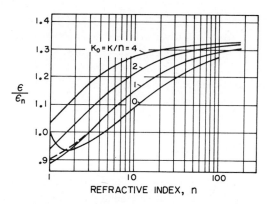

Fig. 2-16   Ratio of the hemispherical to the normal emittance as predicted by electromagnetic theory.

curves are representative of equations (2-52) and (2-53) except for the short, dashed segment for the case of $k_0 = 1$.   Within the range of n and k values that are appropriate to real materials, the ϵ/ϵ_n ratio is greater than one for metals and less than one for dielectrics.   For the latter materials, ϵ is only slightly less than ϵ_n.

The information contained in Fig. 2-16 can be rephrased into a more useful form by noting, in accordance with equation (2-53), that the refractive index n is simply related to the normal emittance ϵ_n for parametric values of k.   Consequently, one may replace n in favor of ϵ_n as abscissa variable in the ϵ/ϵ_n plot.   Such a presentation is made in Fig. 2-17.   For electric nonconductors, the normal emittance ϵ_n is typically greater than 0.6; correspondingly, $0.95 \leqslant \epsilon/\epsilon_n \leqslant 1$.   On the other hand, for highly polished metals, $\epsilon_n \leqslant 0.10$ and $1.1 \leqslant \epsilon/\epsilon_n \leqslant 1.32$.

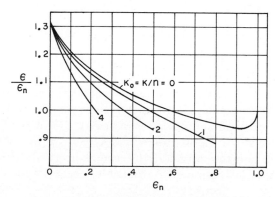

Fig. 2-17   Ratio of the hemispherical to the normal emittance as a function of the normal emittance.

Experimental values of ϵ/ϵ_n for a variety of materials are reported in ref. 10 along with theoretical curves for $k_0 = 0$ and $k_0 = 1$.* These results are also reproduced in the text of Eckert and Drake (ref. 3). The surface finishes of the metals were such that $0.04 \leqslant \epsilon_n \leqslant 0.34$; while for the nonconductors, $0.76 \leqslant \epsilon_n \leqslant 0.96$. The experimental results pertain to total infrared radiation, while the predictions of electromagnetic theory are for monochromatic radiation. Notwithstanding this, the measured values of ϵ/ϵ_n for dielectrics and metals, respectively, fall quite close to the $k_0 = 0$ and $k_0 = 1$ curves of the theory.

It is well to reiterate that the theoretical values of ϵ/ϵ_n for metals apply to oxide-free, optically smooth surfaces. The presence of oxides or roughness will normally decrease the difference between the hemispherical and the normal emittance.

In view of the possible uncertainty or unavailability of the optical constants n and k, there is a motivation to seek predictive expressions for the emittance that involve material properties that are known with greater reliability. Analytical and experimental study along these lines has been primarily concerned with the normal emittance of electric conductors. The simplest representation is the Hagen-Rubens relation† for the monochromatic emittance, which follows from equation (2-43) as

$$\epsilon_n = 0.365\sqrt{\frac{r}{\lambda}} - 0.0464\frac{r}{\lambda} \tag{2-54}$$

where r and λ are the electric resistivity and the wavelength as previously discussed. Schmidt and Eckert have integrated equation (2-54) over all wavelengths, thereby deriving an expression for the *total normal emittance*

$$\epsilon_n = 0.576\sqrt{rT_s} - 0.124rT_s \tag{2-55}$$

in which the numerical constants require that the surface temperature T_s be introduced in degrees Kelvin when r is in ohm-centimeters.

Although the Hagen-Rubens model is strictly valid for very long wavelengths, it has provided accurate results for *some* metals when λ is as low as one or two microns and is believed to be a good approximation for most polished metals for $\lambda > 5\ \mu$ (refs. 5, 6). As far as the aforementioned shorter wavelengths are concerned, this finding is regarded as fortuitous, a state of affairs that has directed attention to

* The $k_0 = 1$ curve appears to fall slightly higher than that of Fig. 2-17.
† The last term is due to Schmidt and Eckert (ref. 10).

other, more sophisticated models. In this connection, recent consideration has been extended to the Drude-Roberts theory, which postulates
the existence of two sets of charge carriers within the metal. Experiments performed with a variety of metals have explored the utility
of this model both for correlating data and for predicting results at
other operating conditions (refs. 34, 35).

REFERENCES

1. R. V. Dunkle, J. T. Gier, and co-workers, Snow characteristics project,
   progress report. University of California, Berkeley, June 1953.

2. W. Sieber, Zusammensetzung der von Werk- und Baustoffen zurückgewor-
   fenen Wärmestrahlung. *Z. Tech. Physik*, **22**, 130–135 (1941).

3. E. R. G. Eckert and R. M. Drake, Jr., *Heat and Mass Transfer*, McGraw-Hill,
   New York, 1959.

4. G. G. Gubareff, J. E. Janssen, and R. H. Torborg, *Thermal Radiation Properties Survey*, Honeywell Research Center, Minneapolis, Minn., 1960.

5. R. V. Dunkle, Thermal radiation characteristics of surfaces. In *Theory and
   Fundamental Research in Heat Transfer* (J. A. Clark, ed.), Pergamon Press,
   New York, 1963.

6. N. W. Snyder, Radiation in metals. *Trans. ASME*, **76**, 541–548 (1954).

7. E. Eckert, Messung der Reflexion von Wärmestrahlen an technischen
   Oberflachen. *Forsch. Gebiete Ingenieurw*, **7**, 265–270 (1936).

8. H. C. Hottel, Normal total emissivity of various surfaces. Table A-23 in
   Heat Transmission (W. H. McAdams, ed.), McGraw-Hill, New York, 1954.

9. J. R. Singham, Tables of emissivity of surfaces. *Intern. J. Heat Mass
   Transfer*, **5**, 67–76 (1962).

10. E. Schmidt and E. Eckert, Über die Richtungsverteilung der Wärmestrah-
    lung. *Forsch. Gebiete Ingenieurw*, **6**, 175–183 (1935).

11. D. K. Edwards and I. Catton, Radiation characteristics of rough and oxidized
    metals. In *Advances in Thermophysical Properties at Extreme Tempera-
    tures and Pressures*, pp. 189–199. The American Society of Mechanical
    Engineers, New York, 1965.

12. R. E. Rolling, A. I. Funai, and J. R. Grammer, Investigation of the effect of
    surface condition on the radiant properties of metals. Technical Report
    No. AFML-TR-64-363, Lockheed Missiles and Space Co., Palo Alto, Calif.,
    November 1964.

13. K. E. Torrance and E. M. Sparrow, Discussion of: Effects of roughness of
    metal surfaces on angular distribution of monochromatic radiation. *J.
    Heat Transfer*, **C87**, 93 (1965).

14. R. V. Dunkle and co-workers, Heated cavity reflectometer for angular
    measurements. In *Progress in International Research on Thermophysical
    and Transport Properties*, pp. 541–562. The American Society of Mechani-
    cal Engineers, New York, 1962.

15. R. C. Birkebak and E. R. G. Eckert, Effects of roughness of metal surfaces on angular distribution of monochromatic radiation. *J. Heat Transfer*, **C87**, 85–94 (1965).

16. K. E. Torrance and E. M. Sparrow, Biangular reflectance of an electric non-conductor as a function of wavelength and surface roughness. *J. Heat Transfer*, **C87**, 283–292 (1965).

17. K. E. Torrance and E. M. Sparrow, Off-specular peaks in the directional distribution of reflected thermal radiation. *J. Heat Transfer*, **C88**, 223–230 (1966).

18. R. V. Dunkle, Emissivity and inter-reflection relationships for infinite parallel specular surfaces. In *Symposium on Thermal Radiation of Solids* (S. Katzoff, ed.), pp. 39–44, *NASA SP-55* (1965).

19. Optics. Section 6 in *American Institute of Physics Handbook* (D. E. Gray, ed.), 2nd ed., McGraw-Hill, New York, 1963.

20. E. E. Bell, The optical constants and their measurement. In *Handbuch der Physik*, vol. 25, part 2, Springer-Verlag, Berlin (in press).

21. Anon., *International Critical Tables*, vol. 5, McGraw-Hill, New York, 1929.

22. P. W. Kruse, L. D. McGlauchlin, and R. B. McQuistan, *Elements of Infrared Technology*, Wiley, New York, 1962.

23. W. S. Martin, Optical constants and spectral emissivities at high temperatures. In *Radiative Heat Transfer from Solid Materials* (H. H. Blau, Jr. and H. Fischer, eds.), Macmillan, New York, 1962.

24. J. R. Beattie and G. K. T. Conn, Optical constants of metals in the infrared. *Phil. Mag.*, **46**, Ser. 7, 989–1001 (1955).

25. S. Roberts, Interpretation of optical constants of metal surfaces. *Phys. Rev.*, **100**, 1667–1671 (1955).

26. S. Roberts, Optical properties of nickel and tungsten and their interpretation according to Drude's formula. *Phys. Rev.*, **114**, 104–115 (1959).

27. S. Roberts, Optical properties of copper. *Phys. Rev.*, **118**, 1509–1518 (1960).

28. H. Davies, The reflection of electromagnetic waves from a rough surface. *Proc. Inst. Elec. Eng.*, **101**, 209–214 (1954).

29. S. O. Rice, Reflection of electromagnetic waves from slightly rough surfaces. *Pure Appl. Math.*, **4**, 351–378 (1951).

30. P. Beckmann and A. Spizzichino, *The Scattering of Electromagnetic Waves from Rough Surfaces*, Macmillan, New York, 1963.

31. H. E. Bennett and J. O. Porteus, Relation between surface roughness and specular reflection at normal incidence. *J. Opt. Soc. Am.*, **51**, 123–129 (1961).

32. H. E. Bennett, Specular reflectance of aluminized ground glass and the height distribution of surface irregularities. *J. Opt. Soc. Am.*, **53**, 1389–1394 (1963).

33. J. O. Porteus, Relation between the height distribution of a rough surface and the reflectance at normal incidence. *J. Opt. Soc. Am.*, **53**, 1394–1402 (1963).

34. R. A. Seban, The emissivity of transition metals in the infrared. *J. Heat Transfer*, **C87**, 173–176 (1965).

35. D. K. Edwards and N. B. deVolo, Useful approximations for spectral and total emissivity of smooth bare metals. In *Advances in Thermophysical Properties at Extreme Temperatures and Pressures*, pp. 174–188, The American Society of Mechanical Engineers, New York, 1965.

36. K. E. Torrance and E. M. Sparrow, Theory for off-specular reflectance from roughened surfaces. *J. Opt. Soc. Am.*, **57**, 1105–1114 (1967).

PROBLEMS

2-1.   For a surface with wavelength-dependent ϵ_λ, discuss whether the maximum of the e_λ distribution occurs at the same λ as the maximum of the $e_{b\lambda}$ distribution.   Both e_λ and $e_{b\lambda}$ correspond to the same temperature T.   Hint: Examine the expression for $de_\lambda/d\lambda$.

2-2.   Plot the distribution of e_λ as a function of λ for anodized aluminum at 350°F.   At which wavelength does e_λ take on its maximum value?   How does this compare with the λ at which $e_{b\lambda}$ is a maximum?   Assume $n = 1$.

2-3.   Two plates, one with a polished aluminum surface and the other painted white, are exposed to solar radiation.   Suppose that each surface is in thermal equilibrium, such that the absorbed solar radiation is exactly balanced by the emitted radiation.   Which surface has a higher equilibrium temperature?   Estimate the numerical value of the equilibrium temperature of each plate.   The incident solar radiation is 400 Btu/hr-ft^2.   Neglect convection.

2-4.   By employing Simpson's rule or some other convenient numerical integration method, compute ϵ for a white painted surface at $T = 500°R$.   Take $e_{b\lambda}$ and ϵ_λ from Table 1-2 and Figure 2-3, respectively.

2-5.   The ϵ_λ distribution for some electric nonconductors can be approximated as

$$\epsilon_\lambda = \epsilon_1, \qquad 0 \leqslant \lambda < \lambda^*$$
$$\epsilon_\lambda = \epsilon_2, \qquad \lambda^* < \lambda \leqslant \infty$$

Suppose that $\lambda^* = 2\ \mu$, $\epsilon_1 = 0.2$, and $\epsilon_2 = 0.95$.   What is the emittance ϵ when the temperature is 100°F?   Is ϵ appreciably different when the temperature is 1000°F?   Hint: Use Table 1-2.

2-6.   Consider a surface for which ϵ is known as a function of temperature.   The ϵ_λ is independent of temperature, but varies with λ.   Describe a computation scheme for determining ϵ_λ.

2-7. By inspection of the ϵ_λ listing of Table 2-1, propose an ϵ value for each material which might serve as a gray-body emittance for infrared radiation at, say, $T = 400°F$. Compare such proposed ϵ values with those listed in Table 2-2.

2-8. Explain why the absorptance α of polished aluminum is markedly different for irradiation by the sun and by a source at 200°F.

2-9. Employ a calculation method similar to that for Problem 2-4 to evaluate α for a white painted surface on which solar radiation is incident. Assume the effective black-body temperature of the sun's surface is 10,000°R. Compare the α thus obtained with the ϵ calculated in Problem 2-4.

2-10. Polished molybdenum at 80°F is irradiated by a gray-body source at 1000°F. Evaluate the absorptance α. Hint: Use equation (2-19) in conjunction with Table 2-2; note that the T_i and T_s of equation (2-19) are absolute temperatures.

2-11. Subdivide the 90-degree range of the angle θ (Fig. 2-5) into nine intervals of 10 degrees each. For the case in which the radiation leaving a surface is diffusely distributed (i = constant), find the radiant flux Φ (per unit time and unit surface area) passing through each such 10-degree interval. In which interval is the largest value of Φ encountered? Hint: Integrate equation (2-24) and recall that $d\omega =$ $\sin \theta\, d\theta\, d\phi$.

2-12. For an isotropic surface, that is, $\epsilon(\theta, \phi) = \epsilon(\theta)$, show that the hemispherical emittance ϵ is given by the following weighted average of $\epsilon(\theta)$

$$\epsilon = \int_0^{\pi/2} \epsilon(\theta) \sin 2\theta\, d\theta$$

2-13. Apply the integral derived in Problem 2-12 to evaluate ϵ for aluminum-bronze and for copper oxide, taking the $\epsilon(\theta)$ from Figs. 2-8 and 2-7 respectively. Use Simpson's rule or any other convenient numerical integration technique to perform the computations. For each material compare the ϵ thus obtained with ϵ_n, the emittance in the direction $\theta = 0°$.

2-14. Show that

$$\rho_{ah}(\psi) = \int_\Omega \rho_{ba}(\psi, \theta, \phi) \cos \theta\, d\omega_r$$

· in which $d\omega_r$ is the solid angle of reflection, and the integration is carried out over the reflection angles θ and ϕ. Hint: The quantity $de_{r,h}$ of equation (2-30) may be expressed as

$$de_{r,h}(\psi) = \int_\Omega d^2e_r(\psi, \theta, \phi)$$

where, once again, the integration is over the reflection angles θ and ϕ.

2-15.   Let the two components of polarization in an incident mono-chromatic beam of radiation be $H_\perp$ and $H_{//}$, such that $H = H_\perp + H_{//}$. Suppose that the incident beam is unpolarized, that is, $H_\perp = H_{//}$. Verify the validity of equation (2-37).

2-16.   Use equation (2-43) to evaluate the normal specular reflectance of an aluminum surface at $200°F$ when $\lambda = 5\ \mu$, $10\ \mu$, and $20\ \mu$.   Hint: First obtain the electrical resistivity from an appropriate reference source.

2-17.   For black-body emission, $i_{b\perp} = i_{b//} = \frac{1}{2}i_b$ and $i_{b\perp} + i_{b//} = i_b$. Also,

$$\epsilon_\perp(\theta) = \frac{i_\perp(\theta)}{i_{b\perp}}, \qquad \epsilon_{//}(\theta) = \frac{i_{//}(\theta)}{i_{b//}}$$

With this information, verify equation (2-46).   Note that $i_\perp + i_{//} = i$.

2-18.   Evaluate equation (2-55) for aluminum, brass, copper, gold, mild steel, platinum, silver, and tin.   Use surface temperatures corresponding to those of Table 2-2.   Compare the ϵ_n thus calculated with the ϵ values of Table 2-2.

PART TWO

RADIANT INTERCHANGE AMONG SURFACES SEPARATED BY RADIATIVELY NONPARTICIPATING MEDIA

In many problems of engineering interest, the intensity of the thermal radiation passing between surfaces is substantially unaffected by the presence of the intervening media. Nonparticipating media include most monatomic and diatomic gases* as well as air* and vacuum. The analysis of radiant interchange among surfaces separated by nonparticipating media was already a subject of study before the twentieth century. However, it is in more recent years, under the impact of the new technologies, that a steady stream of new and more general methods of analysis has evolved.

The next several chapters are devoted to a description of the analysis of radiant interchange between surfaces. In Chapter 3 consideration is given to diffusely emitting and diffusely reflecting surfaces; while in Chapter 4 the angle factors (geometrical factors) that are intimately related to the interchange between such surfaces are treated. Chapter 5 deals with the situation where the participating surfaces have specular as well as diffuse components of the reflectance. Finally, in Chapter 6 attention is directed to various applications, including those where thermal radiation interacts with other modes of heat transfer such as conduction and convection. Interactions of this type are inherently nonlinear, for radiation depends on the fourth power of the temperature, while conduction and convection vary linearly with temperature.

Before proceeding to the next chapter, it is appropriate to mention

---

* At temperature levels below those at which dissociation or ionization occurs.

a remarkably simple, yet general situation that frequently recurs in practice. Consider a region within which there is black-body radiation characterized by a temperature T_e. Typically, such a region may be found within an enclosed space whose walls have a uniform temperature T_e. The radiation properties of the enclosing walls may be arbitrary. Within the enclosure is situated a body at temperature T having arbitrary radiation properties. This body is sufficiently small relative to the size of the enclosure so that its presence does not alter the black-body radiation field.

The radiant energy emitted per unit time and area by the body is $\epsilon\sigma T^4$, while the corresponding absorbed radiant flux is $\alpha\sigma T_e^4$. Consequently, the net rate of radiant outflow Q is

$$\frac{Q}{A} = \epsilon\sigma T^4 - \alpha\sigma T_e^4$$

where A is the surface area of the body. This expression holds regardless of the directional and spectral properties of the body. For the gray-body condition ($\epsilon = \alpha$), there follows

$$\frac{Q}{A} = \epsilon\sigma(T^4 - T_e^4)$$

CHAPTER 3

RADIANT INTERCHANGE AMONG DIFFUSELY EMITTING AND DIFFUSELY REFLECTING SURFACES

For the purposes of computing radiant interchange, engineering surfaces are almost always idealized as diffuse emitters and diffuse reflectors of radiant energy.   In this chapter, methods of analyzing the radiant transport between such surfaces are described.   Black surfaces are also treated here inasmuch as they are perfect diffuse emitters although they do not reflect at all.

In Section 3-1 certain fundamental concepts are described. Radiant interchange among black surfaces is analyzed in Section 3-2. Next, interchange among gray surfaces is treated in Section 3-3.   Then, in Section 3-4, consideration is given to the effect of wavelength-dependent radiation properties (i.e., nongray surfaces).   Finally, in Section 3-5, methods of solving the integral equations of radiant interchange are outlined.

As an alternative to the analytical formulations to be discussed in this chapter, an all-numerical method has recently been devised that appears to have practical utility for the evaluation of radiant interchange in complex systems.   The basic idea of the new method is to follow the probable path of an emitted energy bundle as it reflects and re-reflects among the participating surfaces.   The angular directions defining the path that is followed by a bundle are determined by drawing *random* numbers from an available collection and then interpreting these numbers in accordance with the specified directional distributions of the surface emittance or reflectance.   Because of the fact that the thus-determined path of a bundle depends on the laws of chance (i.e., on the magnitudes of the random numbers that are drawn), the afore-mentioned procedure is often designated as the *Monte Carlo* method. In general, it is necessary to follow the paths of a large number of energy bundles in order to achieve numerical results of satisfactory accuracy. Inasmuch as the Monte Carlo method applies to both diffuse and

non-diffuse surfaces, further discussion will be postponed until
Chapter 5 (specifically, Section 5-4), where the latter class of surfaces is
treated.

3-1   The Enclosure and the Angle Factor

Two concepts are common to all the material that is to be
presented in this chapter.   The first is the idea of the enclosure.   In
calculating the radiant interchange at any surface, it is necessary to
include radiation arriving at that surface from all directions in space.
To make certain that all radiation is fully accounted, one figuratively
constructs an enclosure and specifies the thermal state and the radiation
properties on each surface of the enclosure.   One or more of the surfaces
of the enclosure may not be material surfaces—for instance, an open
window.   Each of such surfaces may be assigned equivalent radiation
properties and an equivalent black-body temperature that corresponds
to the rate at which radiant energy passes through the fictitious surface
into the enclosure.

The second concept that is to be used frequently in the forthcoming
development is the angle factor (alternatively designated as the shape
factor, configuration factor, geometrical factor, or view factor).   The
angle factor provides information on the fraction of the diffusely
distributed radiant energy leaving one surface that arrives at a second
surface.   To illuminate the angle-factor concept, consider a surface i
with area A_i.   The radiant flux leaving surface i, if diffusely distributed,
will stream with uniform intensity away from i into the entire hemispherical space above the surface.   Suppose that in the aforementioned
space above surface i there is another surface j with area A_j.   The
angle factor $F_{A_i - A_j}$ designates the fraction of the radiation leaving i
that arrives at j.   The subscripts appended to the angle factor have the
following meaning: the first subscript denotes the surface from which
the radiation emanates, while the second subscript denotes the surface
receiving the radiation.

The basic characteristics of angle factors and their evaluation will
be treated in depth in Chapter 4.   At this point it is sufficient to state
certain fundamental properties that are needed to facilitate the forthcoming derivations.   These are the so-called reciprocity rules, which
relate the angle factor for radiant energy traveling from surface i to
surface j with the angle factor for radiant energy traveling from surface
j to surface i.   These rules are as follows:

$$A_i F_{A_i - A_j} = A_j F_{A_j - A_i} \tag{3-1a}$$

$$A_i \, dF_{A_i - dA_j} = dA_j \, F_{dA_j - A_i} \tag{3-1b}$$

$$dA_i \, dF_{dA_i - dA_j} = dA_j \, dF_{dA_j - dA_i} \tag{3-1c}$$

It may be noted that when the area of the receiving surface is infinitesimal, the corresponding angle factor is also infinitesimal.

The foregoing reciprocity rules are valid under the following two conditions: (1) the angular distributions of the radiant fluxes leaving the participating surfaces are diffuse, (2) for finite surfaces, the magnitudes of the radiant fluxes do not vary along the respective surfaces. When the leaving radiant fluxes are diffusely and uniformly distributed, it can be shown (Chapter 4) that the angle factors depend only on the geometrical orientation of the participating surfaces.

Another useful property of angle factors can be deduced from the energy conservation principle. The radiant energy leaving any surface i in an enclosure must impinge on the various surfaces of the enclosure; none can be lost. From this, it follows that

$$\sum_{j=1}^{N} F_{A_i - A_j} = 1 \tag{3-2}$$

where N denotes the number of surfaces in the enclosure. The summation includes the term $F_{A_i - A_i}$, which represents the fraction of the radiant energy leaving surface i that is incident on itself. This term is not zero when surface i is concave.

In the presentation that follows, initial consideration is to be given to interchange between black surfaces. Following this, the analysis of radiant transport between nonblack surfaces will be described.

3-2   Interchange among Black Surfaces

Isothermal surfaces.   The first case to be studied is an enclosure consisting of N black surfaces. The geometrical arrangement is shown schematically in Fig. 3-1. The surfaces are each isothermal and have temperatures $T_1, T_2, \ldots, T_N$, respectively; the corresponding areas are $A_1, A_2, \ldots, A_N$. If a given physical surface is not actually isothermal, it is subdivided into smaller surfaces, each of which is essentially isothermal.

The net rate of heat loss Q_i from a typical surface i is the difference between the emitted radiation and the absorbed portion of the incident radiation. Thus, in general,

$$\frac{Q_i}{A_i} = \epsilon_i \sigma T_i^4 - \alpha_i H_i \tag{3-3}$$

where H_i is the radiation incident on surface i per unit time and unit area. For a black surface, the emittance ϵ_i and the absorptance α_i are both unity.

The radiant flux incident on surface i comes from the other surfaces of the enclosure. Consider the radiation coming from any other surface j. Since surface j is a black-body emitter, an energy quantity $\sigma T_j^4 A_j$ streams away from j in all directions. Of this, an amount $\sigma T_j^4 A_j F_{A_j - A_i}$ arrives at surface i. By employing the reciprocity rule as stated in equation (3-1a), it follows that the rate at which energy arrives per unit area at A_i from surface j is $\sigma T_j^4 F_{A_i - A_j}$. Contributions

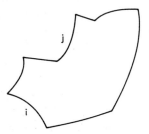

Fig. 3-1   Enclosure consisting of N isothermal surfaces.

such as this arrive at A_i from all the surfaces of the enclosure. Thus H_i is expressible as a sum

$$H_i = \sum_{j=1}^{N} \sigma T_j^4 F_{A_i - A_j} \tag{3-4}$$

With this, and with $\epsilon_i = \alpha_i = 1$, the net heat transfer Q_i follows from equation (3-3) as

$$\frac{Q_i}{A_i} = \sigma T_i^4 - \sum_{j=1}^{N} \sigma T_j^4 F_{A_i - A_j} \tag{3-5}$$

Equation (3-5) holds for any surface $i = 1, 2, \ldots, N$. An alternate form of the heat-loss expression is obtained by using the angle-factor conservation rule as expressed by equation (3-2). Thus

$$\sigma T_i^4 = \sum_{j=1}^{N} \sigma T_i^4 F_{A_i - A_j}$$

Introducing this in equation (3-5) gives

$$\frac{Q_i}{A_i} = \sum_{j=1}^{N} \sigma(T_i{}^4 - T_j{}^4) F_{A_i - A_j} \qquad (3\text{-}6)$$

Thus the heat loss is computed by summing differences in T^4, each weighted by an angle factor.

In some situations the heat flux Q may be prescribed at one or more of the surfaces. For instance, at a no flux or adiabatic surface, Q is zero. Correspondingly, the temperatures of these surfaces are unknown. Consider first the case where the heat transfer rate Q_k is prescribed at one of the surfaces k, while the temperatures are prescribed at all other surfaces. By rearranging equation (3-5), the unknown temperature at surface k is calculable from

$$\sigma T_k{}^4 = \frac{\displaystyle\sum_{j=1}^{N} \sigma T_j{}^4 F_{A_k - A_j} + (Q_k/A_k)}{1 - F_{A_k - A_k}} \qquad (3\text{-}7)$$

in which $j = k$ is specifically excluded from the summation. All quantities appearing on the right-hand side are known, and thus $\sigma T_k{}^4$ can be found. Once $T_k{}^4$ is determined, the heat transfer rates at the other surfaces follow by direct evaluation of equations (3-5) or (3-6).

Suppose next that the heat fluxes Q_k and Q_n are prescribed at surfaces k and n, respectively, along with temperatures $T_1, \ldots, T_N$ at the *other* surfaces of the enclosure. The unknown values of $\sigma T_k{}^4$ and $\sigma T_n{}^4$ are found by simultaneous solution of the following pair of linear algebraic equations:

$$\sigma T_k{}^4(1 - F_{A_k - A_k}) - \sigma T_n{}^4 F_{A_k - A_n} = \sum_{j=1}^{N} \sigma T_j{}^4 F_{A_k - A_j} + \left(\frac{Q_k}{A_k}\right) \qquad (3\text{-}8a)$$

$$-\sigma T_k{}^4 F_{A_n - A_k} + \sigma T_n{}^4(1 - F_{A_n - A_n}) = \sum_{j=1}^{N} \sigma T_j{}^4 F_{A_n - A_j} + \left(\frac{Q_n}{A_n}\right) \qquad (3\text{-}8b)$$

in which the terms $j = k$ and $j = n$ are specifically excluded from the summation. Once these temperatures have been found, the heat fluxes at the other surfaces are immediately calculable from equations (3-5) or (3-6).

Nonisothermal surfaces. In some cases the temperatures may vary substantially across the respective surfaces. To achieve isothermal zones, one may subdivide the surfaces into smaller and smaller elements. The logical limit of this process of subdivision is to produce infinitesimal areas dA. Suppose dA_i is an infinitesimal element on a surface A_i, while

dA_j has a similar relationship to A_j.   The locations of dA_i and dA_j are
specified by position vectors $\mathbf{r}_i$ and $\mathbf{r}_j$ drawn with respect to *any* convenient center of coordinates.   This arrangement is indicated schemati-
cally in Fig. 3-2.

The local rate of heat transfer per unit time and unit area at the
element dA_i on surface i is denoted by q_i.   This quantity is the
difference between the locally emitted and locally absorbed radiant
fluxes, thus

$$q_i(\mathbf{r}_i) = \epsilon_i \sigma T_i{}^4(\mathbf{r}_i) - \alpha_i H_i(\mathbf{r}_i) \qquad (3\text{-}9)$$

in which ϵ_i and α_i are both unity for a black surface.   The functional
dependence of q_i, T_i, and H_i on the position $\mathbf{r}_i$ is shown explicitly.

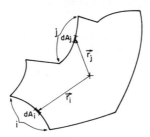

Fig. 3-2   Enclosure consisting of N surfaces, some of which may
be nonisothermal.

The incident radiant flux $H_i(\mathbf{r}_i)$ is found by generalizing the derivation of equation (3-4).   From an element dA_j on surface j, there arrives
per unit time and unit area at dA_i a quantity $\sigma T_j{}^4(\mathbf{r}_j)dF_{dA_i - dA_j}$.
Furthermore, from the entire surface j, there arrives

$$\int_{A_j} \sigma T_j{}^4(\mathbf{r}_j)dF_{dA_i - dA_j}$$

Contributions such as the foregoing are provided by all the surfaces of
the enclosure.   Thus H_i may be expressed as

$$H_i(\mathbf{r}_i) = \sum_{j=1}^{N} \int_{A_j} \sigma T_j{}^4(\mathbf{r}_j)dF_{dA_i - dA_j} \qquad (3\text{-}10)$$

It is also convenient to define

$$K(\mathbf{r}_i, \mathbf{r}_j) = \frac{dF_{dA_i - dA_j}}{dA_j} \qquad (3\text{-}11)$$

The K function is a finite quantity that depends only on the position
and the orientation of the elements dA_i and dA_j.

Upon introducing equations (3-10) and (3-11) into (3-9), one finds (with $\epsilon_i = \alpha_i = 1$)

$$q_i(\mathbf{r}_i) = \sigma T_i{}^4(\mathbf{r}_i) - \sum_{j=1}^{N} \int_{A_j} \sigma T_j{}^4(\mathbf{r}_j) K(\mathbf{r}_i, \mathbf{r}_j)\, dA_j \qquad (3\text{-}12)$$

The summation includes $j = i$. This equation applies at all locations $\mathbf{r}_i$ on surface i. Moreover, it also applies for all surfaces $i = 1, 2, \ldots, N$ in the enclosure. An alternate and fully equivalent way of expressing the incident energy term would be to remove the summation sign and to regard the integration as being extended over all the surfaces of the enclosure. However, the form appearing in equation (3-12) is believed preferable when, for all locations $\mathbf{r}_j$ on a given surface A_j, K is expressible by a single algebraic relationship.

When the temperature distribution $T_j(\mathbf{r}_j)$ is prescribed on all the surfaces j, the determination of the heat flux can be carried out directly. Assuming that the angle factors are available, the integrals appearing on the right-hand side of equation (3-12) involve quantities, all of which are known. The integration can then be carried out either analytically or numerically.

An altogether different state of affairs exists when the heat flux is prescribed on one or more of the surfaces of the enclosure. Then the temperatures of these surfaces are unknown and must be found by analysis. A variety of interesting cases arise, depending on the number of surfaces having prescribed heat flux and on whether or not these surfaces are concave. A few of these cases will be discussed here to illustrate the various types of problems that may arise.

The simplest situation occurs when the heat flux is prescribed along a single *nonconcave* surface k. Then, from equation (3-12), it follows immediately that

$$\sigma T_k{}^4(\mathbf{r}_k) = q_k(\mathbf{r}_k) + \sum_{j=1}^{N} \int_{A_j} \sigma T_j{}^4(\mathbf{r}_j) K(\mathbf{r}_k, \mathbf{r}_j)\, dA_j \qquad (3\text{-}13)$$

in which $j = k$ is specifically excluded from the summation since $K(\mathbf{r}_k, \mathbf{r}_k) = 0$. Once T_k has been found, the heat flux distributions on the other surfaces follow directly by evaluating equation (3-12).

Suppose next that the aforementioned surface k is concave. Thus $dF_{dA_k - dA_k}$ is not zero and, correspondingly, $K(\mathbf{r}_k, \mathbf{r}_k) \neq 0$. The unknown temperature distribution $T_k(\mathbf{r}_k)$ can then be expressed by rephrasing equation (3-12) as

$$\sigma T_k{}^4(\mathbf{r}_k) = q_k(\mathbf{r}_k) + \int_{A_k} \sigma T_k{}^4(\mathbf{r}_k) K(\mathbf{r}_k, \mathbf{r}_k)\, dA_k + \phi_k(\mathbf{r}_k) \qquad (3\text{-}14)$$

in which

$$\phi_k(\mathbf{r}_k) = \sum_{j=1}^{N} \int_{A_j} \sigma T_j^4(\mathbf{r}_j) K(\mathbf{r}_k, \mathbf{r}_j) \, dA_j \qquad (3\text{-}14a)$$

The term $j = k$ is specifically excluded from the summation. Since the temperature distributions appearing in equation (3-14a) are prescribed, ϕ_k is known.

Upon reconsidering equation (3-14), it is seen that the unknown temperature distribution T_k appears both under the integral sign as well as in other terms of the equation. An equation in which the unknown appears under the integral sign is called an integral equation. In the standard terminology of integral equations, K is referred to as the kernel. Since the temperature appears only as T^4, the integral equation may be regarded as linear in T^4.

Integral equations frequently arise in connection with the computation of radiant interchange, particularly for nonblack surfaces. A wide range of methods is available for solving equations of this type. However, a discussion of various solution methods will be deferred until later in this chapter, after the analysis for nonblack surfaces has been developed.

As a final case, which illustrates still another form of the governing equations, consider a situation in which the heat flux is prescribed at two surfaces k and n, both of which are *nonconcave*. The other surfaces have prescribed temperature distributions. Then, by application of equation (3-12) to surfaces k and n, one finds

$$\sigma T_k^4(\mathbf{r}_k) = q_k(\mathbf{r}_k) + \int_{A_n} \sigma T_n^4(\mathbf{r}_n) K(\mathbf{r}_k, \mathbf{r}_n) \, dA_n + \phi_k(\mathbf{r}_k) \qquad (3\text{-}15a)$$

$$\sigma T_n^4(\mathbf{r}_n) = q_n(\mathbf{r}_n) + \int_{A_k} \sigma T_k^4(\mathbf{r}_k) K(\mathbf{r}_n, \mathbf{r}_k) \, dA_k + \phi_n(\mathbf{r}_n) \qquad (3\text{-}15b)$$

The function ϕ_k is represented by equation (3-14a), where both $j = k$ and $j = n$ are now excluded from the summation; ϕ_n is also represented by (3-14a) when k is replaced by n. Both ϕ_k and ϕ_n are known functions of position.

Upon inspection of equations (3-15a) and (3-15b), it may be observed that the two unknown functions appear under the integral signs. So, equations (3-15a) and (3-15b) represent a pair of simultaneous linear integral equations for the temperature distributions T_k^4 and T_n^4. Solution methods for simultaneous equations will also be discussed later.

In modern engineering applications, one sometimes encounters the case in which neither the surface temperature nor the heat flux is prescribed; rather, there is some specified relationship between these quantities. Such a situation is illustrated in Fig. 3-3, which shows an ensemble of radiating fins of rectangular profile mounted on an iso-thermal base plate (temperature T_b). All surfaces are assumed to be black and there is no convection (e.g., vacuum). The fins are very long in the direction into and out of the plane of the figure so that end effects are negligible. Furthermore, the fins are thin ($2t \ll L_2$). Under these conditions, $T = T(y)$ along the height of the fins. More-over, when there are a large number of fins in the ensemble, and uniform conditions exist in the environment, the variations of the temperature and the heat flux are identical along the height of each fin.

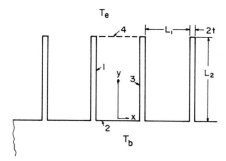

Fig. 3-3   Fins on a cooled wall.

One may envision an enclosure consisting of the surfaces 1, 2, 3, and 4 as indicated in Fig. 3-3. The latter surface, shown dashed, is a fictive surface stretched across the tips of the fins. The temperature of surface 2 is prescribed to be uniform. Similarly, one may assign a uniform black-body temperature T_e to the fictive surface, provided that the radiation in the environment is uniformly and diffusely distributed. Moreover, from symmetry considerations, $T_1(y_1) = T_3(y_3)$ when $y_1 = y_3$. Thus there is only one unknown temperature distribution, $T_1(y_1)$.

By specializing equation (3-12), one can write an integral equation for determining $T_1(y_1)$, thus

$$\sigma T_1{}^4(y_1) = q_1(y_1) + \sigma T_b{}^4 F_{dA_1 - A_2} + \sigma T_e{}^4 F_{dA_1 - A_4}$$

$$+ \int_{A_3} \sigma T_3{}^4(y_3) K(y_1, y_3) \, dA_3 \quad (3\text{-}16)$$

In view of the aforementioned fact that $T_1(y_1) = T_3(y_3)$, it is evident that equation (3-16) contains two unknowns, $T_1(y_1)$ and $q_1(y_1)$. A second relationship connecting these quantities is obtained by applying the Fourier law to the heat conducted along the fin height. From this, there follows

$$q_1(y_1) = kt \frac{d^2 T_1(y_1)}{dy_1{}^2} \tag{3-17}$$

in which k is the thermal conductivity of the fin.

The variations of q and T with y are found by simultaneous solution of equations (3-16) and (3-17). This mathematical system, which includes an integral equation and a differential equation, is called integrodifferential. Since the temperature T appears to both the first and the fourth powers, the system is nonlinear. Solutions to systems of this type will be illustrated in Chapter 6, where consideration is extended to interactions between radiation and other modes of heat transfer.

3-3   Interchange among Gray Surfaces

Engineering calculations of radiant interchange are frequently performed under the assumption that the participating surfaces are gray, that is, $\epsilon = \alpha$. When both the emitted and the incident radiation are confined to the same wavelength range and the spectral emittance is relatively constant within that range, then the gray-body assumption is reasonable. However, if the wavelength ranges corresponding to the emitted and incident radiation are different (e.g., infrared and visible), serious error may result if gray-body conditions are assumed.

The analysis of radiant transport among gray, diffuse surfaces can be formulated at various levels of complexity, depending on the effort that is to be expended in solving the problem. This presentation will begin with a description of the simplest model; this leads to a mathematical problem involving algebraic equations. Later, a more exact model leading to integral equations will be described.

Isothermal surfaces with uniform radiosity.   Consideration is given to an enclosure composed of N finite surfaces such as that pictured in Fig. 3-1. The model is defined by the following five postulates:

(1)   Each surface of the enclosure is isothermal. This condition may be approached by subdividing any nonisothermal surface into smaller sections.

(2)   Each surface is gray.

(3)   The radiation reflected from any surface is diffusely distributed. Under this assumption, all incident radiation is reflected with a uniform

intensity regardless of the direction from which it came. Thus it is clear that contact with a diffusely reflecting surface completely obliterates the past history of the incident radiation. Consequently, there is no need to keep account of specific rays as they interreflect among the surfaces of the enclosure.

(4)   The radiation emitted from any surface is diffusely distributed. This assumed diffuse distribution of the emitted radiation, together with the previously assumed diffuse distribution of the reflected radiation, affords a significant simplification in the analysis. This simplification becomes evident when it is noted that the radiant energy streaming away from a surface is the sum of the emitted radiation and the reflected radiation. If both the emitted and the reflected radiation are diffusely distributed, they are then directionally indistinguishable and there is no need to treat them separately. Indeed, it is advantageous to deal with their sum, which represents the radiant flux streaming away from a surface.

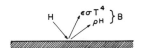

Fig. 3-4   The radiosity B.

The rate at which radiant energy streams away from a surface per unit area is termed the radiosity of the surface and is denoted by the symbol B; the radiosity has units of energy per unit time and unit area. Moreover, the incident radiant energy arriving at a surface per unit time and unit area is to be denoted by the symbol H. Then it follows that (Fig. 3-4)

$$B = \epsilon\sigma T^4 + \rho H \qquad (3\text{-}18)$$

The first term on the right represents the radiation emitted by the surface, while the second term represents the reflected radiation. In accordance with the foregoing assumptions, B is diffusely distributed.

(5)   The radiosity of any surface is constant along that surface. This assumption is necessary in order that the angle factors be independent of the magnitude and surface distribution of the radiant energy flux. From the definition, equation (3-18), it is clear that if B is to be uniform along a surface when T is uniform, it is necessary that the incident radiant flux H be the same at every point on the surface. In general, it is unlikely that this condition will be satisfied in practice.

Consistent with the preceding postulates, an analysis can be devised for calculating the radiant interchange among the surfaces of the enclosure. First, consideration is given to the case where the temperatures are prescribed at all N surfaces. The aim of the analysis is then to determine the corresponding heat transfer rates. The rate of heat transfer at a typical surface i is expressed by equation (3-3).

Upon eliminating the incident flux H between equation (3-3) and equation (3-18), and then specializing to the gray-body condition $\epsilon_i = \alpha_i$, there follows

$$\frac{Q_i}{A_i} = \frac{\epsilon_i}{1 - \epsilon_i}(\sigma T_i{}^4 - B_i) \tag{3-19}$$

Thus, to determine the heat transfer rate at a surface having a prescribed temperature, it is only necessary to determine the radiosity B.

The radiosities are found by applying equation (3-18) at each surface of the enclosure; for instance, at surface i

$$B_i = \epsilon_i \sigma T_i{}^4 + (1 - \epsilon_i)H_i \tag{3-20}$$

in which the gray-body condition $\rho_i = 1 - \epsilon_i$ has been introduced. The next step is to find the incident radiant flux H_i. This is readily accomplished by a simple modification of the expression, equation (3-4), that was previously derived for black surfaces. Indeed, it is only necessary to replace the black-body emissive power $\sigma T_j{}^4$ appearing in equation (3-4) by the radiosity B_j. The same angle factors continue to apply, for both σT^4 and B represent diffusely distributed streams of radiation. Thus

$$H_i = \sum_{j=1}^{N} B_j F_{A_i - A_j} \tag{3-21}$$

and, with this, equation (3-20) becomes

$$B_i = \epsilon_i \sigma T_i{}^4 + (1 - \epsilon_i) \sum_{j=1}^{N} B_j F_{A_i - A_j}, \qquad 1 \leqslant i \leqslant N \tag{3-22}$$

As indicated at the right in equation (3-22), equations such as this are to be written for each of the N surfaces of the enclosure. In this way there are generated N linear, inhomogeneous algebraic equations for the N unknown radiosities $B_1, B_2, \ldots, B_n$.

If there are relatively few surfaces, the corresponding linear algebraic equations can be readily solved by hand methods. However, when there are a large number of surfaces, recourse must be had to an electronic computer. Large-scale electronic computers generally have standard routines in their program libraries for solving hundreds of simultaneous linear algebraic equations. Consequently, the B_i values can be found and, with these, the heat transfer rates Q_i follow directly from equation (3-19).

   When a given physical enclosure must be solved again and again
for different sets of surface temperatures, the solution described in the
previous paragraph becomes cumbersome, even for a large-scale computer.   This solution method requires that the governing equations be
solved anew for each set of prescribed temperatures.   The need to
re-solve a given enclosure many times arises when transient operating
conditions are being analyzed.
   It is possible to devise a general solution method that overcomes
the preceding objection and that, at the same time, provides a direct
relationship between the unknown heat fluxes and the prescribed temperatures.   Indeed, under the new scheme the governing equations
need be solved only once, regardless of how many times the thermal
boundary conditions are altered.
   The analysis begins with a reconsideration of equation (3-22),
which can be rephrased as

$$\sum_{j=1}^{N} X_{ij} B_j = \Omega_i, \qquad 1 \leqslant i \leqslant N \tag{3-23}$$

in which
$$X_{ij} = \frac{\delta_{ij} - (1 - \epsilon_i) F_{A_i - A_j}}{\epsilon_i} \tag{3-23a}$$

and
$$\Omega_i = \sigma T_i^4 \tag{3-23b}$$

The symbol δ_{ij} is the Kronecker delta, which takes on the value 1 when
$i = j$ and the value 0 when $i \neq j$.   The coefficients X_{ij} and the inhomogeneous terms Ω_i are known constants.   Upon evaluating equation
(3-23) for $i = 1$, one finds coefficients $X_{11}, X_{12}, \ldots, X_{1N}$.   Similarly, if
equation (3-23) is evaluated with $i = 2$, one encounters $X_{21}, X_{22}, \ldots, X_{2N}$,
and so forth.   With the coefficients X_{ij}, an N by N array of numbers
can be formed as follows:

$$X = \begin{Vmatrix} X_{11} & X_{12} & \cdots & X_{1N} \\ X_{21} & X_{22} & \cdots & X_{2N} \\ \vdots & \vdots & & \vdots \\ X_{N1} & X_{N2} & \cdots & X_{NN} \end{Vmatrix} \tag{3-24}$$

This array is called the matrix X.
   Modern digital computers, as part of their libraries of standard
programs, have a routine called matrix inversion.   Thus, given a

matrix of numbers such as X, the machine will provide another matrix of numbers ψ, which is called the inverse of X.

$$\psi = X^{-1} = \begin{Vmatrix} \psi_{11} & \psi_{12} & \cdots & \psi_{1N} \\ \psi_{21} & \psi_{22} & \cdots & \psi_{2N} \\ \vdots & \vdots & & \vdots \\ \psi_{N1} & \psi_{N2} & \cdots & \psi_{NN} \end{Vmatrix} \qquad (3\text{-}25)$$

By employing the coefficients ψ_{ij} of the inverse matrix, one can now write the inverse of equation (3-23), thus

$$B_i = \sum_{j=1}^{N} \psi_{ij}\Omega_j, \qquad 1 \leqslant i \leqslant N \qquad (3\text{-}26)$$

Moreover, since $\Omega_j = \sigma T_j{}^4$, there follows

$$B_i = \sum_{j=1}^{N} \psi_{ij}\sigma T_j{}^4, \qquad 1 \leqslant i \leqslant N \qquad (3\text{-}27)$$

It should be noted that as long as the emittances ϵ_i are regarded as constants, then the ψ_{ij} do not depend on the temperatures of the problem. Upon inspecting equation (3-27), it is seen that the radiosities are calculable directly from the given temperatures T_j.

The next step is to calculate the heat flux from equation (3-19). Upon substituting (3-27) into (3-19) and rearranging, there follows

$$\frac{Q_i}{A_i} = \sum_{j=1}^{N} \Lambda_{ij}\sigma T_j{}^4, \qquad 1 \leqslant i \leqslant N \qquad (3\text{-}28)$$

where

$$\Lambda_{ij} = \frac{\epsilon_i}{1 - \epsilon_i}(\delta_{ij} - \psi_{ij}) \qquad (3\text{-}28a)$$

The coefficients Λ_{ij} are known once and for all, provided that the ϵ_i are regarded as constants. Correspondingly, equation (3-28) provides a direct algebraic relation between the unknown heat fluxes of the problem and the prescribed temperatures. Indeed, the heat-flux computation is reduced to a weighted summing of temperatures to the fourth power. This is the same type of computation as was encountered in black enclosures having prescribed surface temperatures.

As was previously discussed in connection with the case of the black-surfaced enclosure, the surface heat flux may be prescribed in lieu of the surface temperature. A similar situation may occur for non-black surfaces. Consideration will now be given to the case in which the temperatures are prescribed at some of the surfaces of the enclosure and the heat fluxes are prescribed at the others. For convenience, suppose that the surfaces are numbered so that those with prescribed temperature are designated as $1 \leqslant i \leqslant N_1$, while those with prescribed heat flux are designated as $(N_1 + 1) \leqslant i \leqslant N$.

For each of the first N_1 surfaces, one can write the radiant flux balances as before:

$$B_i = \epsilon_i \sigma T_i^4 + (1 - \epsilon_i) \sum_{j=1}^{N} B_j F_{A_i - A_j}, \qquad 1 \leqslant i \leqslant N_1 \quad (3\text{-}29)$$

The unknown surface heat flux is related to the known surface temperature by

$$\frac{Q_i}{A_i} = \frac{\epsilon_i}{1 - \epsilon_i} (\sigma T_i^4 - B_i), \qquad 1 \leqslant i \leqslant N_1 \quad (3\text{-}30)$$

For the second group of surfaces $(N_1 + 1) \leqslant i \leqslant N$, equation (3-29) continues to be valid; however, it is not very useful, for the surface temperatures are unknown. Instead, one eliminates σT_i^4 from equations (3-19) and (3-20) and finds

$$\frac{Q_i}{A_i} = B_i - H_i \quad (3\text{-}31)$$

Then, upon substituting for H_i from equation (3-21), there follows

$$B_i = \frac{Q_i}{A_i} + \sum_{j=1}^{N} B_j F_{A_i - A_j}, \qquad (N_1 + 1) \leqslant i \leqslant N \quad (3\text{-}32)$$

Inasmuch as (Q_i/A_i) is prescribed for each of the surfaces $(N_1 + 1) \leqslant i \leqslant N$, then equation (3-32) is the appropriate form of the radiant flux balance for these surfaces. Moreover, the unknown surface temperature is related to the known surface heat flux by

$$\sigma T_i^4 = \frac{1 - \epsilon_i}{\epsilon_i} \frac{Q_i}{A_i} + B_i, \qquad (N_1 + 1) \leqslant i \leqslant N \quad (3\text{-}33)$$

Upon considering equations (3-29) and (3-32), it is seen that there are a total of N unknown B values and that, correspondingly, there are N linear algebraic equations. These equations can be solved by the previously discussed matrix inversion method. The general form, (3-23), which describes the set of N linear equations, continues to be valid, but now the X_{ij} and the Ω_i are given by

$$X_{ij} = \frac{\delta_{ij} - (1 - \epsilon_i) F_{A_i - A_j}}{\epsilon_i}, \qquad 1 \leqslant i \leqslant N_1 \quad (3\text{-}34a)$$

$$X_{ij} = \delta_{ij} - F_{A_i - A_j}, \qquad (N_1 + 1) \leqslant i \leqslant N \quad (3\text{-}34b)$$

$$\Omega_i = \sigma T_i^4, \qquad 1 \leqslant i \leqslant N_1 \quad (3\text{-}34c)$$

$$\Omega_i = \frac{Q_i}{A_i}, \qquad (N_1 + 1) \leqslant i \leqslant N \quad (3\text{-}34d)$$

The X_{ij} and the Ω_i are known constants.

The coefficients X_{ij} can be used to form the X matrix as indicated in equation (3-24), and the inversion of this produces the ψ matrix, equation (3-25). Moreover, the relationship between B_i and Ω_j continues to apply as stated by equation (3-26). The next step is to specialize equation (3-26) to the present case. Upon employing equations (3-34c) and (3-34d), there results

$$B_i = \sum_{j=1}^{N_1} \psi_{ij} \sigma T_j^{\,4} + \sum_{j=(N_1+1)}^{N} \psi_{ij} \frac{Q_j}{A_j}, \qquad 1 \leqslant i \leqslant N \qquad (3\text{-}35)$$

Equation (3-35) applies at any of the surfaces of the enclosure, regardless of whether the temperature or the heat flux is prescribed.

The final step is to compute the unknown surface heat fluxes and surface temperatures. For the surfaces $1 \leqslant i \leqslant N_1$, the substitution of equation (3-35) into equation (3-30) yields

$$\frac{Q_i}{A_i} = \sum_{j=1}^{N_1} \Lambda_{ij} \sigma T_j^{\,4} - \frac{\epsilon_i}{1-\epsilon_i} \sum_{j=(N_1+1)}^{N} \psi_{ij} \frac{Q_j}{A_j}, \qquad 1 \leqslant i \leqslant N_1 \qquad (3\text{-}36)$$

in which the Λ_{ij} retain their definition from equation (3-28a). Similarly, for surfaces $(N_1 + 1) \leqslant i \leqslant N$, the substitution of equation (3-35) into equation (3-33) leads to

$$\sigma T_i^{\,4} = \sum_{j=1}^{N_1} \psi_{ij} \sigma T_j^{\,4} + \sum_{j=(N_1+1)}^{N} \Phi_{ij} \frac{Q_j}{A_j}, \qquad (N_1+1) \leqslant i \leqslant N \qquad (3\text{-}37)$$

where
$$\Phi_{ij} = \frac{1-\epsilon_i}{\epsilon_i} \delta_{ij} + \psi_{ij} \qquad (3\text{-}37\text{a})$$

Thus, by employing equation (3-36) or (3-37), whichever is appropriate, the unknowns of the problem can be determined directly from the prescribed temperatures and heat fluxes.

The electric circuit analogy. The radiant-flux balance equations of the preceding section can be rephrased in a form that suggests an analogy with an electric circuit composed of resistances, voltage sources, and current sources. The analogy was first discussed in detail by A. K. Oppenheim (ref. 1).

For any surface i at which the temperature is prescribed, the radiant flux balance, equation (3-29), can be rewritten as

$$\sum_{j=1}^{N} \frac{B_j - B_i}{(A_i F_{A_i - A_j})^{-1}} + \frac{\sigma T_i^{\,4} - B_i}{(1-\epsilon_i)/\epsilon_i A_i} = 0 \qquad (3\text{-}29\text{a})$$

Similarly, for a surface at which the heat flux is prescribed, the radiant flux balance of equation (3-32) may be rearranged to read

$$\sum_{j=1}^{N} \frac{B_j - B_i}{(A_i F_{A_i - A_j})^{-1}} + Q_i = 0 \qquad (3\text{-}32\text{a})$$

One may associate the radiosities B and the emissive powers σT^4 with potentials and the quantities $1/AF$ and $(1 - \epsilon)/\epsilon A$ with resistances. Then terms such as

$$\frac{B_j - B_i}{(A_i F_{A_i - A_j})^{-1}} \quad \text{and} \quad \frac{\sigma T_i{}^4 - B_i}{(1 - \epsilon_i)/\epsilon_i A_i}$$

are clearly identifiable as currents. Furthermore, Q_i is also a current.

Thus equations (3-29a) and (3-32a) express conservation of current at a node. Each term in the summation represents a current flowing from node j into node i. The term $(\sigma T_i{}^4 - B_i)/[(1 - \epsilon_i)/\epsilon_i A_i]$ is a current flowing into node i through an external connection that links an externally applied voltage $\sigma T_i{}^4$ with node i; the resistance of the connection is $(1 - \epsilon_i)/\epsilon_i A_i$. Q_i represents a current flowing into node i from an external current source.

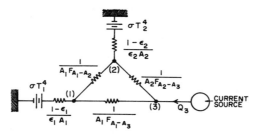

Fig. 3-5   Electric circuit representation of radiant interchange.

In light of the foregoing interpretation, an electric circuit analogy can be constructed from which the radiosities B_i are determinable by measurement of the voltages at the various nodes. Once the radiosities are known, the surface heat fluxes or surface temperatures (whichever are unknown) are calculable from equations (3-30) or (3-33).

To illustrate the analogy, consider an enclosure consisting of three surfaces. Of these, surfaces 1 and 2 have prescribed temperatures, while surface 3 has prescribed heat flux. The circuit analogy for this situation is shown in Fig. 3-5.

Generalized analysis.   In discussing the underlying postulates of the preceding computational method, it was mentioned that the assumption of uniform radiosity on each surface is rarely, if ever, realized. This assumption is no longer necessary if the radiant flux balance is performed for an infinitesimal surface element. Moreover, such an analysis not only permits a continuous variation of the radiosity but also is able to accommodate a continuous variation of the surface temperature and the surface heat flux.

The equations of radiant interchange are readily derived in a manner similar to the foregoing. As a first step, a coordinate system is set up to define the positions of elements dA_i and dA_j on surfaces A_i and A_j (see Fig. 3-2). The radiosity equation at dA_i follows directly from the definition

$$B_i(\mathbf{r}_i) = \epsilon_i \sigma T_i^{\,4}(\mathbf{r}_i) + (1 - \epsilon_i)H_i(\mathbf{r}_i) \tag{3-38}$$

in which the dependence of B_i, T_i, and H_i on the surface coordinate is shown explicitly. The incident flux H_i is most easily derived by modifying the corresponding expression, equation (3-10), for black surfaces. For black surfaces, the energy flux leaving an element dA_j per unit time and unit area is $\sigma T_j^{\,4}(\mathbf{r}_j)$. On the other hand, for a gray, diffuse surface, the corresponding quantity is $B_j(\mathbf{r}_j)$. Thus, upon replacing $\sigma T_j^{\,4}$ by B_j in equation (3-10), one has

$$H_i(\mathbf{r}_i) = \sum_{j=1}^{N} \int_{A_j} B_j(\mathbf{r}_j)\, dF_{dA_i - dA_j} \tag{3-39}$$

The next step is to eliminate H from equation (3-38). At the same time, the differential angle factor $dF_{dA_i - dA_j}$ may be replaced by the kernel function K in accordance with equation (3-11). After these substitutions, there results

$$B_i(\mathbf{r}_i) = \epsilon_i \sigma T_i^{\,4}(\mathbf{r}_i) + (1 - \epsilon_i) \sum_{j=1}^{N} \int_{A_j} B_j(\mathbf{r}_j)K(\mathbf{r}_i, \mathbf{r}_j)\, dA_j, \qquad 1 \leqslant i \leqslant N \tag{3-40}$$

If the temperatures are prescribed at all the surfaces of the enclosure, equation (3-40), when successively applied at each surface, represents N linear integral equations for the N unknown radiosity functions $B_1(\mathbf{r}_1), B_2(\mathbf{r}_2), \ldots, B_N(\mathbf{r}_N)$. Solution methods for equations of this type will be discussed later in this chapter. Once the distributions of B along the various surfaces have been found, the corresponding distributions of the heat flux q follow directly. To compute q, one eliminates the incident flux H between equations (3-9) and (3-38). Upon specializing to the gray-body case, $\epsilon = \alpha$, there results

$$q_i(\mathbf{r}_i) = \frac{\epsilon_i}{1 - \epsilon_i} [\sigma T_i^{\,4}(\mathbf{r}_i) - B_i(\mathbf{r}_i)], \qquad 1 \leqslant i \leqslant N \tag{3-41}$$

Thus, when the temperature distribution is prescribed and B is known from the solution of the governing integral equations, q is calculable by simple arithmetic operations.

In some situations the heat flux distribution on certain surfaces may be prescribed instead of the temperature. Although equation (3-40) is valid for such surfaces, it is not very useful, for both B_i and T_i

would be unknown. As an alternative, one eliminates $\epsilon \sigma T^4$ between equations (3-9) and (3-38) and obtains, after specializing to the gray-body case,

$$B_i(\mathbf{r}_i) = q_i(\mathbf{r}_i) + H_i(\mathbf{r}_i) \qquad (3\text{-}42)$$

Next, H_i is evaluated from equation (3-39), giving

$$B_i(\mathbf{r}_i) = q_i(\mathbf{r}_i) + \sum_{j=1}^{N} \int_{A_j} B_j(\mathbf{r}_j) K(\mathbf{r}_i, \mathbf{r}_j) \, dA_j \qquad (3\text{-}43)$$

Now, consideration may be given to an enclosure in which the temperature is prescribed on some of the surfaces and the heat flux is prescribed on the others. For the former surfaces, the radiant flux balance is applied in the form stated by equation (3-40). On the other hand, for the latter surfaces, the radiant flux balance as expressed by equation (3-43) is utilized. In this way one may write a total of N linear integral equations for the N unknown radiosity distributions. These equations are solved in a manner identical to that employed for the case where all surface temperatures are prescribed; in other words, the problem is in no way made complicated by the fact that some surfaces have prescribed temperature and others have prescribed heat flux.

Once the radiosity equations have been solved, the B distributions can be employed in determining the other unknowns of interest. For those surfaces having prescribed temperature, the distributions of the surface heat flux are found by application of equation (3-41). On the other hand, for those surfaces with prescribed heat flux, the corresponding surface-temperature distributions are calculable from

$$\sigma T_i{}^4(\mathbf{r}_i) = \frac{1 - \epsilon_i}{\epsilon_i} q_i(\mathbf{r}_i) + B_i(\mathbf{r}_i) \qquad (3\text{-}44)$$

A particularly interesting situation arises when the heat flux is prescribed at all the surfaces of the enclosure. Then equation (3-43) is to be applied at all the surfaces. Inspection of equation (3-43) reveals that the radiation properties of the surfaces nowhere appear. Thus the solution for the radiosities is independent of the radiation properties. In addition, these very same integral equations, with B replaced by σT^4, apply for a black-surfaced enclosure wherein the heat fluxes are everywhere prescribed.

The preceding formulation has yielded a set of integral equations for the unknown radiosities. When solved, these equations provide input values to simple algebraic equations from which the unknown heat fluxes or temperatures can be calculated. An alternative (but not necessarily advantageous) formulation yields integral equations that relate the temperature and heat fluxes directly, without recourse to the radiosity. To achieve this alternative formulation, one solves for B from either of equations (3-41) or (3-44) and then introduces this into equation (3-43). Upon rearrangement, there follows

$$\sigma T_i{}^4(\mathbf{r}_i) - \frac{q_i(\mathbf{r}_i)}{\epsilon_i} = \sum_{j=1}^{N} \int_{A_j} \sigma T_j{}^4(\mathbf{r}_j) K(\mathbf{r}_i, \mathbf{r}_j) \, dA_j$$

$$- \sum_{j=1}^{N} \int_{A_j} \frac{1 - \epsilon_j}{\epsilon_j} q_j(\mathbf{r}_j) K(\mathbf{r}_i, \mathbf{r}_j) \, dA_j, \quad 1 \leqslant i \leqslant N \quad (3\text{-}45)$$

It may be observed that equation (3-45) contains a total of $2N$ temperature and heat flux functions. In general, N of these may be prescribed, leaving the remaining N functions to be determined from the N linear integral equations.

It appears to the present authors that equation (3-45) is slightly more cumbersome than the corresponding radiosity equations for cases in which the surface temperatures or the surface heat fluxes are definite prescribed functions. However, equation (3-45) may have some advantage when the thermal boundary condition is specified as a *relationship* between the surface temperature and the surface heat flux as, for instance, is given by the heat conduction equation (3-17).

To help establish these ideas, consideration may be given to the previously discussed case of radiating fins mounted on an isothermal base plate; see Fig. 3-3. The specifics of the geometric configuration have already been described in detail. From that discussion it was concluded that, at most, $T = T(y)$ along the height of the fins. For present purposes the fin and base surfaces will be regarded as gray, diffuse emitters and reflectors. Two situations are to be discussed. First, it will be assumed that the fin conductance is sufficiently high so that the temperature of each fin is uniform and equal to the base surface temperature T_b. In the second case, the fin conductance will be taken as finite and the fin temperature will be nonuniform.

Upon applying equation (3-40) to the enclosure formed by surfaces 1, 2, 3, and 4 (Fig. 3-3), one finds for the case of uniform fin temperature

$$B_1(y_1) = \epsilon_1 \sigma T_b{}^4 + (1 - \epsilon_1) \left[\int_{A_2} B_2(x) K(y_1, x) \, dA_2 \right.$$

$$\left. + \int_{A_3} B_3(y_3) K(y_1, y_3) \, dA_3 + \sigma T_e{}^4 F_{dA_1 - A_4} \right] \quad (3\text{-}46)$$

$$B_2(x) = \epsilon_2 \sigma T_b{}^4 + (1 - \epsilon_2) \left[\int_{A_1} B_1(y_1) K(x, y_1) \, dA_1 \right.$$

$$\left. + \int_{A_3} B_3(y_3) K(x, y_3) \, dA_3 + \sigma T_e{}^4 F_{dA_2 - A_4} \right] \quad (3\text{-}47)$$

in which T_e is the effective black-body temperature of the fictive surface 4 that closes the top of the cavity. Although four surfaces comprise the cavity, only two integral equations need be written in the present situation. Because of the symmetry that results when $\epsilon_1 = \epsilon_3$, it is apparent that

$$B_3(y_3) = B_1(y_1) \qquad \text{when } y_3 = y_1 \qquad (3\text{-}48)$$

This statement replaces the integral equation that would otherwise be written for surface 3. Furthermore, since the fictive surface 4 is regarded as a black radiator at an effective temperature T_e, then

$$B_4 = \sigma T_e{}^4 \qquad (3\text{-}49)$$

This fact has already been utilized in writing equations (3-46) and (3-47).

Thus equations (3-46) and (3-47) constitute a pair of linear integral equations for the unknown radiosity distributions $B_1(y_1)$ and $B_2(x)$. Upon solving these, the corresponding heat flux distribution is calculable from equation (3-41), that is,

$$q_1(y_1) = \frac{\epsilon_1}{1 - \epsilon_1}\left[\sigma T_b{}^4 - B_1(y_1)\right] \qquad q_2(x) = \frac{\epsilon_2}{1 - \epsilon_2}\left[\sigma T_b{}^4 - B_2(x)\right]$$

$$(3\text{-}50)$$

As a result of symmetry, $q_3(y_3) = q_1(y_1)$ when $y_3 = y_1$.

This same problem can be alternatively analyzed by specializing equation (3-45), thus

$$\frac{q_1(y_1)}{\epsilon_1} = \sigma T_b{}^4 - \sigma T_b{}^4 F_{dA_1 - A_2} - \sigma T_b{}^4 F_{dA_1 - A_3} - \sigma T_e{}^4 F_{dA_1 - A_4}$$

$$+ \frac{1 - \epsilon_2}{\epsilon_2}\int_{A_2} q_2(x)K(y_1, x)\,dA_2$$

$$+ \frac{1 - \epsilon_3}{\epsilon_3}\int_{A_3} q_3(y_3)K(y_1, y_3)\,dA_3 \qquad (3\text{-}51)$$

$$\frac{q_2(x)}{\epsilon_2} = \sigma T_b{}^4 - \sigma T_b{}^4 F_{dA_2 - A_1} - \sigma T_b{}^4 F_{dA_2 - A_3} - \sigma T_e{}^4 F_{dA_2 - A_4}$$

$$+ \frac{1 - \epsilon_1}{\epsilon_1}\int_{A_1} q_1(y_1)K(x, y_1)\,dA_1$$

$$+ \frac{1 - \epsilon_3}{\epsilon_3}\int_{A_3} q_3(y_3)K(x, y_3)\,dA_3 \qquad (3\text{-}52)$$

In addition, as already noted, $q_3(y_3) = q_1(y_1)$ when $y_3 = y_1$. Equations (3-51) and (3-52) represent a pair of linear integral equations for the two unknown heat flux distributions $q_1(y_1)$ and $q_2(x)$.

It is interesting to compare the pair of integral equations for the radiosity, (3-46) and (3-47), with the pair of integral equations for the heat flux, (3-51) and (3-52). Upon inspection, it is seen that the major difference lies in the appearance of additional angle factors such as $F_{dA_1 - A_2}$, $F_{dA_1 - A_3}$ in the latter. These angle factors are functions of position which may be regarded as known. Thus, when employing the integral equations for the heat flux, one usually deals with a somewhat more complex inhomogeneous term than when employing the integral equations for the radiosity. This is not a decisive factor when the solutions are carried out with the aid of a digital computer.

Next, consider the case where the conductance of the fins is finite. Correspondingly, the temperature varies along the height of each fin. The problem will first be analyzed in terms of the radiosities. Equations (3-46) and (3-47) continue to apply, but now the term $\epsilon_1 \sigma T_b{}^4$ that formerly appeared on the right-hand side of equation (3-46) is replaced by $\epsilon_1 \sigma T_1{}^4(y_1)$. Moreover, owing to symmetry, $T_3(y_3) = T_1(y_1)$. Since one more dependent variable has been added to the problem, an additional equation must be provided. The additional relationship is the heat conduction equation (3-17), thus

$$q_1(y_1) = kt \frac{d^2 T_1(y_1)}{dy_1{}^2} = \frac{\epsilon_1}{1 - \epsilon_1} [\sigma T_1{}^4(y_1) - B_1(y_1)] \qquad (3\text{-}53)$$

in which equation (3-41) has been also employed. An appropriate pair of temperature boundary conditions must also be stated for the differential equation. Upon considering the thus-modified equations (3-46) and (3-47), together with equation (3-53), it is seen that the mathematical system is now integrodifferential and nonlinear.

The problem may also be formulated directly in terms of heat fluxes and temperatures. Considering equations (3-51) and (3-52), terms such as $\sigma T_b{}^4 F_{dA_1 - A_3}$, $\sigma T_b{}^4 F_{dA_2 - A_1}$, and $\sigma T_b{}^4 F_{dA_2 - A_3}$ are deleted and respectively replaced by

$$\int_{A_3} \sigma T_3{}^4(y_3) K(y_1, y_3)\, dA_3, \qquad \int_{A_1} \sigma T_1{}^4(y_1) K(x, y_1)\, dA_1, \text{ etc.} \qquad (3\text{-}54)$$

Moreover, from equation (3-17)

$$q_1(y_1) = kt \frac{d^2 T_1(y_1)}{dy_1{}^2} \qquad (3\text{-}55)$$

Thus the resulting mathematical system is also integrodifferential and nonlinear.

The question of whether the radiosity formulation has distinct advantages over the heat flux formulation or vice versa rests on which leads to an easier solution. Undoubtedly, in view of the complexity of the problem, all such solutions would have to be carried out numerically. The structure of the numerical method chosen for the task would determine which formulation would be advantageous. If a direct numerical iteration were to be employed, it appears to the authors that the formulation in terms of the heat flux might well offer some advantages.

3-4 Wavelength-Dependent Properties

It has already been noted that the gray-body model approaches reality when the radiant energy participating in the interchange is confined to a limited range of wavelengths within which the spectral emittance is essentially uniform. The gray-body model becomes prone to significant error when the radiant energy spans a wide wavelength range that encompasses the visible as well as the infrared. Such a situation may occur, for example, when a surface emits infrared radiation and is simultaneously irradiated from a high-temperature source such as the sun. The gray-body assumption also becomes tenuous when the range of active wavelengths contains strong emission (or absorption) bands.

An exact accounting of the wavelength dependence of the radiation properties can, in principle, be achieved by performing the radiant interchange calculations monochromatically and then integrating over the entire range of wavelengths. Such a procedure will be described briefly in the forthcoming paragraphs. However, the detailed spectral information needed to perform such calculations is not now available for the vast majority of engineering materials. Moreover, even if the requisite data were available, computations of this type are so lengthy as to be unrealistic except in special cases. Consequently, in lieu of a monochromatic analysis, it appears more practical to subdivide the wavelength range of interest into several finite bands and to assume that the participating surfaces are gray within each band. The analysis of radiant interchange based on a band model will be described in subsequent paragraphs.

Monochromatic analysis. Consideration will be given here to surfaces that emit and reflect diffusely at each wavelength. Furthermore, to simplify the presentation, it will be assumed that each of the participating surfaces is isothermal and that the radiosity is uniform along any given surface.   In principle, there is no difficulty in including surface variations of temperature and of radiosity.   The equations of radiant interchange are to be applied at each wavelength;* overall effects are found by integrating the monochromatic results over all active wavelengths.

The first case to be studied is that in which the temperatures are prescribed at all of the surfaces of the enclosure.   The analysis closely parallels that described on pp. 86–90 of Section 3-3.   As a logical modification of equation (3-18), the monochromatic radiosity B_λ is defined as

$$B_\lambda = \epsilon_\lambda e_{b\lambda}(T) + \rho_\lambda H_\lambda \tag{3-56}$$

in which ϵ_λ and ρ_λ denote the spectral emittance and the spectral reflectance, respectively.   In addition, $e_{b\lambda}(T)$ is the spectral emissive power of a black body as given by Planck's law:

$$e_{b\lambda}(T) = \frac{c_1}{\lambda^5(e^{c_2/\lambda T} - 1)} \tag{3-57}$$

Numerical values of the constants c_1 and c_2 have already been stated in Chapter 1.

Upon following through the analysis for an enclosure consisting of N finite surfaces as in Section 3-3, one finds

$$B_{i\lambda} = \epsilon_{i\lambda} e_{b\lambda}(T_i) + (1 - \epsilon_{i\lambda}) \sum_{j=1}^{N} B_{j\lambda} F_{A_i - A_j}, \qquad 1 \leqslant i \leqslant N \tag{3-58}$$

or alternatively

$$\sum_{j=1}^{N} X_{ij}(\lambda) B_{j\lambda} = \Omega_i(\lambda), \qquad 1 \leqslant i \leqslant N \tag{3-59}$$

where

$$X_{ij}(\lambda) = \frac{\delta_{ij} - (1 - \epsilon_{i\lambda}) F_{A_i - A_j}}{\epsilon_{i\lambda}} \tag{3-59a}$$

and

$$\Omega_i(\lambda) = e_{b\lambda}(T_i) \tag{3-59b}$$

Equation (3-59) is clearly analogous to equation (3-23) of Section 3-3. Therefore one can employ the matrix solution of the latter equation, from which there follows

$$B_{i\lambda} = \sum_{j=1}^{N} \psi_{ij}(\lambda) e_{b\lambda}(T_j), \qquad 1 \leqslant i \leqslant N \tag{3-60}$$

The ψ_{ij} are the elements of the inverse matrix $\psi = X^{-1}$.

* More precisely, within an infinitesimal band $d\lambda$.

The expression for the monochromatic heat flux is deduced by analogy with equations (3-28) and (3-28a), thus

$$\frac{Q_{i\lambda}}{A_i} = \sum_{j=1}^{N} \Lambda_{ij}(\lambda) e_{b\lambda}(T_j), \qquad 1 \leqslant i \leqslant N \qquad (3\text{-}61)$$

$$\Lambda_{ij}(\lambda) = \frac{\epsilon_{i\lambda}}{1 - \epsilon_{i\lambda}} [\delta_{ij} - \psi_{ij}(\lambda)] \qquad (3\text{-}61a)$$

The heat flux Q_i corresponding to radiant interchange over all wavelengths is then obtained by integration of equation (3-61).

$$Q_i = \int_{\lambda=0}^{\infty} Q_{i\lambda} \, d\lambda \qquad (3\text{-}62)$$

In practice, the integration would be restricted to a finite range, for $Q_{i\lambda}$ would be essentially zero outside that range.

Thus, for the case wherein the surface temperatures are prescribed, the monochromatic analysis emerges as a logical generalization of what has gone before. Moreover, it would appear that there is a real possibility that the necessary numerical computations could be carried out with the aid of a digital computer.

A slightly more complicated state of affairs occurs when the surface heat flux is prescribed. The difficulty arises from the fact that it is the integrated Q_i rather than the monochromatic $Q_{i\lambda}$ that would be prescribed. To illustrate the nature of the problem, suppose that the heat flux Q_N is prescribed at surface N, while the temperatures T_i are prescribed at all the other surfaces $i = 1, 2, \ldots, (N - 1)$ of the enclosure. A formal solution of the problem can be written by specializing equation (3-61). In particular, for $i = N$,

$$\frac{Q_{N\lambda}}{A_N} = \sum_{j=1}^{N-1} \Lambda_{Nj}(\lambda) e_{b\lambda}(T_j) + \Lambda_{NN}(\lambda) e_{b\lambda}(T_N) \qquad (3\text{-}63)$$

As it stands, this equation contains two unknowns: $Q_{N\lambda}$ and T_N. Upon integrating over all wavelengths, one finds

$$\frac{Q_N}{A_N} - \int_{\lambda=0}^{\infty} \sum_{j=1}^{N-1} \Lambda_{Nj}(\lambda) e_{b\lambda}(T_j) \, d\lambda = \int_{\lambda=0}^{\infty} \Lambda_{NN}(\lambda) e_{b\lambda}(T_N) \, d\lambda \qquad (3\text{-}63a)$$

Since Q_N and T_j $(1 \leqslant j \leqslant N - 1)$ are prescribed, the left-hand side of this equation can be regarded as known. The temperature T_N appearing on the right-hand side can thus be determined, very likely by a trial and error procedure. Once T_N has been found, the $Q_{i\lambda}$ at the other surfaces can be computed from equation (3-61) and, in turn, the Q_i are obtained from equation (3-62) by integration.

The generalization to cases where the heat flux is prescribed at two or more surfaces follows in a straightforward manner.

Band model. As a compromise between the monochromatic and the gray-body analyses, one can subdivide the range of active wavelengths into finite bands within which the radiation properties are assumed gray. To illustrate this approach, consider the case where two such bands, $\Delta\lambda_{\mathrm{I}}$ and $\Delta\lambda_{\mathrm{II}}$, are employed. Again, to simplify the presentation, the discussion will be related to an enclosure with N finite surfaces, each having uniform temperature and uniform radiosity.

To begin, one notes that the black-body emissive powers for the two bands $\Delta\lambda_{\mathrm{I}}$ and $\Delta\lambda_{\mathrm{II}}$ are

$$e_b(\Delta\lambda_{\mathrm{I}},\, T) = \int_{\Delta\lambda_{\mathrm{I}}} e_{b\lambda}(T)\, d\lambda \qquad e_b(\Delta\lambda_{\mathrm{II}},\, T) = \int_{\Delta\lambda_{\mathrm{II}}} e_{b\lambda}(T)\, d\lambda \quad (3\text{-}64)$$

The numerical values of these quantities are most conveniently determined from Table 1-2 in Chapter 1. The corresponding emissive powers for the nonblack surfaces are then expressed as

$$\epsilon(\Delta\lambda_{\mathrm{I}})e_b(\Delta\lambda_{\mathrm{I}},\, T) \qquad \epsilon(\Delta\lambda_{\mathrm{II}})e_b(\Delta\lambda_{\mathrm{II}},\, T) \qquad\qquad (3\text{-}65)$$

The quantities $\epsilon(\Delta\lambda_{\mathrm{I}})$ and $\epsilon(\Delta\lambda_{\mathrm{II}})$ are regarded as gray-body emittances for the respective bands $\Delta\lambda_{\mathrm{I}}$ and $\Delta\lambda_{\mathrm{II}}$.

Suppose first that the temperatures are prescribed at all the surfaces of the enclosure. Then the equations of radiant interchange for the two bands follow as a direct generalization of the foregoing monochromatic analysis. Thus, for the band $\Delta\lambda_{\mathrm{I}}$,

$$\sum_{j=1}^{N} X_{ij}(\Delta\lambda_{\mathrm{I}}) B_j(\Delta\lambda_{\mathrm{I}}) = \Omega_i(\Delta\lambda_{\mathrm{I}}), \qquad 1 \leqslant i \leqslant N \qquad (3\text{-}66)$$

and for the band $\Delta\lambda_{\mathrm{II}}$

$$\sum_{j=1}^{N} X_{ij}(\Delta\lambda_{\mathrm{II}}) B_j(\Delta\lambda_{\mathrm{II}}) = \Omega_i(\Delta\lambda_{\mathrm{II}}), \qquad 1 \leqslant i \leqslant N \qquad (3\text{-}67)$$

The coefficients X_{ij} and the inhomogeneous terms Ω_i are given by equations (3-59a) and (3-59b), after suitable changes of notation for the black-body emissive power and the emittance in accordance with the definitions of equations (3-64) and (3-65). The two sets of equations represented by (3-66) and (3-67) are independent of one another and the corresponding solutions can be obtained separately.

The heat fluxes for the separate bands follow by analogy with equation (3-61) as

$$\frac{Q_i(\Delta\lambda_{\mathrm{I}})}{A_i} = \sum_{j=1}^{N} \Lambda_{ij}(\Delta\lambda_{\mathrm{I}})e_b(\Delta\lambda_{\mathrm{I}}, T_j), \qquad 1 \leqslant i \leqslant N \qquad (3\text{-}68)$$

$$\frac{Q_i(\Delta\lambda_{\mathrm{II}})}{A_i} = \sum_{j=1}^{N} \Lambda_{ij}(\Delta\lambda_{\mathrm{II}})e_b(\Delta\lambda_{\mathrm{II}}, T_j), \qquad 1 \leqslant i \leqslant N \qquad (3\text{-}69)$$

in which the Λ_{ij} have a definition similar to that given in equation (3-61a). Then the overall heat transfer rate at any surface i is obtained by summing the separate contributions from each band.

$$Q_i = Q_i(\Delta\lambda_{\mathrm{I}}) + Q_i(\Delta\lambda_{\mathrm{II}}) \qquad (3\text{-}70)$$

In light of the foregoing development, it is apparent that the solution of a two-band model will require about twice as much time as the solution of a gray-body model.

When the heat flux is prescribed instead of the surface temperature, minor complications arise as in the case of the monochromatic analysis. For purposes of illustration, suppose that the heat flux Q_N is prescribed at surface N, while the temperatures T_i are prescribed at all the other surfaces $i = 1, 2, \ldots, (N-1)$ of the enclosure. By specializing equations (3-68), (3-69), and (3-70) to surface N, one finds

$$\frac{Q_N}{A_N} - \sum_{j=1}^{N-1} [\Lambda_{Nj}(\Delta\lambda_{\mathrm{I}})e_b(\Delta\lambda_{\mathrm{I}}, T_j) + \Lambda_{Nj}(\Delta\lambda_{\mathrm{II}})e_b(\Delta\lambda_{\mathrm{II}}, T_j)]$$

$$= \Lambda_{NN}(\Delta\lambda_{\mathrm{I}})e_b(\Delta\lambda_{\mathrm{I}}, T_N) + \Lambda_{NN}(\Delta\lambda_{\mathrm{II}})e_b(\Delta\lambda_{\mathrm{II}}, T_N) \quad (3\text{-}71)$$

The left-hand side of this equation contains only prescribed quantities and may therefore be regarded as known. The only unknown on the right-hand side is the temperature T_N. This can be determined with the aid of Table 1-2, probably by a trial and error procedure. Once T_N has been found, the results for the other surfaces follow by application of equations (3-68), (3-69), and (3-70).

The generalization of the band model to include three or more bands is straightforward. Additionally, there is, in principle, no difficulty in applying the band model to the previously described integral equation formulation.

3-5 Solution Methods for the Integral Equations of Radiant Interchange

It has been shown by experience that exact analytical solutions of the integral equations of radiant interchange are possible only in very special geometrical configurations; these will be discussed later. For more general problems, one must have recourse either to numerical solutions or to approximate analytical methods. In view of the wide prevalence of high-speed digital computers, the numerical approach appears to have the greater likelihood of application to problems of direct practical interest. However, approximate analytical techniques are attractive, for they frequently yield closed-form solutions that facilitate investigation of the effect of changes in the values of the geometrical and radiation parameters.

The various approaches just mentioned will be discussed in greater detail in the paragraphs that follow. Consideration will first be given to numerical procedures, then to approximate analytical methods, and, finally, to exact solutions.

Numerical methods. Consider first the case of integral equations that are linear in σT^4, B, or q; in other words, there is no nonlinear coupling such as that afforded by the heat conduction equation (3-17). Typically, the linear integral equations under consideration may be represented by any of equations (3-14), (3-15), (3-40), (3-43), and (3-45). In any of these problems, an unknown integral term has the form

$$\int_{A_j} f(\mathbf{r}_j) K(\mathbf{r}_i, \mathbf{r}_j)\, dA_j \tag{3-72}$$

where $f(\mathbf{r}_j)$ represents any of the unknown dependent variables.

One numerical approach is to reduce the linear integral equations to a corresponding set of linear algebraic equations. Once this reduction has been effected, advantage can be taken of standard computer routines for solving linear algebraic equations. If the integrals are approximated by the trapezoidal rule, the resulting algebraic equations are identical to those that would be derived by applying radiant-flux balances directly to finite surfaces in the enclosure. Perhaps the only distinction between the trapezoidal-rule representation of the integral equations and the finite-surface formulation is that in the former case a special effort would be made to use a sufficient number of mesh points to ensure close approximation of the integrals.

In most cases, a better finite-difference approximation to an integral is achieved by employing Simpson's rule. The resulting algebraic equations are still linear in the unknowns, but these equations differ from those obtained from the direct finite-surface formulation. Perlmutter and Siegel (refs. 2 to 4) have successfully employed the Simpson rule representation for solving radiant interchange problems in cylindrical tubes.

Higher-order, finite-difference approximations (four, five, etc. point formulas) of the integral terms may be employed; such representations are available in standard textbooks on numerical analysis. In all cases the resulting algebraic equations are linear, and these are solvable using the same computer routines as would be employed for solving the algebraic equations deduced from the simple trapezoidal-rule approximation.

A second numerical approach is that of successive iterations. This method can be advantageously described by referring to a specific example; for instance, equations (3-46) and (3-47), which pertain to gray, diffuse interchange in the enclosure pictured in Fig. 3-3. The first step is to guess the radiosity distribution $B_1(y_1)$. Moreover, since $B_3(y_3) = B_1(y_1)$ when $y_3 = y_1$, this guess also fixes the distribution $B_3(y_3)$. Then, turning our attention to equation (3-47), it may be observed that for a specific x, the kernel $K(x, y)$ is a function of y alone. Thus, with the guessed distributions of B_1 and B_3 and a given x, the integrations appearing on the right-hand side of equation (3-47) can be performed numerically. This operation gives a value of B_2 at the preselected x. Next, another location x is chosen and the integration is repeated, and a value of B_2 is obtained at this second x. Proceeding in this way, a distribution $B_2(x)$ is generated along the entire surface A_2.

With this $B_2(x)$ and with the initially guessed distribution $B_3(y_3)$, one can then turn to equation (3-46). For a preselected y_1, the integrals on the right-hand side of the equation can be evaluated numerically, which yields a value of B_1 at that y_1. By assigning successive y_1, a new distribution $B_1(y_1)$ can be generated. The new distribution of B_1 is then compared with that initially guessed. If the agreement is not satisfactory, the procedure is repeated again and again until the B_1 distributions from successive outputs are the same to within some preselected tolerance.

The foregoing numerical integrations can be carried out by any number of methods. These are so well documented in standard texts on numerical analysis that repetition here is superfluous. Experience with the solution of radiation problems has indicated that the direct iteration procedure will converge without difficulty when the governing integral equations are linear. The iterative procedure has been employed successfully to solve a wide range of problems relating to cavities; refs. 5 and 6 describe representative solutions.

The linearity of the mathematical problem is destroyed when there is a coupling between radiation and conduction [e.g., equation (3-17)] or convection. However, the numerical schemes previously discussed continue to be applicable, provided that appropriate modifications are made. The finite-difference representation, when applied both to integral and derivative terms, yields a set of nonlinear algebraic equations. A solution method for such systems is described in general by Scarborough (ref. 7) and has been specialized to nonlinear problems of radiant interchange by Ness (ref. 8). The finite-difference method of solution has been utilized for a problem of combined radiation and conduction in ref. 9.

Care must be exercised when applying the iterative method to nonlinear systems. A direct iteration (i.e., using the output of one iteration as input to the next) may well diverge. This difficulty can almost always be circumvented by using weighted inputs. By this procedure, one weights a given output with the prior input and uses this as input to the next cycle of the iteration. Such an iterative procedure was successfully employed in an analysis of radiating-conducting fins (ref. 10).

Approximate analytical methods. A variety of approximate analytical techniques are available for solving the integral equations of radiant interchange. A total of five such methods will be discussed here.

Two of the methods aim at reducing the governing integral equations to differential equations. The first is the approximate-kernel method whereby the actual angle factors are replaced by approximate, but more manageable functions. Thus far the method appears to show promise only for angle factors of the form

$$K(x, y) = \text{function of } |x - y| \tag{3-73}$$

Angle factors of this type occur in cylindrical cavities and tubes and in plane-walled cavities and passages.

The actual angle factors are approximated by exponentials

$$K(x, y) = Ae^{-a|x-y|} + Be^{-b|x-y|} + \cdots \tag{3-74}$$

Consider first the case where one exponential is employed. A typical integral term, such as that given by equation (3-72), then becomes

$$I(y) = A \int_{x_1}^{x_2} f(x)e^{-a|x-y|} \, dx \tag{3-75}$$

If I is twice differentiated with respect to y and proper cognizance taken of the absolute magnitude signs, one obtains

$$\frac{d^2I}{dy^2} = a^2 A \int_{x_1}^{x_2} f(x)e^{-a|x-y|} \, dx - 2aAf(y) \tag{3-76}$$

The last term on the right-hand side is present only when $x_1 \leqslant y \leqslant x_2$, as has been the case in all previously published applications.

Thus the integral term I is restored by the double differentiation. Then the integral terms can be eliminated between the undifferentiated and the twice-differentiated forms of the integral equation. These operations lead to a second-order *differential* equation, the solution of which can be undertaken by either analytical or numerical means, whichever appears suitable.

When two exponential terms are employed, as exemplified by equation (3-74), it can be shown that a typical integral term in the governing equations is restored by a combination of quadruple and double differentiation. Consequently, each integral equation can be transformed into a fourth-order differential equation.

The foregoing approximate-kernel method seems first to have been employed in 1927 by Buckley (ref. 11) for interchange in a cylindrical cavity. Later it was applied to a series of radiant interchange problems involving the cylindrical tube (refs. 2 to 4, 12) and the rectangular-groove cavity (ref. 13). As a recent innovation, Perlmutter and Siegel (ref. 4) have devised a correction procedure to improve the accuracy of the solutions, the accuracy of the uncorrected solutions having proved insufficient for long tubes.

A second approach that reduces the governing integral equations to differential equations is the Taylor's series method. For purposes of illustration, this method will be applied here to the relatively simple integral equation

$$f(y) = G(y) + \beta \int_{x_1}^{x_2} f(x) K(y, x)\, dx \qquad (3\text{-}77)$$

where $x_1 \leqslant y \leqslant x_2$ and β is any constant. However, in principle, there is no reason why the method cannot be applied to systems involving more than one equation, for instance, equations (3-46) and (3-47), or to cases where nonlinear coupling occurs.

The essential idea of the method is to expand the unknown function appearing in the integrand in terms of a Taylor's series as follows:

$$f(x) = f(y) + (x - y)\left[\frac{d}{dy} f(y)\right]_y + \frac{1}{2}(x - y)^2 \left[\frac{d^2}{dy^2} f(y)\right]_y + \cdots \quad (3\text{-}78)$$

Upon introducing this into the integral equation (3-77) and rearranging, one finds

$$\frac{d^2 f}{dy^2}\left[\frac{1}{2}\int_{x_1}^{x_2} (x - y)^2 K(y, x)\, dx\right] + \frac{df}{dy}\left[\int_{x_1}^{x_2} (x - y) K(y, x)\, dx\right]$$

$$+ f\left[\int_{x_1}^{x_2} K(y, x)\, dx - \frac{1}{\beta}\right] + \frac{G}{\beta} = 0 \quad (3\text{-}79)$$

After the integration in x has been performed, the bracketed quantities are functions of y alone.   Thus a second-order, linear, ordinary differential equation with variable coefficients is derived for $f(y)$.   Depending on the nature of the coefficients, such an equation may be solved either by analytical or by numerical methods.

Whenever an integral equation is transformed into a differential equation, it is necessary to provide an appropriate number of boundary conditions.   In general, sufficient boundary conditions may be generated by requiring that the original integral equation be satisfied at selected points.   Alternative boundary conditions may be obtained by using whatever symmetries exist in the problem.   Further details may be found in ref. 4, where the Taylor's series method was successfully employed in solving for the radiant interchange in a tube.

An altogether different point of view is employed in another pair of approximate analytical methods.   In these, one proposes a solution for the unknowns in the form of a linear sum of preselected functions g_i, thus

$$f(\mathbf{r}) = c_1 g_1(\mathbf{r}) + c_2 g_2(\mathbf{r}) + c_3 g_3(\mathbf{r}) + \cdots \qquad (3\text{-}80)$$

The c_i are constants that remain to be found.   One approach to the determination of the constants may be termed point matching.   In utilizing this method, the governing integral equations are applied at a specific set of points whose number is the same as the number of unknown constants.   The proposed solutions, as embodied by equation (3-80), are substituted into the integral equations and the indicated integrations performed.   This leaves a set of linear algebraic equations that is just sufficient to determine the unknowns c_i.   Alternatively, the governing equations may be applied at a set of points whose number exceeds the number of unknown constants.   Under these circumstances, least-squares techniques can be employed.

A second approach to finding the c_i is to employ variational methods.   It can be shown that, corresponding to the governing equations of radiant interchange, there is a variational expression that has a very special property; namely, that when functions are found which make the variational expression an extremum, those same functions are solutions of the original integral equations.   To find the functions that provide an absolute extremum is extremely difficult. Instead, one attempts to come as close as possible to the extremum and in this way to get a close approximation to the solution of the radiation problem.

In practice, proposed solutions in the form of equation (3-80) are substituted into the variational expression, and constants c_i are found that provide a relative extremum corresponding to the preselected form of solution. The variational method is described in detail in ref. 14. Applications are presented in refs. 4, 12, and 15.

The last of the approximate analytical approaches to be discussed here is the method of successive substitutions. This is the analytical analog of the previously discussed numerical method of successive iterations. In this analytical approach, one selects a simple first approximation for each of the unknown functions and then substitutes these into the integral terms appearing on the right-hand sides of the governing equations. The integrals are carried out analytically, thus yielding a new analytical expression for each unknown function. In turn, these new expressions are substituted into the integral terms and the required integrations performed. This operation provides yet another set of expressions for the unknowns, and so forth. In practice, the required integrations become so complex that only one or two stages of the successive substitutions can be carried out. The method is well-documented in the text of Hildebrand (ref. 16).

Exact solutions. At least two geometrical configurations seem to permit exact solutions of the governing integral equations. One of these is the spherical cavity pictured in Fig. 3-6. It can be readily shown that the angle factor for diffuse interchange between any two surface elements within the spherical cavity is a constant, that is (ref. 17),

$$dF_{dA_1 - dA_2} = \frac{dA_2}{4\pi R^2} \tag{3-81}$$

where R is the radius of the spherical shell. Correspondingly, the radiant flux balance at a particular location (ϕ_0, θ_0) becomes

$$B(\phi_0, \theta_0) = \epsilon\sigma T^4(\phi_0, \theta_0) + \rho\left[H^*(\phi_0, \theta_0) + \frac{1}{4\pi R^2}\int_A B(\phi, \theta)\, dA\right] \tag{3-82}$$

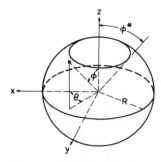

Fig. 3-6 Spherical cavity.

in which $H^*(\phi_0, \theta_0)$ is the incident radiation at (ϕ_0, θ_0) that enters the cavity through its opening and A is the total surface area of the cavity wall.

A general solution for B that is valid for any prescribed distributions of T and H^* is

$$B(\phi, \theta) = G(\phi, \theta) + \frac{\dfrac{(1 - \epsilon)}{4\pi R^2} \displaystyle\int_A G \, dA}{1 - \dfrac{(1 - \epsilon)A}{4\pi R^2}} \qquad (3\text{-}83)$$

in which

$$G = \epsilon\sigma T^4 + (1 - \epsilon)H^* \qquad (3\text{-}83a)$$

If the prescribed distributions of T and H^* cannot be analytically integrated, a numerical quadrature is necessary. A closed-form solution similar to equation (3-83) can also be derived for the case of prescribed surface heat flux.

The second configuration that invites analytical treatment is the cylindrical arc cavity pictured in Fig. 3-7. The extension of the surface normal to the plane of the figure is very large. For this case, a typical integral term in the governing equation has the form

$$I(\alpha) = \int_0^{\theta_0} f(\theta) \sin \frac{1}{2}|\theta - \alpha| \, d\theta \qquad (3\text{-}84)$$

in which $0 \leqslant \alpha \leqslant \theta_0$. If this integral is twice differentiated with respect to α (with proper accounting of the absolute magnitude signs), it is found that

$$\frac{d^2 I}{d\alpha^2} = f(\alpha) - \frac{1}{4} \int_0^{\theta_0} f(\theta) \sin \frac{1}{2}|\theta - \alpha| \, d\theta \qquad (3\text{-}85)$$

that is, the integral is restored after the two differentiations. Thus the integrals can be eliminated between the undifferentiated and twice-differentiated forms of the governing equation. This operation gives

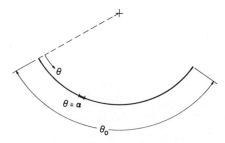

Fig. 3-7   Circular-arc cavity.

rise to a second-order differential equation that may be solved analytically for certain types of boundary conditions. Additional details relating to this problem are available in ref. 18.

The authors are unaware of other configurations that permit analytical solutions of the type just described for the spherical cavity and the circular-arc cavity.

REFERENCES

1. A. K. Oppenheim, Radiation analysis by the network method. *Trans. ASME*, **78**, 725–735 (1956).

2. M. Perlmutter and R. Siegel, Heat transfer by combined forced convection and thermal radiation in a heated tube. *J. Heat Transfer*, **C84**, 301–311 (1962).

3. R. Siegel and M. Perlmutter, Convective and radiant heat transfer for flow of a transparent gas in a tube with a gray wall. *Intern. J. Heat Mass Transfer*, **5**, 639–660 (1962).

4. M. Perlmutter and R. Siegel, Effect of specularly reflecting gray surface on thermal radiation through a tube and from its heated wall. *J. Heat Transfer*, **C85**, 55–62 (1963).

5. E. M. Sparrow, L. U. Albers, and E. R. G. Eckert, Thermal radiation characteristics of cylindrical enclosures. *J. Heat Transfer*, **C84**, 73–81 (1962).

6. E. M. Sparrow and V. K. Jonsson, Radiant emission characteristics of diffuse conical cavities. *J. Opt. Soc. Am.*, **53**, 816–821 (1963).

7. J. B. Scarborough, *Numerical Analysis*, 4th ed., The Johns Hopkins Press, Baltimore, Md., 1958.

8. A. J. Ness, Solution of equations of a thermal network on a digital computer. *Solar Energy*, **3**, 37 (1960).

9. E. M. Sparrow, G. B. Miller, and V. K. Jonsson, Radiating effectiveness of annular-finned space radiators, including mutual irradiation between radiator elements. *J. Aerospace Sci.*, **29**, 1291–1299 (1962).

10. E. M. Sparrow, E. R. G. Eckert, and T. F. Irvine, Jr., The effectiveness of radiating fins with mutual irradiation. *J. Aerospace Sci.*, **28**, 763–772 (1961).

11. H. Buckley, On the radiation from the inside of a circular cylinder. *Phil. Mag.*, (7) **4** (23), 753–762 (1927).

12. C. M. Usiskin and R. Siegel, Thermal radiation from a cylindrical enclosure with specified wall heat flux. *J. Heat Transfer*, **C82**, 369–374 (1960).

13. E. M. Sparrow and V. K. Jonsson, Thermal radiation absorption in rectangular-groove cavities. *J. Appl. Mech.*, **E30**, 237–244 (1963).

14. E. M. Sparrow and A. Haji-Sheikh, A generalized variational method for calculating radiant interchange between surfaces. *J. Heat Transfer*, **C87**, 103–109 (1965).

15. E. M. Sparrow, Application of variational methods to radiation heat transfer calculations. *J. Heat Transfer*, **C82**, 375–380 (1960).

16. F. B. Hildebrand, *Methods of Applied Mathematics*, Prentice-Hall, Englewood Cliffs, N.J., pp. 421–427, 1952.

17. E. M. Sparrow and V. K. Jonsson, Absorption and emission characteristics of diffuse spherical enclosures. *J. Heat Transfer*, **C84**, 188–189 (1962).

18. E. M. Sparrow, Radiant absorption characteristics of concave cylindrical surfaces. *J. Heat Transfer*, **C84**, 283–293 (1962).

PROBLEMS

3-1.   A 15-foot length of steam pipe passes through a room whose walls are at 70°F.   The pipe is insulated, the outer diameter of the insulation being 2 inches.   The outer surface of the insulation is at 105°F.   Calculate the heat loss by radiation from the pipe to the room, assuming that the surface of the insulation has an emittance ϵ of 0.95.   For comparison purposes, the heat loss by natural convection is 250 Btu/hr.

3-2.   Consider two geometrically identical black enclosures.   For the first enclosure, the temperatures of the surfaces are $T_1, T_2, \ldots, T_N$. For the second enclosure, the temperatures are $(T_1{}^4 - C)^{\frac{1}{4}}, (T_2{}^4 - C)^{\frac{1}{4}}, \ldots, (T_N{}^4 - C)^{\frac{1}{4}}$, where C is a constant.   Show that the heat transfer rate Q_i at any surface i is the same for both enclosures, regardless of the value of C.

3-3.   Prove that the sum of the $Q_1, Q_2, \ldots, Q_N$ given by equation (3-5) is zero.

3-4.   An enclosure consists of four plane black surfaces, 1, 2, 3, and 4. Each surface is isothermal.   For the case in which surfaces 1 and 3 are adiabatic (i.e., $Q_1 = Q_3 = 0$), derive expressions for $T_1{}^4$ and $T_3{}^4$ in terms of the known temperatures of surfaces 2 and 4.   Assume that all angle factors are known.

3-5.   For the conditions of Problem 3-4, how are Q_2 and Q_4 related?

3-6.   The top and bottom walls of a black cubical enclosure are at a temperature T_a and the four side walls are at a temperature T_b.   Derive an expression for the heat transfer rate Q at the bottom wall.   The numerical value of all of the participating angle factors is 0.2.

3-7.   A V-groove cavity is shown schematically in diagram (d) of

Fig. 6-1, page 174. The walls are very long in the direction into and out of the plane of the page. The angle factor for radiant interchange between the cavity walls is $(1 - \sin \phi)$. Assume that both walls are black and are at temperature T. Neglect any radiation entering the cavity from the external environment. Show that the radiation leaving the cavity is $2h\sigma T^4$ per unit length normal to the plane of the figure.

3-8. Two parallel walls separated by a distance h are pictured in the figure below. The walls are very long in the direction into and out of the plane of the page. The temperature of the lower wall (wall 1) is a uniform value T_1. At the upper wall (wall 2), the heat flux q_2 is uniform. Both walls are black. The radiation in the external environment is black-body radiation at temperature T_e. The angle factors between the strips dA_1 and dA_2 are

$$\frac{dF_{dA_1 - dA_2}}{dx_2} = \frac{dF_{dA_2 - dA_1}}{dx_1} = \frac{h^2/2}{[(x_2 - x_1)^2 + h^2]^{3/2}}$$

Show that the temperature distribution along wall 2 is

$$\sigma T_2^4(x_2) - \sigma T_e^4 = q_2 + (\sigma T_1^4 - \sigma T_e^4) \int_0^L dF_{dA_2 - dA_1}$$

Carry out the integral and obtain the explicit dependence of T_2^4 on x_2. Hint:

$$\int_{A_3} dF_{dA_2 - dA_3} + \int_{A_4} dF_{dA_2 - dA_4} = 1 - \int_{A_1} dF_{dA_2 - dA_1}$$

where A_3 and A_4 are fictitious surfaces stretched across the openings at $x = 0$ and $x = L$.

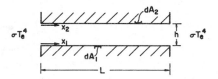

3-9. For Problem 3-8, derive an expression for $q_1(x_1)$.

3-10. Consider a gray diffuse enclosure, the surfaces of which have temperatures $T_1, T_2, \ldots, T_N$. The radiosities are to be assumed uniform along the respective surfaces. In an identical enclosure, the temperatures are $(T_1^4 - C)^{1/4}, (T_2^4 - C)^{1/4}, \ldots, (T_N^4 - C)^{1/4}$. Show that the radiosities for the latter case are $B_1^I - \sigma C, B_2^I - \sigma C, \ldots,$

$B_N{}^I - \sigma C$, where $B_1{}^I, B_2{}^I, \ldots, B_N{}^I$ are the radiosities for the first case. Then, prove that Q_i is identical for the two cases.

3-11.   Prove that the sum of the Q_i for all the surfaces of a gray diffuse enclosure is zero.   Hint: Start with the expression $Q_i/A_i = B_i - H_i$ and then evaluate H_i from equation (3-21).

3-12.   Two parallel plates of identical surface area are situated so that the spacing between the plates is very small compared with both the length and the width of the surfaces.   Let plate 1 have temperature T_1 and emittance ϵ_1, while plate 2 has temperature T_2 and emittance ϵ_2.   Both plates are gray diffuse emitters and reflectors of radiant energy.   Derive an expression for Q_1/A_1.   How does Q_1/A_1 compare with Q_2/A_2?   Hint: The pertinent angle factors are unity.

3-13.   Reconsider Problem 3-7.   Let the walls of the cavity be gray diffuse emitters and reflectors of radiant energy.   Both walls have identical emittances ϵ and temperatures T.   Formulate the problem under the assumption that the radiosities are uniform along the respective walls.   As in Problem 3-7, neglect the radiation entering the cavity from the external environment.   Show that the radiation emerging from the opening of the cavity is

$$Q = \frac{2h\epsilon\sigma T^4}{\epsilon + (1 - \epsilon)\sin\phi}$$

per unit length in the direction normal to the plane of the figure.

3-14.   Generalize Problem 3-13.   Let the radiation in the external environment be black radiation at temperature T_e.   Show that the expression for Q, derived in Problem 3-13, continues to be valid provided that T^4 is replaced by $T^4 - T_e{}^4$.

3-15.   Draw the electric circuit representation for Problem 3-14.

3-16.   Reconsider Problem 3-6.   All conditions remain unchanged, but now let the top and the bottom walls have an emittance ϵ_a, while each side wall has an emittance ϵ_b.   All walls are gray diffuse emitters and reflectors.   Assume that the radiosity is uniform along each one of the surfaces.   Derive an expression for the heat transfer rate at the bottom wall.

3-17.   Draw the electric circuit representation for Problem 3-16.

3-18.   In Problem 3-16, how does the heat transfer rate at any one of the side walls compare with the heat transfer rate at the bottom wall?

3-19.   Refer to the figure in Problem 3-8.   Suppose now that wall 1 has uniform temperature T_1 and emittance ϵ_1, while wall 2 has uniform temperature T_2 and emittance ϵ_2.   Both walls are gray diffuse emitters and reflectors.   The radiation in the external environment is blackbody radiation at temperature T_e.   Write the integral equations for $B_1(x_1)$ and $B_2(x_2)$.   Use the expressions for the angle factors as well as the hint given in Problem 3-8.

Answer:

$$\hat{B}_1(x_1) = \epsilon_1 \sigma \hat{T}_1{}^4 + (1 - \epsilon_1) \frac{h^2}{2} \int_0^L \frac{\hat{B}_2(x_2)\, dx_2}{[(x_2 - x_1)^2 + h^2]^{3/2}}$$

$$\hat{B}_2(x_2) = \epsilon_2 \sigma \hat{T}_2{}^4 + (1 - \epsilon_2) \frac{h^2}{2} \int_0^L \frac{\hat{B}_1(x_1)\, dx_1}{[(x_2 - x_1)^2 + h^2]^{3/2}}$$

where $\hat{B} = B - \sigma T_e{}^4$, $\hat{T}^4 = T^4 - T_e{}^4$.

3-20.   By introducing dimensionless coordinates $X = x/h$, the angle factor appearing in the B_1 equation of Problem 3-19 takes the form

$$\frac{dF_{dA_1 - dA_2}}{dX_2} = \frac{\frac{1}{2}}{[(X_2 - X_1)^2 + 1]^{3/2}} \equiv K(X_1, X_2)$$

It is proposed to approximate K by $\frac{1}{2} \exp[-|X_2 - X_1|]$.   Examine the accuracy of the approximation.   (A better approximation is given in Fig. 6 of reference 6, cited at the end of Chapter 6.)

3-21.   Use the approximate angle factor of Problem 3-20 to convert the integral equations of Problem 3-19 into differential equations.   Hint: Use equation (3-76).

3-22.   Derive equation (3-83).

CHAPTER 4

ANGLE FACTORS FOR DIFFUSE INTERCHANGE

The analysis of radiant interchange among diffusely emitting and diffusely reflecting surfaces depends heavily on a knowledge of angle factors. This chapter is concerned with the analytical representation of angle factors and with methods by which numerical values of these quantities can be determined. In Section 4-1 the expressions that relate the angle factor to the shape and the orientation of the participating surfaces are derived. The application of these expressions is illustrated. Then, in Section 4-2, it is demonstrated that the basic equations of definition, which involve area integrals, can be rephrased in alternative and more convenient forms, which involve contour integrals. Certain shorthand methods of determining angle factors are described in Section 4-3. Lastly, Section 4-4 presents in graphical form a catalogue of available angle-factor information. When carrying out computations that require angle factors, it is advisable to consult such a catalogue before attempting a direct evaluation of these quantities.

In addition to the analytical methods to be discussed, certain graphical and analogy methods may also be applied to the determination of angle factors. These are well-documented in refs. 1 to 4 and will not be described here.

4-1 The Defining Equations and Their Application

The equations of definition. Consideration is given first to interchange between a pair of infinitesimal surfaces dA_i and dA_j as illustrated in Fig. 4-1. The normals to the elements are denoted, respectively, by unit vectors $\hat{n}_i$ and $\hat{n}_j$. The angles β_i and β_j are formed by the respective normals and the connecting line between the elements. The length of the connecting line is r.

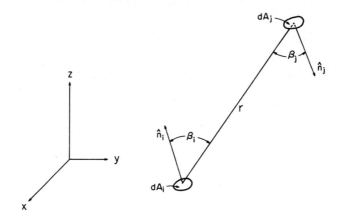

Fig. 4-1   Configuration for interchange between two infinitesimal elements.

Suppose that B_i represents the radiosity at the element dA_i; it is assumed that the radiosity is diffusely distributed. As previously noted in Chapter 1, equation (1-9), the intensity corresponding to any diffusely distributed flux of radiant energy Φ is uniform and equal to Φ/π. Therefore the intensity corresponding to B_i is

$$i_i = \frac{B_i}{\pi} \qquad (4\text{-}1)$$

The radiant energy leaving dA_i in the direction of the element dA_j may be written, in accordance with equation (2-24), as

$$i_i \, dA_i \cos \beta_i \, d\omega \qquad (4\text{-}2)$$

in which $d\omega$ is the solid angle subtended by dA_j when viewed from dA_i. From the geometry of the situation, it follows that $d\omega = dA_j \cos \beta_j / r^2$. Introducing this into expression (4-2) and eliminating i_i in favor of B_i, one obtains

$$\frac{B_i \cos \beta_i \cos \beta_j \, dA_i \, dA_j}{\pi r^2} \qquad (4\text{-}3)$$

It may also be noted that the radiant energy leaving dA_i in all directions is

$$B_i \, dA_i \qquad (4\text{-}4)$$

The ratio of expression (4-3) to expression (4-4) represents the fraction of the radiant energy leaving dA_i that is incident on dA_j.

This is precisely the definition of the angle factor $dF_{dA_i - dA_j}$, thus

$$dF_{dA_i - dA_j} = \frac{\cos \beta_i \cos \beta_j \, dA_j}{\pi r^2} \qquad (4\text{-}5)$$

It is evident that this angle factor is of infinitesimal order, for it is proportional to the infinitesimal area element dA_j. As noted in Chapter 3, the first subscript appended to the angle factor identifies the surface from which the radiation emanates, while the second subscript identifies the surface on which the radiation is incident.

By retracing the derivation of the previous paragraphs, it is readily shown that the angle factor for diffusely distributed radiant energy leaving the element dA_j and arriving at dA_i is

$$dF_{dA_j - dA_i} = \frac{\cos \beta_i \cos \beta_j \, dA_i}{\pi r^2} \qquad (4\text{-}6)$$

Upon comparing equations (4-5) and (4-6), one sees that

$$dA_i \, dF_{dA_i - dA_j} = dA_j \, dF_{dA_j - dA_i} \qquad (4\text{-}7)$$

This is the reciprocity rule that was stated in equation (3-1c) of Chapter 3.

Next, consideration is given to the angle factors for diffuse interchange between an infinitesimal element dA_i and a finite surface A_j as shown in Fig. 4-2. The radiant energy passing from dA_i to a typical

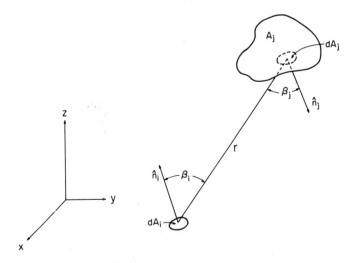

Fig. 4-2   Configuration for interchange between an infinitesimal element and a finite surface.

area element dA_j on surface j is represented by expression (4-3). Then, the radiation passing from dA_i to all such elements comprising A_j is obtained by integration, thus

$$B_i \, dA_i \int_{A_j} \frac{\cos \beta_i \cos \beta_j \, dA_j}{\pi r^2} \qquad (4\text{-}8)$$

The factor $B_i \, dA_i$ is independent of A_j and has therefore been removed from under the integral sign. The range of integration is extended over those parts of A_j that are directly visible from dA_i.

The rate at which radiant energy leaves dA_i in all directions has already been stated in expression (4-4). From the ratio of expression (4-8) to expression (4-4), there is obtained the fraction of the radiation leaving the element dA_i that arrives at the finite surface A_j. This is the angle factor $F_{dA_i - A_j}$.

$$F_{dA_i - A_j} = \int_{A_j} \frac{\cos \beta_i \cos \beta_j \, dA_j}{\pi r^2} \qquad (4\text{-}9)$$

The angle factor for radiation traveling from A_j to dA_i may now be derived. First, the radiant energy passing from a typical element dA_j of A_j to dA_i may be stated by interchanging subscripts i and j in expression (4-3), thus

$$\frac{B_j \cos \beta_i \cos \beta_j \, dA_i \, dA_j}{\pi r^2} \qquad (4\text{-}10)$$

By integrating the foregoing over all parts of A_j that are directly visible from dA_i, one has

$$dA_i \int_{A_j} \frac{B_j \cos \beta_i \cos \beta_j \, dA_j}{\pi r^2} \qquad (4\text{-}11)$$

This represents the radiation passing from A_j to dA_i. The radiant energy leaving A_j in all directions is

$$\int_{A_j} B_j \, dA_j \qquad (4\text{-}12)$$

where the integral is extended over the same area as in expression (4-11). The angle factor $dF_{A_j - dA_i}$ then follows as the ratio of (4-11) to (4-12).

$$dF_{A_j - dA_i} = \frac{dA_i \int_{A_j} B_j \cos \beta_i \cos \beta_j \, dA_j / \pi r^2}{\int_{A_j} B_j \, dA_j} \qquad (4\text{-}13)$$

As it stands, the foregoing angle factor depends on the magnitude and the surface distribution of the radiosity B_j. In the treatment of radiant interchange, it is customary to assume that the radiosity is uniformly distributed along the participating surfaces. This condition is precisely fulfilled for black, isothermal surfaces. However, as discussed in Chapter 3, it is violated to various degrees when the surfaces are not black. When B_j is uniform, it follows that $\int_{A_j} B_j \, dA_j = B_j A_j$. With this, equation (4-13) becomes

$$dF_{A_j - dA_i} = \frac{dA_i}{A_j} \int_{A_j} \frac{\cos \beta_i \cos \beta_j \, dA_j}{\pi r^2} \qquad (4\text{-}14)$$

Owing to the simplification provided by the assumption of uniform radiosity, the resulting angle factor depends only on the geometry of the system. Since dA_i appears as a multiplier on the right-hand side, it is clear that the angle factor is properly designated as infinitesimal.

By comparing equations (4-9) and (4-14), one finds

$$A_j \, dF_{A_j - dA_i} = dA_i F_{dA_i - A_j} \qquad (4\text{-}15)$$

This is the reciprocity rule as previously stated by equation (3-1b) of Chapter 3. It should be emphasized that reciprocity is strictly valid only when the radiant flux leaving the participating finite surface is uniformly distributed along that surface.

Finally, there is the angle factor for interchange between two finite surfaces A_i and A_j as is illustrated in Fig. 4-3. The radiant energy passing from an element dA_i of surface A_i to the surface A_j is given by expression (4-8). The radiation passing from A_i to A_j then follows as

$$\int_{A_i} \int_{A_j} \frac{B_i \cos \beta_i \cos \beta_j \, dA_i \, dA_j}{\pi r^2} \qquad (4\text{-}16)$$

Correspondingly, the energy leaving A_i in all directions is

$$\int_{A_i} B_i \, dA_i \qquad (4\text{-}17)$$

The ratio of the foregoing provides the desired angle factor

$$F_{A_i - A_j} = \frac{\displaystyle\int_{A_i} \int_{A_j} B_i \cos \beta_i \cos \beta_j \, dA_i \, dA_j / \pi r^2}{\displaystyle\int_{A_i} B_i \, dA_i} \qquad (4\text{-}18)$$

To render the angle factor independent of the magnitude and the

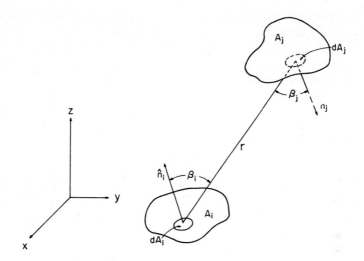

Fig. 4-3   Configuration for interchange between two finite surfaces.

surface distribution of the radiosity, it is necessary to assume that the latter quantity is uniform over A_i.   With this assumption, there follows

$$F_{A_i - A_j} = \frac{1}{A_i} \int_{A_i} \int_{A_j} \frac{\cos \beta_i \cos \beta_j \, dA_i \, dA_j}{\pi r^2} \qquad (4\text{-}19)$$

A corresponding derivation for radiant energy traveling from A_j toward A_i leads to an equation identical to (4-18) except that the subscripts i and j are interchanged.   In analogy with the foregoing, it is again necessary to assume that B_j is uniform over A_j in order to achieve a general representation.   The result is

$$F_{A_j - A_i} = \frac{1}{A_j} \int_{A_i} \int_{A_j} \frac{\cos \beta_i \cos \beta_j \, dA_i \, dA_j}{\pi r^2} \qquad (4\text{-}20)$$

By comparing equations (4-19) and (4-20), the reciprocity rule previously stated in equation (3-1a) of Chapter 3 emerges:

$$A_i F_{A_i - A_j} = A_j F_{A_j - A_i} \qquad (4\text{-}21)$$

It should be emphasized that the reciprocity strictly applies when the respective radiosity distributions are uniform along each of the partici- pating surfaces.

Evaluation of the equations of definition.   In order to use the foregoing equations of definition for computational purposes, it is necessary to evaluate $\cos \beta_i$, $\cos \beta_j$, and r in terms of the coordinates and other quantities that stipulate the shape of the participating surfaces. In general, the shape of a surface can be described by specifying the direction of the normal at all points on the surface. In turn, the direction of the normal is generally expressed in terms of the direction cosines l, m, and n. The direction cosine l is the cosine of the angle between the surface normal and the x direction, the direction cosine m is the cosine of the angle between the surface normal and the y direction, and so forth. The local normal to a surface can then be expressed as

$$\hat{n} = \hat{i}l + \hat{j}m + \hat{k}n \qquad (4\text{-}22)$$

in which $\hat{i}$, $\hat{j}$, and $\hat{k}$ are the standard unit vectors lying along the coordinate axes and l, m, and n are the local direction cosines.

The length r of the connecting line between two elements dA_i and dA_j is readily expressed in terms of the coordinates (x_i, y_i, z_i) of the former and (x_j, y_j, z_j) of the latter.

$$r^2 = (x_i - x_j)^2 + (y_i - y_j)^2 + (z_i - z_j)^2 \qquad (4\text{-}23)$$

It is also convenient to define unit vectors $\hat{r}_{ij}$ and $\hat{r}_{ji}$ that lie along the connecting line and are directed, respectively, from dA_i to dA_j, and vice versa.

$$\hat{r}_{ij} = \frac{1}{r}[(x_j - x_i)\hat{i} + (y_j - y_i)\hat{j} + (z_j - z_i)\hat{k}] \qquad (4\text{-}24a)$$

$$\hat{r}_{ji} = \frac{1}{r}[(x_i - x_j)\hat{i} + (y_i - y_j)\hat{j} + (z_i - z_j)\hat{k}] \qquad (4\text{-}24b)$$

The quantities $\cos \beta_i$ and $\cos \beta_j$ then follow directly as the scalar products $\hat{n}_i \cdot \hat{r}_{ij}$ and $\hat{n}_j \cdot \hat{r}_{ji}$, respectively. Upon evaluating these with the aid of equations (4-22) and (4-24), one obtains

$$\cos \beta_i = \frac{l_i(x_j - x_i) + m_i(y_j - y_i) + n_i(z_j - z_i)}{r} \qquad (4\text{-}25a)$$

$$\cos \beta_j = \frac{l_j(x_i - x_j) + m_j(y_i - y_j) + n_j(z_i - z_j)}{r} \qquad (4\text{-}25b)$$

Equations (4-23) and (4-25) relate the geometric quantities that appear in the angle-factor equations to the coordinates and to the direction cosines of the participating surfaces. For the case of interchange between infinitesimal surfaces, substitution of equations (4-23) and (4-25) into equations (4-5) and (4-6) leads to a complete specification of the angle factor.

When finite surfaces are involved, equations (4-23) and (4-25) are substituted into the integrands, say, in equations (4-9) or (4-14). To proceed, it is necessary to specify the variation of the direction cosines (l, m, n) as a function of the surface coordinates. For interchange between an infinitesimal element dA_i and a finite surface A_j, the quantities (l_i, m_i, n_i) are constants with respect to the integration over A_j, but (l_j, m_j, n_j) must be prescribed as a function of x_j, y_j, and z_j. For interchange between a pair of finite surfaces, the direction cosines of each surface must be prescribed as functions of the corresponding coordinates.

Once the shape of the surface has been specified in terms of the direction cosines, the indicated integrations must still be performed. For exchange between an infinitesimal and a finite surface, an area integral (usually a double integral) must be carried out. When two finite surfaces are involved, a double area integral (usually a quadruple integral) must be performed.

The integrals just described have been analytically evaluated in the literature for a number of simple geometrical configurations, and the corresponding results for many of these are presented in an angle-factor catalogue in Appendix A. When the form of the participating surfaces becomes moderately complex, the likelihood of determining the angle factor by purely analytical means is not very great, although one or two stages of the integrations may be performable analytically. In view of the general availability of high-speed computing equipment, the numerical evaluation of analytically intractable integrals poses no difficulties.

To illustrate the evaluation of the basic equations of definition, consider the case of interchange between two plane surfaces sharing a common edge and oriented at right angles to one another. The configuration is shown schematically in Fig. 4-4. For the surface A_i, the direction of the normal is constant with position and is specified by $l_i = m_i = 0$, $n_i = 1$. For surface A_j, the direction of the normal is also constant and is specified by $l_j = 0$, $m_j = 1$, $n_j = 0$.

Then, upon specializing equations (4-25a) and (4-25b), one obtains

$$\cos \beta_i = \frac{z_j}{r} \qquad \cos \beta_j = \frac{y_i}{r} \qquad (4\text{-}26)$$

Moreover, equation (4-23) becomes

$$r^2 = (x_i - x_j)^2 + y_i^2 + z_j^2 \qquad (4\text{-}27)$$

These quantities may be introduced into the equation of definition, (4-19), for the angle factor $F_{A_i - A_j}$, from which there follows

$$bc F_{A_i - A_j} = \int_0^b dx_j \int_0^b dx_i \int_0^a z_j \, dz_j \int_0^c \frac{y_i \, dy_i}{\pi[(x_i - x_j)^2 + y_i^2 + z_j^2]^2} \qquad (4\text{-}28)$$

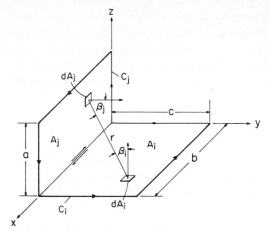

Fig. 4-4   Plane rectangular surfaces sharing a common edge.

The integration of equation (4-28) can be carried through without difficulty with the aid of a table of integrals, although the actual execution is quite lengthy.   Some of the detailed steps are shown in ref. 4.   The end result is tabulated in the angle-factor catalogue in Appendix A (configuration 2, $\Phi = 90°$).

4-2   Contour-Integral Representation

As derived in the preceding section, the angle factors for diffuse interchange are represented in terms of area integrals when one or both of the participating surfaces are finite.   An alternate form of the defining equations for angle factors can be achieved by replacing the area integrals by contour (line) integrals.   The new representation affords a significant advantage in that the order of the integration is reduced; the former double integral is reduced to a single integral and the former quadruple integral is reduced to a double integral.

The basic mathematical tool needed in carrying out the transformation is Stokes' theorem, which takes an area integral into a contour integral.   Stokes' theorem states that, given a surface with area A having a boundary curve C, there follows

$$\oint_C P\,dx + Q\,dy + R\,dz$$

$$= \int_A \left[l\left(\frac{\partial R}{\partial y} - \frac{\partial Q}{\partial z}\right) + m\left(\frac{\partial P}{\partial z} - \frac{\partial R}{\partial x}\right) + n\left(\frac{\partial Q}{\partial x} - \frac{\partial P}{\partial y}\right) \right] dA \quad (4\text{-}29)$$

where P, Q, and R are twice-differentiable functions and l, m, and n are the local direction cosines of dA.

   Angle factor between infinitesimal and finite areas. Stokes' theorem will first be applied to the reformulation of the angle factor for interchange between an infinitesimal and a finite surface. By introducing equations (4-23) and (4-25) into equation (4-9) and rearranging, it can be seen that the angle factor $F_{dA_i - A_j}$ has the same general form as the right-hand side of Stokes' theorem, that is,

$$F_{dA_i - A_j} = \int_{A_j} \{l_j[(x_i - x_j)f] + m_j[(y_i - y_j)f] + n_j[(z_i - z_j)f]\} \, dA_j$$

$$(4\text{-}30)$$

in which f is an abbreviation for

$$f = \frac{l_i(x_j - x_i) + m_i(y_j - y_i) + n_i(z_j - z_i)}{\pi r^4} \qquad (4\text{-}31)$$

   A precise correspondence between equation (4-30) and the right-hand side of Stokes' theorem is obtained by equating corresponding terms

$$\frac{\partial R}{\partial y_j} - \frac{\partial Q}{\partial z_j} = (x_i - x_j)f$$

$$\frac{\partial P}{\partial z_j} - \frac{\partial R}{\partial x_j} = (y_i - y_j)f$$

$$\frac{\partial Q}{\partial x_j} - \frac{\partial P}{\partial y_j} = (z_i - z_j)f \qquad (4\text{-}32)$$

Moreover, the foregoing set of partial differential equations can be solved to yield

$$2\pi r^2 P = -m_i(z_j - z_i) + n_i(y_j - y_i) \qquad (4\text{-}33\text{a})$$

$$2\pi r^2 Q = l_i(z_j - z_i) - n_i(x_j - x_i) \qquad (4\text{-}33\text{b})$$

$$2\pi r^2 R = -l_i(y_j - y_i) + m_i(x_j - x_i) \qquad (4\text{-}33\text{c})$$

   Inasmuch as a precise correspondence between $F_{dA_i - A_j}$ and the right-hand side of Stokes' theorem has been achieved, an identical correspondence also exists with the left-hand side of Stokes' theorem. Thus

$$F_{dA_i - A_j} = \oint_{C_j} P \, dx + Q \, dy + R \, dz \qquad (4\text{-}34)$$

Upon substituting P, Q, and R from equation (4-33) and rearranging, one arrives at

$$F_{dA_i - A_j} = l_i \oint_{C_j} \frac{(z_j - z_i)\, dy_j - (y_j - y_i)\, dz_j}{2\pi r^2}$$

$$+ m_i \oint_{C_j} \frac{(x_j - x_i)\, dz_j - (z_j - z_i)\, dx_j}{2\pi r^2}$$

$$+ n_i \oint_{C_j} \frac{(y_j - y_i)\, dx_j - (x_j - x_i)\, dy_j}{2\pi r^2} \qquad (4\text{-}35)$$

This is the contour-integral representation of the angle factor for interchange between an infinitesimal and a finite surface. It may be observed that, in contrast to the double (i.e., area) integral appearing in the standard equation of definition (4-9), only single integrals appear in the contour-integral representation.

The structure of equation (4-35) will now be discussed. The integration is performed along the boundary curve C_j of the area A_j; (x_j, y_j, z_j) are coordinates representing any point on the contour C_j and these may vary during the integration. The direction of movement along C_j is the same as that of a traveler who walks around the contour with his body aligned with the normal and keeps the interior of A_j at his left. The distance between dA_i and any point on C_j is r; this quantity may also vary during the integration. On the other hand, the quantities (x_i, y_i, z_i) and (l_i, m_i, n_i), respectively representing the position and the direction cosines of dA_i, are fixed constants during the integration.

In practice, it is convenient to orient the coordinate axes so that the normal to dA_i lies precisely along a coordinate line. Then two of the three direction cosines are identically zero, and, correspondingly, two of the three integrals appearing in equation (4-35) may be deleted. Therefore, if the axes are so aligned, it is only necessary to perform one contour integration to determine the desired angle factor. Additional simplifications occur when one or more of the coordinates are constant along any part of the contour C_j; then the corresponding differentials are zero and part of the integrand is eliminated.

To illustrate the facility with which the contour-integral method can be applied, consideration is given to the case of diffuse interchange between an off-axis element and a circular segment. The configuration is shown schematically in the lower portion of Fig. 4-5. Upon inspecting the figure, it is seen that $l_i = m_i = 0$, $n_i = 1$; $x_i = a$, $y_i = z_i = 0$, $z_j = c$. With these facts, equation (4-35) reduces to

$$F_{dA_i - A_j} = \oint_{C_j} \frac{y_j\, dx_j - (x_j - a)\, dy_j}{2\pi[(x_j - a)^2 + y_j^2 + c^2]} \qquad (4\text{-}36)$$

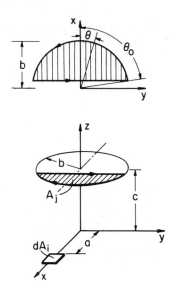

Fig. 4-5   An off-axis element and a circular segment.

The contour C_j that bounds the area A_j is composed of two parts: an arc running from $-\theta_0 \leqslant \theta \leqslant \theta_0$ and a straight line running from $-b \sin \theta_0 \leqslant y_j \leqslant b \sin \theta_0$.   On   the   arc, $x_j = b \cos \theta$, $y_j = b \sin \theta$; while on the straight line, $x_j = b \cos \theta_0$, $dx_j = 0$.   With these, equation (4-36) becomes

$$F_{dA_i - A_j} = \frac{2}{2\pi} \int_{\theta_0}^{0} \frac{ab \cos \theta - b^2}{a^2 + b^2 + c^2 - 2ab \cos \theta} \, d\theta$$

$$- \frac{2}{2\pi} (b \cos \theta_0 - a) \int_{0}^{b \sin \theta_0} \frac{dy_j}{(b \cos \theta_0 - a)^2 + c^2 + y_j^2} \qquad (4\text{-}37)$$

The integrals appearing in equation (4-37) can be found directly in standard integral tables.   The final result for the angle factor is stated in the angle-factor catalogue in Appendix A (see configuration no. 12).

Upon further consideration of equation (4-35), it is seen that the contour-integral representation separates itself naturally into three parts.   Each of the three contour integrals can be evaluated independently of the others and of the values of l_i, m_i, and n_i.   Indeed, each one of the contour integrals is, in itself, an angle factor between a specially arranged element at (x_i, y_i, z_i) and the surface A_j.

In particular, the absolute value of the first contour integral represents the angle factor for an element at (x_i, y_i, z_i) having direction cosines $l_i = \pm 1$, $m_i = n_i = 0$.   Either the plus or the minus sign is appropriate depending on whether the normal to the aforementioned element lies along the $+x$ or the $-x$ axis as the element views A_j. Moreover, in some instances an element with $l_i = +1, m_i = n_i = 0$ sees only a portion of A_j, and an element with $l_i = -1, m_i = n_i = 0$ sees the other portion of A_j. In such cases the first contour integral of equation (4-35) represents the sum of the angle factors from each of these elements to A_j. In light of this interpretation, one can define

$$\left| \oint_{C_j} \frac{(z_j - z_i) \, dy_j - (y_j - y_i) \, dz_j}{2\pi r^2} \right| = F_{dA_i - A_j}(\pm 1, 0, 0) \quad (4\text{-}38\text{a})$$

Next, attention may be directed to the second contour integral of equation (4-35), the absolute value of which is the angle factor for an element at (x_i, y_i, z_i) having direction cosines $l_i = 0, m_i = \pm 1, n_i = 0$. The sign is selected in a manner analogous to that just discussed; namely, the plus sign applies when the element normal lies along the $+y$ axis as the element views A_j, while the minus sign applies when the element normal lies along the $-y$ axis as it views A_j. Moreover, in some cases only part of A_j is viewed by the element $l_i = 0, m_i = +1, n_i = 0$, and the other part is viewed by the element $l_i = 0, m_i = -1, n_i = 0$. Then the second contour integral represents (in magnitude) the sum of the individual angle factors between these elements and A_j. Consistent with this interpretation, one defines

$$\left| \oint_{C_j} \frac{(x_j - x_i) \, dz_j - (z_j - z_i) \, dx_j}{2\pi r^2} \right| = F_{dA_i - A_j}(0, \pm 1, 0) \quad (4\text{-}38\text{b})$$

The third contour integral appearing in equation (4-35) has an interpretation that is completely analogous to the foregoing. Correspondingly,

$$\left| \oint_{C_j} \frac{(y_j - y_i) \, dx_j - (x_j - x_i) \, dy_j}{2\pi r^2} \right| = F_{dA_i - A_j}(0, 0, \pm 1) \quad (4\text{-}38\text{c})$$

The foregoing definitions can then be introduced into equation (4-35), giving

$$F_{dA_i - A_j} = |l_i| F_{dA_i - A_j}(\pm 1, 0, 0) + |m_i| F_{dA_i - A_j}(0, \pm 1, 0)$$
$$+ |n_i| F_{dA_i - A_j}(0, 0, \pm 1) \quad (4\text{-}39)$$

Equation (4-39) can be recognized as stating a superposition principle; namely, that the angle factor between an element dA_i with arbitrary direction cosines (l_i, m_i, n_i) and a surface A_j is expressible as a linear sum of three basic angle factors. The weighting factors in the superposition are the direction cosines of the element. The superposition principle serves to generalize results for basic configurations to more complex configurations.

To illustrate the principle of superposition, consider the calculation of the angle factor between an element with direction cosines $l_i = 0$, m_i, n_i and a circular disk as shown in the upper part of Fig. 4-6. The catalogue in Appendix A gives the angle factors for the basic configurations (nos. 10 and 11) shown in the lower part of the figure. Then, in terms of these known results, the desired angle factor becomes

$$F_{dA_i - A_j} = |m_i| F_{dA_i - A_j}(0, -1, 0) + |n_i| F_{dA_i - A_j}(0, 0, 1) \qquad (4\text{-}40)$$

It would appear that the contour-integral representation (4-35) provides some advantages relative to the standard definition, equation

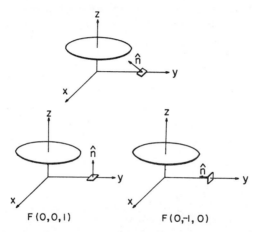

F (0,0,1)   F (0,-1, 0)

Fig. 4-6   Illustration of the superposition principle.

(4-9). In particular, the order of the integration is reduced from double to single. This simplifies the numerical as well as the analytical determination of angle factors. The contour-integral representation also leads to a superposition principle that extends the utility of angle-factor information that is available for basic configurations.

Angle factor between two finite areas. The defining equation (4-19) for the angle factor between two finite surfaces can also be transformed to a contour-integral representation by employing Stokes' theorem. In carrying out the transformation, it is convenient to take advantage of the results of the foregoing derivation for the case of interchange between an infinitesimal and a finite surface. Thus the area integral over A_j in equation (4-19) may be regarded as being transformed to a contour integral in accordance with equation (4-35). Picking up at this point and rearranging, one finds

$$2\pi A_i F_{A_i - A_j} = \oint_{C_j} dx_j \int_{A_i} \frac{-m_i(z_j - z_i) + n_i(y_j - y_i)}{r^2} \, dA_i$$

$$+ \oint_{C_j} dy_j \int_{A_i} \frac{l_i(z_j - z_i) - n_i(x_j - x_i)}{r^2} \, dA_i$$

$$+ \oint_{C_j} dz_j \int_{A_i} \frac{-l_i(y_j - y_i) + m_i(x_j - x_i)}{r^2} \, dA_i \quad (4\text{-}41)$$

Equation (4-41) naturally divides itself into three parts, and it is convenient to treat each part separately. In particular, Stokes' theorem will be applied separately to each part. Upon comparing the first area integral appearing in equation (4-41) with the right-hand side of Stokes' theorem, equation (4-29), there follows

$$\frac{\partial R}{\partial y_i} - \frac{\partial Q}{\partial z_i} = 0$$

$$\frac{\partial P}{\partial z_i} - \frac{\partial R}{\partial x_i} = -\frac{z_j - z_i}{r^2}$$

$$\frac{\partial Q}{\partial x_i} - \frac{\partial P}{\partial y_i} = \frac{y_j - y_i}{r^2} \quad (4\text{-}42)$$

It is readily verified that this set of partial-differential equations is satisfied by

$$P = \ln r, \quad Q = 0, \quad R = 0 \quad (4\text{-}43)$$

Therefore, in accordance with Stokes' theorem, the first area integral of equation (4-41) may be replaced by

$$\oint_{C_i} \ln r \, dx_i \quad (4\text{-}44a)$$

In a similar manner, the second and the third area integrals appearing in equation (4-41) can be respectively transformed into the following:

$$\oint_{C_i} \ln r \, dy_i \quad \text{and} \quad \oint_{C_i} \ln r \, dz_i \quad (4\text{-}44b)$$

In light of equations (4-44a) and (4-44b), the area integrals can be eliminated from equation (4-41), giving

$$F_{A_i - A_j} = \frac{1}{2\pi A_i} \oint_{C_i} \oint_{C_j} \left(\ln r \, dx_i \, dx_j + \ln r \, dy_i \, dy_j + \ln r \, dz_i \, dz_j \right)$$

$$(4\text{-}45)$$

This is the contour-integral representation of the angle factor for interchange between two finite surfaces. It is seen to involve a double integration in contrast to the quadruple integration that appears in the conventional angle-factor representation.

As indicated in equation (4-45), the angle factor is determined by integrating around the contours C_i and C_j of the participating areas A_i and A_j; r is the distance between points on the respective contours. If any of the coordinates are constant over some portion of either contour, the corresponding differentials are zero and the integration is thus simplified.

The application of the contour-integral representation, equation (4-45), will now be illustrated. For this purpose, consider the pair of perpendicular plane surfaces as shown in Fig. 4-4; this same configuration was previously used to illustrate the application of the conventional area-integral representation, equation (4-19). First, the contour integral around C_i will be evaluated. It may be noted that on all segments of C_i, $dz_i = 0$ and that either $dx_i = 0$ or $dy_i = 0$. Also, $dy_j = 0$ on all segments of C_j. Then, upon specializing equation (4-45), one finds

$$2\pi bc\, F_{A_i - A_j} = \oint_{C_j} dx_j \left\{ \int_0^b \ln\left[(x_i - x_j)^2 + z_j^2\right]^{1/2} dx_i \right.$$

$$\left. + \int_b^0 \ln\left[(x_i - x_j)^2 + c^2 + z_j^2\right]^{1/2} dx_i \right\} \quad (4\text{-}46)$$

The next step is to specialize the contour integral around C_j.

$$2\pi bc\, F_{A_i - A_j} = \int_0^b dx_j \left\{ \int_0^b \ln\left[(x_i - x_j)^2 + a^2\right]^{1/2} dx_i \right.$$

$$\left. + \int_b^0 \ln\left[(x_i - x_j)^2 + c^2 + a^2\right]^{1/2} dx_i \right\}$$

$$+ \int_b^0 dx_j \left\{ \int_0^b \ln\left[(x_i - x_j)^2\right]^{1/2} dx_i \right.$$

$$\left. + \int_b^0 \ln\left[(x_i - x_j)^2 + c^2\right]^{1/2} dx_i \right\} \quad (4\text{-}47)$$

Upon comparing equation (4-47) with equation (4-28), the reduction in the order of the integration from quadruple to double is clearly seen. The required double integrals are readily carried out with the aid of standard tables. The final result for the angle factor is included in the catalogue at the end of this chapter.

As a final remark, it may be noted that the reduction in the order of the integration afforded by the contour-integral representation may be even more advantageous for numerical than for analytical determination of angle factors.

4-3 Shorthand Methods

In this section, procedures for bypassing the evaluation of integrals in the determination of angle factors are discussed.   One such procedure is termed angle-factor algebra, whereby an unknown angle factor is computed by adding or subtracting known angle factors for related configurations.   A second group of procedures applies to surfaces that are greatly elongated in one of their dimensions.

Angle-factor algebra.   The basic principle employed in angle-factor algebra is that energy is conserved.   Thus, if radiant energy leaves a surface A_i and arrives at a surface A_j that is subdivided into parts A'_j and A''_j $(A_j = A'_j + A''_j)$, then

$$A_i F_{A_i - A_j} = A_i F_{A_i - A'_j} + A_i F_{A_i - A''_j} \qquad (4\text{-}48)$$

The left-hand side represents the energy passing from A_i to A_j, and this is represented on the right-hand side as the sum of the energies respectively passing from A_i to A'_j and A''_j.   This basic idea, even when applied in a straightforward manner, is extremely useful in generalizing available angle-factor information.   However, when applied with ingenuity, this procedure permits the computation of angle factors for configurations that, at first glance, look quite different from those employed in the synthesis.

As a first illustration of angle-factor algebra, consider the surfaces

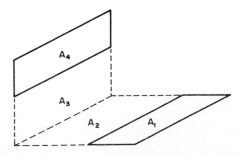

Fig. 4-7   An illustration of the application of angle-factor algebra.

A_1 and A_4 as shown in Fig. 4-7; it is desired to determine $F_{A_1-A_4}$. As indicated in the figure, the intervening areas are denoted by A_2 and A_3; in addition, $A_1 + A_2 \equiv A_{12}$ and $A_3 + A_4 \equiv A_{34}$. The application of energy conservation gives

$$A_1 F_{A_1-A_4} = A_{12} F_{A_{12}-A_4} - A_2 F_{A_2-A_4} \tag{4-49}$$

Since neither of the angle factors appearing on the right-hand side is a commonly tabulated quantity, one proceeds by further application of energy conservation, thus

$$A_{12} F_{A_{12}-A_4} = A_{12} F_{A_{12}-A_{34}} - A_{12} F_{A_{12}-A_3} \tag{4-50a}$$

$$A_2 F_{A_2-A_4} = A_2 F_{A_2-A_{34}} - A_2 F_{A_2-A_3} \tag{4-50b}$$

Upon substituting these into equation (4-49), one obtains

$$F_{A_1-A_4} = \frac{1}{A_1} (A_{12} F_{A_{12}-A_{34}} + A_2 F_{A_2-A_3} - A_{12} F_{A_{12}-A_3} - A_2 F_{A_2-A_{34}})$$
$$\tag{4-51}$$

All the angle factors appearing on the right-hand side pertain to pairs of perpendicular plane surfaces. Information for such surfaces is given in Appendix A (configuration no. 2, $\Phi = 90°$). Thus all quantities on the right may be regarded as known and, correspondingly, $F_{A_1-A_4}$ is readily computed.

A somewhat more interesting application of angle-factor algebra may now be discussed with the aid of Fig. 4-8, which shows two plane parallel rectangles of different size, A_1 and A_7, respectively. Other surface areas that are required in the derivation are labeled in the figure. As in the prior illustration, the following notation will be employed:

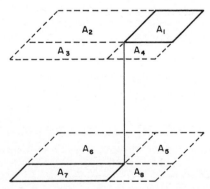

Fig. 4-8   A second illustration of angle-factor algebra.

$A_{12} \equiv A_1 + A_2$, $A_{1234} \equiv A_1 + A_2 + A_3 + A_4$, and so forth. Let the problem be the computation of $F_{A_1 - A_7}$.

To begin the derivation, certain additional relationships between the participating angle factors may be noted. If one successively specializes the basic equation of definition, equation (4-19), to interchange between surfaces 1 and 6 and to interchange between surfaces 2 and 5, it is found by inspection of the integrals (without executing the integration) that

$$A_1 F_{A_1 - A_6} = A_2 F_{A_2 - A_5} \qquad (4\text{-}52a)$$

By a similar procedure, one finds

$$A_1 F_{A_1 - A_8} = A_4 F_{A_4 - A_5}, \quad A_4 F_{A_4 - A_7} = A_3 F_{A_3 - A_8}, \text{ etc. } \quad (4\text{-}52b)$$

Other useful relationships follow from energy conservation. For instance,

$$A_{12} F_{A_{12} - A_{56}} = A_1 F_{A_1 - A_5} + A_1 F_{A_1 - A_6} + A_2 F_{A_2 - A_6} + A_2 F_{A_2 - A_5} \qquad (4\text{-}53)$$

Upon substitution of equation (4-52a) in the foregoing and rearrangement, one finds

$$A_1 F_{A_1 - A_6} = \frac{1}{2} \left(A_{12} F_{A_{12} - A_{56}} - A_1 F_{A_1 - A_5} - A_2 F_{A_2 - A_6} \right) \quad (4\text{-}54)$$

It may be observed that all the angle factors appearing on the right-hand side represent configurations consisting of pairs of congruent rectangles, situated one above the other. Such angle factors are widely available; for instance, see configuration 1 in the angle-factor catalogue. Relationships such as equation (4-54) can be readily derived for all the participating surfaces.

Another group of angle-factor relationships is needed before the final determination of $F_{A_1 - A_7}$ can be made. By successively applying the equation of definition, equation (4-19), to interchange between surfaces 1 and 7 and to interchange between surfaces 2 and 8, one finds

$$A_1 F_{A_1 - A_7} = A_2 F_{A_2 - A_8} \qquad (4\text{-}55)$$

This follows directly from an inspection of the integrals. Furthermore, by employing angle-factor reciprocity in conjunction with equation (4-55), one gets

$$A_1 F_{A_1 - A_7} = A_3 F_{A_3 - A_5} = A_4 F_{A_4 - A_6} \qquad (4\text{-}56)$$

Now, consideration can be given to the determination of $F_{A_1 - A_7}$. On the basis of energy conservation, one can write

$$A_1 F_{A_1 - A_7} = A_{1234} F_{A_{1234} - A_{5678}}$$

$$- (A_1 F_{A_1 - A_5} + A_1 F_{A_1 - A_6} + A_1 F_{A_1 - A_8})$$

$$- (A_2 F_{A_2 - A_5} + A_2 F_{A_2 - A_6} + A_2 F_{A_2 - A_7} + A_2 F_{A_2 - A_8})$$

$$- (A_3 F_{A_3 - A_5} + A_3 F_{A_3 - A_6} + A_3 F_{A_3 - A_7} + A_3 F_{A_3 - A_8})$$

$$- (A_4 F_{A_4 - A_5} + A_4 F_{A_4 - A_6} + A_4 F_{A_4 - A_7} + A_4 F_{A_4 - A_8})$$

$$(4\text{-}57)$$

Angle factors such as $A_1 F_{A_1 - A_6}$, $A_1 F_{A_1 - A_8}$, $A_2 F_{A_2 - A_5}$, $A_2 F_{A_2 - A_7}$, $A_3 F_{A_3 - A_6}$, $A_3 F_{A_3 - A_8}$, $A_4 F_{A_4 - A_5}$, and $A_4 F_{A_4 - A_7}$ can be eliminated by applying relationships such as equation (4-54). Furthermore, $A_2 F_{A_2 - A_8}$, $A_3 F_{A_3 - A_5}$, and $A_4 F_{A_4 - A_6}$ are equal to $A_1 F_{A_1 - A_7}$ in accordance with equations (4-55) and (4-56). After making these substitutions and rearranging, there results

$$F_{A_1 - A_7} = \frac{1}{4A_1} (A_{1234} F_{A_{1234} - A_{5678}} + A_1 F_{A_1 - A_5} + A_2 F_{A_2 - A_6}$$

$$+ A_3 F_{A_3 - A_7} + A_4 F_{A_4 - A_8})$$

$$- \frac{1}{4A_1} (A_{12} F_{A_{12} - A_{56}} + A_{14} F_{A_{14} - A_{58}}$$

$$+ A_{34} F_{A_{34} - A_{78}} + A_{23} F_{A_{23} - A_{67}}) \quad (4\text{-}58)$$

By inspecting the right-hand side of equation (4-58), it is seen that all the angle factors appearing there represent interchange between pairs of congruent rectangles, one situated above the other. As previously noted, angle factors of this type are widely available (e.g., configuration no. 1 of the angle-factor catalogue). Thus $F_{A_1 - A_7}$ is readily determined by summing the known values of the angle factors appearing on the right-hand side.

The preceding discussion of angle-factor algebra dealt with interchange between finite surfaces. However, this approach can also be advantageously employed for interchange between infinitesimal surfaces. Angle factors between such surfaces are required in the integral-equation formulation that was discussed in Chapter 3.

The application of angle-factor algebra to infinitesimal surfaces may be illustrated by referring to the case of two ring elements dA_i and dA_j as shown in Fig. 4-9. The elements may be regarded as lying on the wall of a tube of radius R; the separation distance is $x_j - x_i$.

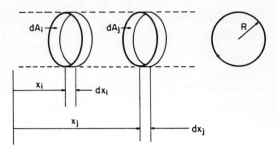

Fig. 4-9 Application of angle-factor algebra to determining infinitesimal angle factors.

To begin the derivation, consider the radiant interchange between circular disks stretched across the tube cross section. Let a disk at x_i be denoted by $D(x_i)$, a disk at $(x_i + dx_i)$ be denoted by $D(x_i + dx_i)$, and so forth. In accordance with the energy conservation principle, one can write

$$dF_{D(x_i)-dA_j} = F_{D(x_i)-D(x_j)} - F_{D(x_i)-D(x_j+dx_j)} \qquad (4\text{-}59)$$

$$= -\frac{\partial F_{D(x_i)-D(x_j)}}{\partial x_j}\, dx_j \qquad (4\text{-}59a)$$

Moreover, by applying the reciprocity rule, there follows

$$F_{dA_j-D(x_i)} = -\frac{R}{2}\frac{\partial F_{D(x_i)-D(x_j)}}{\partial x_j} \qquad (4\text{-}60)$$

A further application of conservation of energy gives

$$dF_{dA_j-dA_i} = F_{dA_j-D(x_i+dx_i)} - F_{dA_j-D(x_i)} \qquad (4\text{-}61)$$

$$= \frac{\partial F_{dA_j-D(x_i)}}{\partial x_i}\, dx_i \qquad (4\text{-}61a)$$

$$= -\frac{R}{2}\frac{\partial^2 F_{D(x_i)-D(x_j)}}{\partial x_j\, \partial x_i}\, dx_i \qquad (4\text{-}61b)$$

where equation (4-60) has been employed. Upon inspection of equation (4-61b), it is seen that the ring-to-ring angle factor is derivable from a double differentiation of the disk-to-disk angle factor. The latter is commonly available in the literature; for instance, configuration no. 3 in Appendix A. When the differentiation operation is executed, there results

$$dF_{dA_j-dA_i} = \left\{1 - |X_i - X_j|\frac{2(X_i - X_j)^2 + 3}{2[(X_i - X_j)^2 + 1]^{\frac{3}{2}}}\right\} dX_i \qquad (4\text{-}62)$$

where $X = x/2R$.

Additional applications of angle-factor algebra may be found in ref. 4.

Elongated surfaces.   A remarkably simple method for determining angle factors can be employed when each of the participating surfaces is greatly elongated in one of its dimensions.   For instance, the surfaces pictured in Fig. 4-10 are envisioned as being very long (in effect, infinite) in the direction into and out of the plane of the page.   Under this condition, the angle factor for interchange between two infinitesimal elements dA_i and dA_j (left-hand sketch of Fig. 4-10) can be computed from

$$dF_{dA_i - dA_j} = \frac{1}{2} d(\sin \theta) \tag{4-63}$$

The angle θ is formed by the normal at dA_i and the connecting line between dA_i and dA_j.

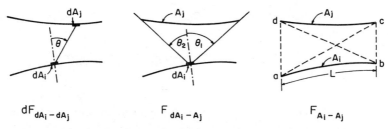

Fig. 4-10   Angle-factor determination for surfaces that are elongated in the direction normal to the plane of the page.

The angle factor for interchange between an infinitesimal surface dA_i and a finite surface A_j (center sketch, Fig. 4-10) follows directly from equation (4-63) as

$$F_{dA_i - A_j} = \frac{1}{2} (\sin \theta_1 + \sin \theta_2) \tag{4-64}$$

As illustrated in the figure, the angles θ_1 and θ_2 are formed by the normal at dA_i and the connecting lines drawn to the extremities of A_j. Positive values of θ_1 are measured clockwise with respect to the normal; on the other hand, positive values of θ_2 are measured counterclockwise with respect to the normal.

The angle factor for interchange between a pair of finite surfaces (right-hand sketch, Fig. 4-10) is most easily computed by applying Hottel's string method (ref. 5).   According to this procedure, strings are imagined to be tightly stretched between the endpoints of A_i and A_j. The angle factor $F_{A_i - A_j}$ is then determined by adding the lengths of the crossed strings, subtracting the lengths of the uncrossed strings, and then dividing by twice the length of A_i. In terms of the nomenclature of the figure, these operations are expressible as

$$F_{A_i - A_j} = \frac{1}{2L} [(\overline{ac} + \overline{bd}) - (\overline{ad} + \overline{bc})] \qquad (4\text{-}65)$$

4-4 Angle-Factor Catalogue

Angle-factor information for a wide range of surface configurations has appeared in various sources in the literature. Results for those configurations that recur most frequently have been compiled here to form an angle-factor catalogue. The presentation of such results in terms of algebraic expressions and in graphs is quite space consuming; consequently, it was decided to place this information in Appendix A. However, the layout of the appendix will be discussed here.

Figures A-1 and A-2 display the configurations for which angle-factor information is presented. The first of these figures pertains to interchange between pairs of finite surfaces, while the second figure pertains to interchange between an infinitesimal element and a finite surface. These figures contain the dimensional nomenclature that is employed in the forthcoming formulas and graphs.

Following Figs. A-1 and A-2 is a listing of the algebraic equations that express the angle factors for the various configurations. Next, a graphical presentation of angle-factor values is made in Figs. A-3 through A-10 for those configurations for which numerical results are available.

Other extensive compilations of angle-factor information are found in refs. 4 and 6. The latter is primarily concerned with cylindrical assemblies such as configuration no. 5.

REFERENCES

1. E. R. G. Eckert and R. M. Drake, *Heat and Mass Transfer*, McGraw-Hill, New York, 1959.

2. M. Jakob, *Heat Transfer*, vol. 2., Wiley, New York, 1957.

3. F. Kreith, *Radiation Heat Transfer*, International Textbook, Scranton, Penn., 1962.

4. D. C. Hamilton and W. R. Morgan, Radiant interchange configuration factors. *NACA TN 2836* (1952).

5. H. C. Hottel, Radiant heat transmission. In *Heat Transmission* (W. H. McAdams, ed.), 3rd ed., ch. 3, McGraw-Hill, New York, 1954.

6. H. Leuenberger and R. A. Pearson, Compilation of radiation shape factors for cylindrical assemblies. ASME Annual Meeting, New York, 1956. *ASME Paper No. 56-A-144* (1956).

PROBLEMS

4-1. Verify the following relations between angle factors

$$F_{A_i - A_j} = \frac{1}{A_i} \int_{A_i} F_{dA_i - A_j} \, dA_i$$

$$F_{A_j - A_i} = \int_{A_i} dF_{A_j - dA_i}$$

4-2. Specialize the defining equation for $F_{dA_i - A_j}$, equation (4-9), to configuration 8 of page 341. Carry out the integration and verify the expression given on page 343.

4-3. Deduce a simplified form, appropriate to large a/c, for the angle factor for configuration 8. Using the thus-derived expression, check the horizontal portions of the curves of Fig. A-7, page 348.

4-4. Consider a modification of configuration 8 of page 341. Suppose that the element dA_1, instead of being directly beneath the upper right corner of the area A_2, is relocated to a more forward position. Specifically, dA_1 still lies directly beneath the right-most boundary of area A_2, but is now situated a distance ξ ($0 \leqslant \xi \leqslant b$) forward of its initial location. Utilizing the angle factor expression for configuration 8 as given on page 343, derive an expression for $F_{dA_1 - A_2}$ corresponding to the new location of dA_1.

4-5. Demonstrate that the angle factor $F_{dA_1 - A_2}$ for configuration 14 of page 341 is the average, taken over $0 \leqslant \xi \leqslant b$, of the angle factor derived in Problem 4-4.

4-6. Repeat Problem 4-2 for configuration 11 on page 341 and verify the expression for $F_{dA_1 - A_2}$ given on page 344.

4-7. The horizontal portions of the curves of Figs. A-5 and A-9 are identical. Explain why.

4-8. Specialize the defining equation for $F_{A_i - A_j}$, equation (4-19), to configuration 1 of page 340.   Attempt to perform the integrations with a view to verifying the expression given on page 339.

4-9.   Simplify the $F_{A_1 - A_2}$ expression for configuration 1 for large a/c and compare the numerical results with those of Fig. A-3.

4-10.   Verify that equation (4-30) follows from the substitution of equations (4-23) and (4-25) into equation (4-9).

4-11.   By direct substitution, show that the $P, Q,$ and R of equations (4-33) satisfy any one of equations (4-32).

4-12.   Solve Problem 4-2 by the contour integral method and compare the effort involved in obtaining the angle factor with that expended in Problem 4-2.

4-13.   Apply the contour integral method to find $F_{dA_1 - A_2}$ for configuration 11 of page 341.   Compare the effort involved in obtaining the result with that encountered in Problem 4-6.

4-14.   Verify that the $P, Q,$ and R of equation (4-43) satisfy equation (4-42).

4-15.   Apply the contour integral method to configuration 1 of page 301.   Comment on the relative ease or difficulty in determining $F_{A_1 - A_2}$ compared with the direct application of the defining equation (4-19).   (See Problem 4-8.)

4-16.   Refer to Figs. 4-8 and use angle-factor algebra to derive an expression for $F_{A_5 - A_2}$ in terms of known angle factors (specifically, configuration 1, page 340).

4-17.   Derive the expression for the angle factor $dF_{dA_1 - dA_2}$ that was stated in Problem 3-8.   Hint: Apply equation (4-63) and evaluate $\sin \theta$ in terms of the nomenclature of the figure accompanying Problem 3-8. Then differentiate

$$d (\sin \theta) = \frac{d (\sin \theta)}{dx_2} \, dx_2$$

4-18.   Find $F_{dA_1 - A_2}$ for the configuration of Problem 3-8.   Perform the derivation both by direct application of equation (4-64) and by integrating $dF_{dA_1 - dA_2}$.

4-19.   Apply equation (4-65) to derive $F_{A_1 - A_2}$ for the configuration of Problem 3-8.   The resulting expression should give values in agreement with the $b/c = \infty$ curve of Fig. A-3.

CHAPTER 5

RADIANT INTERCHANGE AMONG SURFACES HAVING SPECULAR COMPONENTS OF REFLECTION

The calculation of radiant interchange is commonly performed under the assumption that the participating surfaces are diffuse emitters and diffuse reflectors of radiant energy. There is, however, considerable experimental evidence that real surfaces depart by various degrees from this model. Unfortunately, the analysis of the radiant interchange on the basis of nonidealized directional distributions leads to an exceedingly complex computational undertaking. Moreover, detailed directional distributions of emittance and reflectance are generally unavailable for real engineering materials.

As an alternative to the diffuse model, it is logical to consider other limiting cases that can be analyzed and evaluated without excessive complication. One such limit is the case of specular reflection. Indeed, as will be demonstrated later in this chapter, a model including diffuse emission and specular reflection lends itself to analytical formulation, especially if the participating surfaces are plane. A second model that may be somewhat closer to reality includes a subdivision of the hemispherical reflectance into specular and diffuse components, thus

$$\rho = \rho^s + \rho^d \tag{5-1}$$

This representation, coupled with the assumption of diffuse emission, also lends itself to analytical treatment.

In both of the foregoing cases, the analysis of the radiant interchange is facilitated by the use of the exchange-factor concept. This quantity plays a role similar to that played by the angle factor in the case of interchange among diffuse surfaces. The exchange factor is introduced and illustrated in Section 5-1. Then, in Section 5-2, the analysis of radiant interchange in enclosures containing both purely specular reflecting surfaces and purely diffuse reflecting surfaces is described. Next, in Section 5-3, the analysis is generalized to include

145

the case in which each of the participating surfaces may possess both specular and diffuse reflectance components. In all cases to be discussed in this chapter, the radiation emitted by the participating surfaces is diffusely distributed.

Results obtained from the application of the aforementioned analytical methods are referenced at appropriate places throughout this chapter. Furthermore, selected numerical results are presented in the forthcoming Chapter 6.

A purely numerical computation procedure that appears to hold promise for nonelementary directional distributions of emission and reflection is the Monte Carlo method. According to this procedure, one follows the *probable* path of a discrete bundle of energy until its absorption somewhere in the system. By employing a large number of such bundles, the net heat transfer can be determined. The Monte Carlo method was first applied to radiation problems by Howell and Perlmutter (ref. 1). These authors were concerned with the case of a radiatively participating medium. The method has recently been used for problems of surface-to-surface radiant transfer. A somewhat fuller discussion of the Monte Carlo method is given in Section 5-4.

5-1   The Exchange Factor

General properties of the exchange factor.   A basic requirement in the analysis of radiant interchange is the knowledge of what fraction of the radiant energy leaving one surface arrives at a second surface. For the case of diffusely emitting and diffusely reflecting surfaces, this information is embodied in the angle factor. When specularly reflecting surfaces are involved, an analogous quantity may be employed. This quantity may be termed the exchange factor.

Consider a surface i away from which there streams a *diffusely distributed* flux of thermal radiation. The fraction of this radiation that arrives at a second surface j both *directly* and by all possible *intervening specular reflections* is called the exchange factor $E_{A_i - A_j}$. From the foregoing definition, it is clear that the exchange factor is expressible in the form of a series as follows:

$$E_{A_i - A_j} = f_0 + \rho_{11}{}^s f_1 + \rho_{12}{}^s \rho_{22}{}^s f_2 + \rho_{13}{}^s \rho_{23}{}^s \rho_{33}{}^s f_3 + \cdots \quad (5\text{-}2)$$

The first term of the series, f_0, represents the direct transport of radiant energy from surface i to surface j. Inasmuch as the radiant flux under consideration is specified as being diffuse, it is evident that

$$f_0 = F_{A_i - A_j} \tag{5-3}$$

where F is the diffuse angle factor.

The succeeding terms of the series represent the transport due to specular interreflections. In general, a term such as $\rho_{1n}{}^s \, \rho_{2n}{}^s \cdots \rho_{nn}{}^s f_n$ represents the fraction of the radiant energy that leaves surface i and arrives at surface j after having sustained n intervening specular reflections. $\rho_{1n}{}^s$ is the specular reflectance of the surface on which the first of the n specular reflections occurs; $\rho_{2n}{}^s$ is the specular reflectance of the surface on which the second of the n specular reflections occurs, and so forth. Therefore, referring to equation (5-2), the term $\rho_{11}{}^s f_1$ takes account of one intervening specular reflection, the term $\rho_{12}{}^s \rho_{22}{}^s f_2$ takes account of two intervening specular reflections, and so on.

In some situations there may be more than one path by which radiation can travel from A_i to A_j while experiencing a common number of specular reflections. In such cases the appropriate term of equation (5-2) is evaluated along each such path.

The quantities $f_1, f_2, \ldots$ also lend themselves to interpretation. f_1 is the fraction of the diffusely distributed radiant energy that leaves surface i and arrives at surface j with one intervening *ideal* ($\rho^s = 1$) specular reflection. Similarly, f_2 is the fraction of the diffusely distributed radiant energy that leaves surface i and arrives at surface j with two intervening *ideal* ($\rho^s = 1$) specular reflections, and so forth. The $f_1, f_2, \ldots$ are purely geometrical quantities. Indeed, as will be demonstrated later, they are diffuse angle factors.

A representation such as equation (5-2) can be written for radiant energy traveling from an element dA_i to a surface A_j; the corresponding exchange factor is designated as $E_{dA_i - A_j}$. Similarly, an expression of the same form would apply for the exchange factor $dE_{dA_i - dA_j}$ for exchange between elements dA_i and dA_j. Here, the quantities $f_0, f_1, f_2, \ldots$ represent infinitesimal angle factors.

A set of reciprocity relationships can be derived for exchange factors that are similar to those for angle factors, that is,

$$A_i E_{A_i - A_j} = A_j E_{A_j - A_i} \tag{5-4a}$$

$$A_i \, dE_{A_i - dA_j} = dA_j E_{dA_j - A_i} \tag{5-4b}$$

$$dA_i \, dE_{dA_i - dA_j} = dA_j \, dE_{dA_j - dA_i} \tag{5-4c}$$

The foregoing reciprocity rules are strictly valid only when the participating surfaces are gray.

The determination of exchange factors is relatively straightforward when the surfaces having specular components of reflection are plane.

This will be demonstrated in the succeeding paragraphs. However, when the specular surfaces are curved, the task of finding the exchange factor becomes very formidable. The case of curved specular surfaces will be discussed later.

Plane specularly reflecting surfaces, the image method. When the specularly reflecting surfaces are plane, it is convenient to use a basic property of plane mirrors; namely, that the radiant energy (or light) reflected from a plane mirror appears to come from an image located behind the mirror. The distance between the image and the mirror is identical to the distance at which the object is placed in front of the mirror.

With this concept, attention may be directed to diffusely distributed radiant energy leaving a surface i, reflecting specularly on a mirror M, and arriving at a surface j. To an observer located on surface j, the radiant energy appears to originate not at surface i but, instead, at the image of surface i as formed in the mirror M. This image may be denoted by $i(M)$. Thus the fraction of the diffusely distributed radiation leaving surface i and arriving at surface j via one ideal* specular reflection is equal to the angle factor corresponding to interchange between the image $i(M)$ and j. If the mirror is not a perfect reflector, that is, if $\rho^s \neq 1$, the aforementioned angle factor is multiplied by the value of ρ^s appropriate to the specular surface.

These ideas may first be applied to the simple enclosure pictured in Fig. 5-1. The surface 3 is either a pure specular reflector with reflectance ρ^s or has a specular component of reflection ρ^s (plus a diffuse component ρ^d). The surfaces 1, 2, and 4 are diffuse reflectors of radiant energy. The shapes of the latter surfaces may be arbitrary; however, for simplicity, they are indicated as planes. All four surfaces are diffuse emitters of radiant energy. The extension of the surfaces in the

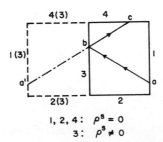

1, 2, 4: $\rho^s = 0$
3: $\rho^s \neq 0$

Fig. 5-1   Enclosure with one specular surface.

* $\rho^s = 1$.

direction into and out of the plane of the page is sufficiently large so that end effects can be neglected.

The first step is to construct the images as they are formed by the mirror, surface 3. These images are indicated in Fig. 5-1 by dashed lines. With these, the exchange factors may now be determined.

For diffusely distributed radiation leaving surface 1, there are two paths by which surface 4 is reached. First, there is the direct transfer, which is characterized by the diffuse angle factor $F_{A_1-A_4}$. Second, there is the indirect transfer via an intervening specular reflection at surface 3. This is illustrated by the ray abc in Fig. 5-1. To an observer located at point c, the radiant energy that actually originates from point a on surface 1 *appears* to originate at point a' on the image surface. A similar state of affairs applies for all radiation passing from surface 1 to surface 4 via the mirror 3; that is, the image surface 1(3) appears to be a source of diffusely distributed radiant energy. If the specular reflectance of surface 3 is $\rho_3{}^s$, then the effective radiosity of the image surface 1(3) is $\rho_3{}^s$ times the radiosity of surface 1.

In view of the foregoing, it is evident that the fraction of the radiation leaving surface 1 that arrives at surface 4 with one intervening specular reflection is $\rho_3{}^s F_{A_{1(3)}-A_4}$, where the angle factor corresponds to diffuse interchange between surfaces 1(3) and 4. Then, by summing the contributions of the direct and the indirect transfer, one has

$$E_{A_1-A_4} = F_{A_1-A_4} + \rho_3{}^s F_{A_{1(3)}-A_4} \qquad (5\text{-}5)$$

It is readily verified that equation (5-5) has exactly the same form as the general exchange-factor representation, equation (5-2).

The exchange factors corresponding to other pairs of surfaces can be written without difficulty. For instance,

$$E_{A_2-A_4} = F_{A_2-A_4} + \rho_3{}^s F_{A_{2(3)}-A_4} \qquad (5\text{-}6)$$

and so forth. One interesting fact is that surface 1, although plane, now transfers energy to itself, that is,

$$E_{A_1-A_1} = \rho_3{}^s F_{A_{1(3)}-A_1} \qquad (5\text{-}7)$$

A more interesting situation occurs when there are multiple specular reflections, as is illustrated by the enclosure shown in Fig. 5-2. The shape and orientation of the surfaces are the same as those of the preceding example, except that now both surfaces 2 and 3 have specular components of reflection. The image surfaces formed by these mirrors are shown as dashed lines in the figure. The notation on the images indicates the reflection pattern. Thus the image 1(3, 2) is the image

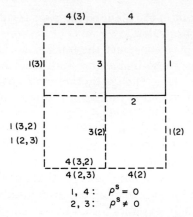

I, 4 : $\rho^s = 0$

2, 3 : $\rho^s \neq 0$

Fig. 5-2   Enclosure with two adjacent specular surfaces.

of surface 1 formed by successive reflections in the mirrors 3 and 2, and so forth.   It may be noted that a pair of images may occupy the same spacial position although each image is created by a different reflection pattern.

Let the initial task be the determination of the exchange factor $E_{A_1 - A_4}$.   The first contribution is from the direct transport $F_{A_1 - A_4}$. A second contribution is given by radiation from surface 1 that strikes the mirror 3 and is specularly reflected to surface 4; this quantity is $\rho_3{}^s F_{A_{1(3)} - A_4}$.   Another portion of the radiation from surface 1 that reflects in mirror 3 may, in turn, impinge on the second mirror 2.   It is thus necessary to determine whether any of this re-reflected radiation from mirror 2 can fall on surface 4.   The re-reflected energy appears to originate at the image surface 1(3, 2).   An observer situated on surface 4 and looking through the mirror 2 (the last mirror in the reflection sequence) is unable to see any part of the image surface 1(3, 2).   Hence none of this doubly reflected energy from surface 1 reaches surface 4.

However, there are additional paths by which radiant energy leaving surface 1 may reach surface 4.   Radiation from 1 may impinge directly on the mirror 2, and a fraction $\rho_2{}^s F_{A_{1(2)} - A_4}$ of 1's output arrives at surface 4.   Moreover, a portion of the radiation reflected from mirror 2 may re-reflect in mirror 3.   This re-reflected energy appears to come from the image 1(2, 3).   An observer stationed on surface 4 and looking through mirror 3 is able to see the image 1(2, 3). Therefore, via this path of double specular reflection, there arrives at surface 4 a fraction $\rho_2{}^s \rho_3{}^s F_{A_{1(2,3)} - A_4}$ of the output of surface 1.

It can be verified that there are no additional paths of interchange between surfaces 1 and 4 that involve specular reflections.   Next, the various contributions discussed above can be brought together and summed, giving

$$E_{A_1 - A_4} = F_{A_1 - A_4} + \rho_3{}^s F_{A_{1(3)} - A_4} + \rho_2{}^s F_{A_{1(2)} - A_4} + \rho_2{}^s \rho_3{}^s F_{A_{1(2,3)} - A_4}$$

$$(5\text{-}8)$$

Equation (5-8) is of the same form as the general exchange-factor representation, equation (5-2). An interesting feature of equation (5-8) is that it indicates two different paths by which radiation can travel from A_1 to A_4 while experiencing one specular reflection. The possible existence of such multiple paths was mentioned previously in the discussion following equation (5-2).

It is also of interest to derive the exchange factor, $E_{A_4 - A_4}$, for surface 4 to itself. First, there is the contribution $\rho_2{}^s F_{A_{4(2)} - A_4}$ that results from radiation that leaves surface 4, specularly reflects at mirror 2, and returns to surface 4. Another portion of the energy that is reflected in mirror 2 impinges on mirror 3 and is re-reflected, the radiation appearing to come from the image 4(2, 3). An observer on surface 4 looking through mirror 3 can see a large part of the image 4(2, 3) if he stands near the extreme left-hand end of surface 4, but he is able to see only a small part of the image if he stands near the extreme right. Thus energy from surface 4 that undergoes successive reflections in mirrors 2 and 3 can return to 4, but account must be taken of the just-discussed partial view of the image surface. This contribution can be expressed as $\rho_2{}^s \rho_3{}^s F^*_{A_{4(2,3)} - A_4}$, where the asterisk is affixed to identify the partial view.

In addition, radiation leaving surface 4 may reflect in mirror 3, then re-reflect in mirror 2, and return to surface 4. This contributes $\rho_3{}^s \rho_2{}^s F^*_{A_{4(3,2)} - A_4}$, where there is once again a partial view as discussed above.

After verifying that there are no further contributions to $E_{A_4 - A_4}$, one can bring together the foregoing and write

$$E_{A_4 - A_4} = \rho_2{}^s F_{A_{4(2)} - A_4} + \rho_2{}^s \rho_3{}^s (F^*_{A_{4(2,3)} - A_4} + F^*_{A_{4(3,2)} - A_4}) \quad (5\text{-}9)$$

By considering the geometry of the situation, it can be shown that

$$F^*_{A_{4(2,3)} - A_4} + F^*_{A_{4(3,2)} - A_4} = F_{A_{4(2,3)} - A_4} = F_{A_{4(3,2)} - A_4} \quad (5\text{-}10)$$

Thus, in the present situation, the two partial views merge into a total view.

Generally, however, it may be necessary to derive the angle factor corresponding to a partial view. To illustrate how this is done, consider the angle factor $F^*_{A_{4(2,3)}-A_4}$. First, an element $dA_{4(2,3)}$ on the image surface is selected, and it is noted which portion of A_4 is seen from that element as one looks through the mirror A_3. The corresponding angle factor is then determined. This operation is performed for all $dA_{4(2,3)}$ on the image surface. The angle factors thus determined are averaged, and this gives $F^*_{A_{4(2,3)}-A_4}$.

When a pair of specularly reflecting plane surfaces are placed parallel to one another, the expression for the exchange factor becomes an infinite series. Other interesting reflection patterns result when a pair of specularly reflecting surfaces is arranged obliquely (i.e., neither parallel nor perpendicular to one another).

Although the image method has been illustrated here by applications involving interchange between finite surfaces, the method applies equally well to the determination of the exchange factor for infinitesimal surfaces.

The image method has been employed in refs. 2 to 5 to facilitate the computation of radiant interchange in enclosures containing specularly reflecting, plane surfaces. Even though the exchange factor was not used as such therein; still, in essence, it was evaluated. These references include some applications of the image method additional to those described here.

Curved specularly reflecting surfaces. When the specularly reflecting surfaces are curved and have an arbitrary shape, there appears to be little hope of formulating a general analytical procedure for determining the exchange factors. However, for axisymmetric configurations, there exists a systematic method by which exchange factors can, at least in principle, be evaluated. This method will be described with the aid of Fig. 5-3. For the sake of concreteness, this figure represents a definite physical situation; namely, the radiant interchange *within* a cone. However, no aspect of the method is specific to this con-

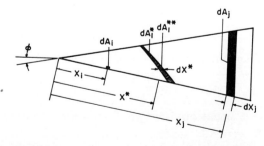

Fig. 5-3 Schematic of an axisymmetric, specularly reflecting surface configuration.

figuration; rather, the description applies to any axisymmetric situation.

Consider the diffusely distributed radiant energy leaving the element dA_i and arriving at the darkened ring element dA_j, either directly or by all possible intervening specular reflections. The exchange factor for this process is expressed by equation (5-2) except that now $E_{A_i - A_j}$ is replaced by $dE_{dA_i - dA_j}$; also, the f_n are infinitesimal quantities. Attention may be focused on any typical term of this expression, say the term corresponding to transport with n intervening specular reflections. This term contains the quantity f_n, the determination of which is the goal of the forthcoming discussion. Once the f_n have been found (for all possible n), then the exchange factor is, in essence, determined.

As noted earlier, f_n represents the fraction of the diffusely distributed radiation from dA_i that arrives at dA_j with n intervening specular reflections. Suppose that for those rays making n such surface contacts, the first surface contact is made within a band dA_1^* as is illustrated in Fig. 5-3. In general, the band will have a nonelementary shape and a nonuniform axial extension dx^*. The fraction of the diffusely distributed radiation leaving dA_i that strikes dA_1^* is $dF_{dA_i - dA_1^*}$. Correspondingly,

$$f_n = dF_{dA_i - dA_1^*} \qquad (5\text{-}11)$$

Thus the f_n can be determined if one can find the band in which rays make the first of their n specular contacts with the surface.

By studying the specular reflection pattern of a typical ray, it is possible (at least in principle) to derive an expression relating the axial location of the point of origination x_i, the axial location of the receiving point x_j, and the axial location x^* of the first point of contact. It is evident from the figure that x^* will depend on the angular displacement between x_i and x^*. If this angular displacement is denoted by $\Delta\theta$, one can write

$$x^* = g(x_i, x_j, \Delta\theta) \qquad (5\text{-}12)$$

in which g represents a functional relationship.

Moreover, the bandwidth at x^* corresponding to the bandwidth dx_j of the receiving element is

$$dx^* = \frac{\partial g}{\partial x_j} dx_j \qquad (5\text{-}13)$$

Since $\partial g/\partial x_j$ will, in general, depend on the angular displacement $\Delta\theta$, so also will the bandwidth dx^*. This dependence is illustrated in the figure.

It is convenient to denote the darkened subelement of dA_1^* as dA_1^{**}. Under the assumption that the functional relationship expressed by equation (5-12) is known, then the subelement is completely determined once x_i, x_j, and $\Delta\theta$ are specified. Correspondingly, the angle factor $dF_{dA_i - dA_1^{**}}$ can be computed. Once this has been found, the desired angle factor $dF_{dA_i - dA_1^*}$ is obtained by integrating over $\Delta\theta$ from 0 to 2π. With this, and in light of equation (5-11), one can write

$$f_n = \int_{\Delta\theta = 0}^{2\pi} dF_{dA_i - dA_1^{**}} \tag{5-14}$$

It remains to discuss the determination of the functional relationship, equation (5-12), which connects the coordinates of the radiating and receiving elements with the location of the first of n specular reflections. For this purpose, it is necessary to consider the specular reflection pattern of a typical ray. The law of specular reflection requires that the incident and reflected rays be equally inclined to, and coplanar with, the surface normal at the point of contact. If the direction of an incident ray and its point of impingement are prescribed, then the location of the next point of contact with the surface is determined by solving the following system of equations:

(a)   The equation of a plane that contains the incident ray and the normal at the point of impingement

$$C_x x + C_y y + C_z z = 1 \tag{5-15a}$$

(b)   The equation of the surface

$$h(x, y, z) = 0 \tag{5-15b}$$

(c)   The equation expressing the equality of the angle of incidence α_i and the angle of reflection α_r

$$\alpha_i = \alpha_r \quad \text{or} \quad \cos \alpha_i = \cos \alpha_r \tag{5-15c}$$

The coefficients C_x, C_y, and C_z are conveniently determined by successively substituting into equation (5-15a) the coordinates of the point of impingement, the coordinates of the point from which the incident ray originated, and the coordinates of a point on the surface normal. Then, equations (5-15a) and (5-15b) may be combined to give the line of intersection of the plane and the surface. The reflected ray must strike the surface somewhere along this line, the precise location being fixed by equation (5-15c). Once the locations of three points along the path of a ray have been interrelated by the solution of equations (5-15), then, at least in principle, the reflection pattern is known and the g function of equation (5-12) can be found.

With this, the description of the procedure for evaluating the exchange factor for axisymmetric configurations may be regarded as complete. However, the application of the method will, in general, be a formidable undertaking, the aspect of greatest difficulty being the determination of the g function. The method has been successfully applied in ref. 6 for specular interchange in a conical cavity. The details of that analysis are quite complex and do not lend themselves to exposition here. Instead, consideration will now be directed to the case of the circular cylindrical tube, which leads to a simpler form of the analysis while, at the same time, illustrating the essential features of the method.

The discussion of the specularly reflecting cylindrical tube is facilitated by reference to Fig. 5-4. The upper sketch of the figure is a

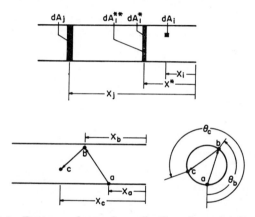

Fig. 5-4   Pattern of specular reflections in a circular tube.

schematic of such a tube showing an emitting element dA_i and a receiving ring element dA_j. The problem is to evaluate the exchange factor $dE_{dA_i - dA_j}$. In accordance with the method just described, the key to the problem is the determination of the quantity f_n; this has been identified as the angle factor for diffusely distributed radiation leaving dA_i and arriving at an element dA_1^* where the first of n specular surface contacts is achieved. The element dA_1^* is shown in the figure, along with the subelement dA_1^{**}.

The derivation of f_n necessitates a knowledge of the reflection pattern of a typical ray as it reflects and re-reflects on the walls of the tube. The reflection pattern may be derived by specializing equations (5-15a), (5-15b), and (5-15c) to the configuration under study. Considering three typical points a, b, and c that represent three successive surface contacts, one may evaluate equations (5-15a) and (5-15b) with the aid of the lower sketch of Fig. 5-4. Upon combining these equations, there results

$$(x_c - x_b) \sin \theta_b = (x_b - x_a) \sin (\theta_c - \theta_b) \qquad (5\text{-}16)$$

In addition, the equality of the angles of incidence and reflection at point b can be evaluated as

$$\frac{1 - \cos (\theta_c - \theta_b)}{\{2[1 - \cos (\theta_c - \theta_b)] + (x_c - x_b)^2\}^{1/2}} = \frac{1 - \cos \theta_b}{[2(1 - \cos \theta_b) + (x_b - x_a)^2]^{1/2}} \qquad (5\text{-}17)$$

A simultaneous solution of the foregoing yields

$$\theta_c - \theta_b = \theta_b \equiv \Delta\theta \qquad (5\text{-}18\text{a})$$

$$x_c - x_b = x_b - x_a \qquad (5\text{-}18\text{b})$$

Inspection of equations (5-18) indicates that as a ray reflects and re-reflects, it undergoes both a uniform angular displacement and a uniform linear displacement between successive points of contact.

Consequently, it is evident that a ray, leaving dA_i and making n intervening specular reflections before striking dA_j, will make its initial contact at the point

$$x^* = x_i + \frac{x_j - x_i}{n + 1} \qquad (5\text{-}19)$$

Equation (5-19) is the specific form of the functional relationship (5-12) that pertains to the reflection pattern in a circular tube. Moreover, by differentiating in accordance with equation (5-13), there follows

$$dx^* = \frac{dx_j}{n + 1} \qquad (5\text{-}20)$$

Both x^* and dx^* are independent of the angular displacement $\Delta\theta$. Correspondingly, dA_1^* is indicated as a ring of uniform thickness in the upper sketch of Fig. 5-4.

By specifying $\Delta\theta$ and employing equations (5-19) and (5-20), the position and the axial extension of the darkened subelement dA_1^{**} are fixed. The diffuse angle factor for interchange between dA_i and dA_1^{**} can be derived without difficulty. Then, in accordance with equation (5-14), this angle factor is integrated over the range from $\Delta\theta = 0$ to $\Delta\theta = 2\pi$, from which there results

$$f_n = \left\{ 1 - \frac{\left[\dfrac{\Delta x}{2(n + 1)}\right]^3 + \dfrac{3}{2}\dfrac{\Delta x}{2(n + 1)}}{\left[1 + \left(\dfrac{\Delta x}{2(n + 1)}\right)^2\right]^{1/2}} \right\} d\left[\frac{x_j}{2(n + 1)}\right] \qquad (5\text{-}21)$$

in which x has been made dimensionless by the radius of the tube and

$$\Delta x = |x_j - x_i| \qquad \text{(5-21a)}$$

Moreover, when $n = 0$, equation (5-21) reduces identically to the diffuse angle factor for direct transfer between dA_i and dA_j.

To complete the evaluation of the exchange factor, the f_n thus derived may be substituted into a series such as that of equation (5-2). For the special case in which the specular component of the reflectance is the same at all points along the tube wall, this becomes

$$dE_{dA_i - dA_j} = \sum_{n=0}^{\infty} (\rho^s)^n f_n \qquad \text{(5-22)}$$

It may be observed that the exchange factor just evaluated for the cylindrical tube can also be derived by a less formal approach, as in ref. 7. This alternative approach is successful for this configuration because of the particularly simple pattern of specular interreflections. For other axisymmetric configurations, one is not so fortunate as to find a reflection pattern as simple as that for the tube. For instance, if one end of the tube were closed, the reflection pattern for radiation passing from an element on the closing disk to an element on the tube wall would be quite complex. This latter situation has been successfully analyzed in ref. 6. It appears that the formalism described in the foregoing paragraphs represents the most general, presently available analytical approach for determining exchange factors in curved-surface configurations.

5-2 Radiant Interchange in Enclosures Containing Specularly Reflecting Surfaces and Diffusely Reflecting Surfaces

Consideration is now given to an enclosure made up of N finite surfaces, some of which are pure specular reflectors while the remainder are pure diffuse reflectors. All surfaces are gray bodies and emit diffusely. For convenience, suppose that the surfaces are numbered so that $1, 2, \ldots, N_d$ are diffusely reflecting and $(N_d + 1), (N_d + 2), \ldots, N$ are specularly reflecting. Initially, it will be assumed that the temperatures are prescribed at all the surfaces of the enclosure.

The radiant energy that is transported between the surfaces by specular reflection is taken fully into account by the exchange factors. Thus, once the exchange factors have been determined, no further consideration need be given to the specular reflections.

A radiant flux balance may now be made at a typical diffusely reflecting surface i, $1 \leqslant i \leqslant N_d$. From equation (3-20) of Chapter 3, there follows

$$B_i = \epsilon_i \sigma T_i{}^4 + (1 - \epsilon_i) H_i \tag{5-23}$$

in which B is the radiosity and H is the incident radiation per unit time and area. The contribution to H_i from the other diffusely reflecting surfaces in the enclosure can be represented in accordance with equation (3-21), except that now the angle factor F is replaced by the exchange factor E, that is,

$$\sum_{j=1}^{N_d} B_j E_{A_i - A_j} \tag{5-24}$$

In formulating the contribution of the specularly reflecting surfaces to H_i, cognizance has to be taken of the fact that radiation reflected from these surfaces is included in the exchange factors. Hence, it is only necessary to take account of radiation that is emitted at the specular surfaces and reaches i both directly and by all possible intervening specular reflections. In analogy with the foregoing expression (5-24), this contribution to H_i is

$$\sum_{j=(N_d+1)}^{N} \epsilon_j \sigma T_j{}^4 E_{A_i - A_j} \tag{5-25}$$

After combining the foregoing components of H_i, introducing them into equation (5-23), and rearranging, there follows

$$\sum_{j=1}^{N_d} X_{ij} B_j = \Omega_i, \qquad 1 \leqslant i \leqslant N_d \tag{5-26}$$

in which
$$X_{ij} = \frac{\delta_{ij} - (1 - \epsilon_i) E_{A_i - A_j}}{\epsilon_i} \tag{5-26a}$$

and
$$\Omega_i = \sigma T_i{}^4 + \frac{1 - \epsilon_i}{\epsilon_i} \sum_{j=(N_d+1)}^{N} \epsilon_j \sigma T_j{}^4 E_{A_i - A_j} \tag{5-26b}$$

Equation (5-26) may be applied at each of the diffusely reflecting surfaces in the enclosure. Consequently, there is a total of N_d equations; correspondingly, there are N_d unknown values of B. Thus, although there is a total of N surfaces in the enclosure, those that are specularly reflecting do not contribute unknowns to the problem. From this point of view, the present computational task is somewhat lightened compared with that for the case of an enclosure in which all surfaces are diffuse.

The algebraic system defined by equations (5-26) is identical in form to that defined by equations (3-23) of Chapter 3. Therefore the matrix method of solution developed in Chapter 3 continues to apply here without modification. In particular, if ψ_{ij} are the elements of the inverse matrix, then

$$B_i = \sum_{j=1}^{N_d} \psi_{ij}\Omega_j, \qquad 1 \leqslant i \leqslant N_d \qquad (5\text{-}27)$$

The heat transfer rate at any of the diffusely reflecting surfaces follows from equation (3-19) as

$$\frac{Q_i}{A_i} = \frac{\epsilon_i}{1-\epsilon_i}(\sigma T_i^{\,4} - B_i), \qquad 1 \leqslant i \leqslant N_d \qquad (5\text{-}28)$$

On the other hand, the heat transfer rate at any of the specularly reflecting surfaces is evaluated from the defining equation (with $\alpha = \epsilon$ for gray surfaces) as

$$\frac{Q_i}{A_i} = \epsilon_i(\sigma T_i^{\,4} - H_i), \qquad (N_d + 1) \leqslant i \leqslant N \qquad (5\text{-}29)$$

in which

$$H_i = \sum_{j=1}^{N_d} B_j E_{A_i - A_j} + \sum_{j=(N_d+1)}^{N} \epsilon_j \sigma T_j^{\,4} E_{A_i - A_j} \qquad (5\text{-}30)$$

If the heat transfer rate Q is prescribed at any of the diffusely reflecting surfaces in lieu of the temperature, the generalization described in Chapter 3, p. 91, can be incorporated into the foregoing formulation without difficulty. However, if Q is prescribed at any one of the specularly reflecting surfaces, a somewhat modified solution method is appropriate.

Consider first the case where the heat transfer rate is prescribed at only one of the specular surfaces, say surface k. The analysis begins as before, and the B_i for the diffusely reflecting surfaces are represented in accordance with equation (5-27). However, one or more of the Ω_j may contain the unknown $\sigma T_k^{\,4}$; correspondingly, one or more of the B_i may include linear terms in $\sigma T_k^{\,4}$. Next, equations (5-29) and (5-30) are specialized for $i = k$. In light of the foregoing, it is evident that either or both summations of equation (5-30) may contain terms linear in $\sigma T_k^{\,4}$. Then, upon combining the thus-specialized equations (5-29) and (5-30) and noting that Q_k/A_k is known, it is evident that one must solve a linear algebraic equation for $\sigma T_k^{\,4}$. Once this quantity has been determined, all other results of interest may be evaluated as previously discussed.

If the heat transfer rate is prescribed at two of the specular surfaces, say k and n, two simultaneous algebraic equations must be solved for $\sigma T_k{}^4$ and $\sigma T_n{}^4$, and so forth.

The preceding analysis tacitly involves the simplifying assumption that the radiosity is uniform along each of the diffusely reflecting surfaces. This limitation can be lifted, and, at the same time, consideration can be extended to surface variations of temperature and heat flux by performing the radiant flux balance at infinitesimal elements rather than at finite surfaces. For this purpose, infinitesimal exchange factors dE are employed. The treatment is analogous to that described on p. 98 of Chapter 3 for purely diffuse surfaces.

Upon reconsidering the analysis that has been described here for radiant interchange in enclosures containing both specularly and diffusely reflecting surfaces, it is seen that there are close parallels with the case of purely diffuse enclosures. Thus, superficially, it appears that the presence of specular reflectors does not lead to significant difficulties. However, the actual use of the analytical method requires a knowledge of the pertinent exchange factors. As was previously noted, the determination of exchange factors is a formidable undertaking except when the participating surfaces are plane or in certain elementary axisymmetric, curved-surface configurations.

Numerical solutions of radiant interchange in enclosures containing specularly reflecting surfaces are reported in refs. 2 to 7. Interchange among finite surfaces is considered in refs. 2 to 5, while interchange among infinitesimal surfaces is treated in refs. 6 to 7. Some of these results will be discussed in Chapter 6.

5-3   Radiant Interchange in Enclosures Containing Surfaces Having Both Specular and Diffuse Reflectance Components

It is uncommon to find surfaces that are either perfect diffuse reflectors or perfect specular reflectors. A model that may be somewhat closer to reality and that still lends itself to analytical treatment consists of subdividing the hemispherical reflectance into a diffuse and a specular component, as in equation (5-1). For a gray surface, this becomes

$$\rho = 1 - \epsilon = \rho^s + \rho^d \tag{5-31}$$

It will be assumed that all the participating surfaces are gray, diffuse emitters of radiant energy.

The analysis of the radiant interchange among surfaces having the foregoing reflectance characteristic, equation (5-31), is formulated as a direct generalization of the method of Section 5-2. Consideration is given to an enclosure containing N surfaces. In general, each surface may have a reflectance distribution given by equation (5-31). For some surfaces, ρ^s may be zero; for other surfaces, ρ^d may be zero. However, the general analysis covers these cases and no special provision need be made for them. As an initial assumption, the temperature of each surface in the enclosure is regarded as known.

Consider a typical surface i in the enclosure. The radiosity B may still be employed to describe the flux of diffusely distributed radiation leaving surface i, thus

$$B_i = \epsilon_i \sigma T_i{}^4 + \rho_i{}^d H_i, \qquad 1 \leqslant i \leqslant N \qquad (5\text{-}32)$$

in which ρ^d is the diffuse reflectance. H_i represents the flux of radiant energy arriving at surface i due to emission, diffuse reflection, and specular reflection at the other surfaces of the enclosure.

In evaluating H, cognizance is taken of the fact that specularly reflected radiation is fully accounted for by the exchange factor. Thus, once the exchange factors have been evaluated, no further accounting need be made of the specularly reflected radiation. Hence the incident radiation H_i can be expressed in analogy to equation (3-21) of Chapter 3 as

$$H_i = \sum_{j=1}^{N} B_j E_{A_i - A_j} \qquad (5\text{-}33)$$

Upon introducing this into equation (5-32) and rearranging, there follows

$$\sum_{j=1}^{N} X_{ij} B_j = \Omega_i, \qquad 1 \leqslant i \leqslant N \qquad (5\text{-}34)$$

where

$$X_{ij} = \frac{\delta_{ij} - \rho_i{}^d E_{A_i - A_j}}{\epsilon_i} \qquad (5\text{-}34a)$$

and

$$\Omega_i = \sigma T_i{}^4 \qquad (5\text{-}34b)$$

A comparison reveals that equations (5-34) are, in essence, identical to equations (3-23), the only differences being that $\rho_i{}^d$ and $E_{A_i - A_j}$ appear in place of $(1 - \epsilon_i)$ and $F_{A_i - A_j}$. Therefore the matrix method of solution continues to apply. In particular, in terms of the elements of the inverse matrix ψ_{ij}, the solution for B_i can be expressed as

$$B_i = \sum_{j=1}^{N} \psi_{ij} \sigma T_j{}^4, \qquad 1 \leqslant i \leqslant N \qquad (5\text{-}35)$$

The heat transfer rate at any surface i can be evaluated from the defining equation

$$\frac{Q_i}{A_i} = \epsilon_i(\sigma T_i^4 - H_i) \tag{5-36}$$

in which the gray-body assumption has been employed. For all surfaces for which $\rho^d \neq 0$, H_i can be eliminated from the foregoing with the aid of equation (5-32). After rearrangement, there results

$$\frac{Q_i}{A_i} = \frac{\epsilon_i}{\rho_i^d}[(1 - \rho_i^s)\sigma T_i^4 - B_i], \qquad \begin{array}{c} 1 \leqslant i \leqslant N \\ \rho^d \neq 0 \end{array} \tag{5-37}$$

where equation (5-31) has been employed. When $\rho^s = 0$, this heat transfer expression reduces identically to equation (3-19), which applies for diffusely reflecting surfaces.

For those surfaces for which $\rho^d = 0$, the heat transfer is evaluated by eliminating H_i between equations (5-33) and (5-36), giving

$$\frac{Q_i}{A_i} = \epsilon_i\left(\sigma T_i^4 - \sum_{j=1}^{N} B_j E_{A_i - A_j}\right) \tag{5-38}$$

It may also be noted that $B_j = \epsilon_j \sigma T_j^4$ for a surface at which $\rho^d = 0$.

The extension of the analysis to include the case in which the heat flux is prescribed at some of the surfaces is carried out along lines previously described on p. 91 in Chapter 3. The only difference is that equation (5-32) is replaced by

$$B_i = \frac{Q_i}{A_i} + (1 - \rho_i^s)H_i \tag{5-39}$$

for the surfaces at which the heat flux is prescribed, and corresponding changes are made in the X_{ij} and Ω_i of equations (5-34a) and (5-34b). No special provisions need be made for the case of $\rho^d = 0$.

The model of the reflectance distribution represented by equation (5-31) also lends itself to analysis of interchange between infinitesimal surfaces. To illustrate the use of the method in such cases as well as to illuminate the preceding discussion for interchange between finite surfaces, consideration is given to the radiant interchange in an isothermal cylindrical cavity. Such a cavity is pictured schematically in Fig. 5-5. For purposes of simplicity, the cavity is assumed to be, in effect, of infinite depth.

The radiation properties ϵ, ρ^s, and ρ^d are constant along the cavity wall. The wall temperature is designated as T_w. A diffusely distributed radiant flux, characterized by an effective black-body temperature T_∞, streams inward through the cavity opening ($x = 0$) from the external environment. Since the governing equation is linear in

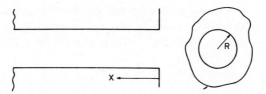

Fig. 5-5   Circular cylindrical cavity.

T^4, the problem just described is fully equivalent to an alternate problem in which the wall temperature is $(T_w{}^4 - T_\infty{}^4)^{1/4}$ and the external environment is at absolute zero.   The latter problem is somewhat simpler and is therefore the one analyzed here.

The flux of diffusely distributed radiant energy leaving a typical location x on the cavity wall may be expressed in a form analogous to equation (5-32), thus

$$B(x) = \epsilon\sigma(T_w{}^4 - T_\infty{}^4) + \rho^d H(x) \qquad (5\text{-}40)$$

As in the just-discussed analysis for interchange between finite surfaces, the specularly reflected radiation is fully included in the exchange factor. Correspondingly, the radiant flux incident at x may be written in a form similar to that of equation (5-33), except that now the summation is replaced by an integral

$$H(x) = \int_{\xi=0}^{\infty} B(\xi)\, dE_{dA_x - dA_\xi} \qquad (5\text{-}41)$$

in which ξ is the running variable of integration.   The exchange factor appearing in the integrand is exactly that represented by equations (5-21) and (5-22).

By combining equations (5-40) and (5-41), one gets

$$B(x) = \epsilon\sigma(T_w{}^4 - T_\infty{}^4) + \rho^d \int_{\xi=0}^{\infty} B(\xi)\, dE_{dA_x - dA_\xi} \qquad (5\text{-}42)$$

This integral equation is of the same type as that previously encountered in Chapter 3 for the analysis of diffuse interchange.   The various solution methods discussed there continue to apply.   It may be noted that two independent parameters,* ϵ and ρ^d, influence the solution of equation (5-42).

Once the radiosity has been determined by solution of equation (5-42), the local heat flux per unit area follows as

$$q(x) = \frac{\epsilon}{\rho^d}[(1 - \rho^s)\sigma(T_w{}^4 - T_\infty{}^4) - B(x)] \qquad (5\text{-}43)$$

* ρ^s also appears in dE, but this is related to ϵ and ρ^d in accordance with equation (5-31).

This expression is derived in a manner similar to equation (5-37). It may be noted that, on the basis of physical reasoning, q will approach zero as x approaches infinity. Correspondingly, $B(x)$ approaches $(1 - \rho^s)\sigma(T_w{}^4 - T_\infty{}^4)$ at large x.

The integral equation (5-42) for the radiosity was solved numerically in ref. 8 by the method of successive iterations. In carrying out the solutions, the upper limit on the integral was taken as a finite value ξ^*. The magnitude of ξ^* was selected so that $B(x)$ approached very closely to the aforementioned limiting value. Numerical results based on these solutions will be discussed in the forthcoming chapter.

As a final remark, it might be mentioned that contributions to the analysis of interchange between surfaces having both specular and diffuse reflectance components have been made by several authors (refs. 8 to 10). These references also include numerical results for various special cases.

5-4   The Monte Carlo Method

The Monte Carlo method provides a systematic numerical approach for taking account of nonelementary directional distributions of emission and reflection. The method derives its name from the fact that the laws of probability (i.e., chance) are employed in determining the direction of travel of energy bundles and in deciding if a bundle is absorbed or reflected.

The use of the Monte Carlo method for solving radiation heat transfer problems was first accomplished in a series of papers by Howell and Perlmutter (refs. 1, 11, 12). These papers were primarily concerned with the effects of radiatively participating gases, and the bounding surfaces of the gas body were taken to be diffuse emitters and reflectors having elementary geometrical configuration (parallel plates, concentric cylinders). More recently, several application-oriented problems involving participating media were solved by the Monte Carlo method (refs. 13, 14, and 15). The utility of the Monte Carlo method for solving problems of surface-to-surface radiation has also been recognized (refs. 16 to 20). A valuable survey article setting forth the application of Monte Carlo to heat transfer problems in general (and, in particular, to radiation) has been prepared by Howell (ref. 21).

The general ideas of the Monte Carlo method will now be discussed. Details may be found in the aforementioned references, where flow charts that demonstrate the successive computer operations are also

presented.   First, a surface location is selected at which emitted radiation is to be studied.   The emitted energy is regarded as being carried
in discrete bundles.   The number N of such bundles is then selected.
Typical values of N used in practical computation have ranged from
10,000 to 50,000.

The next step is to set up equations that express the probability
that an energy bundle will be emitted from the surface in a given direction (see refs. 20 and 21).   It is convenient to express the direction of
bundle emission in terms of the "cone" angle θ and the polar angle ϕ of
a spherical coordinate system centered at the emitting location.   In
principle, any directional distribution of the emittance can be accom-
modated.   The probability function can be normalized so that it takes
on values between zero and unity.   For instance, for a diffusely emitting
surface, the probability distribution functions for the angles θ and ϕ
are $\sin^2 \theta$ and $\phi/2\pi$.

Once the probability functions are established, the path of the
first energy bundle can be traced.   The direction of departure (θ, ϕ)
is found by drawing a pair of random numbers R_θ and R_ϕ from a
uniformly distributed set between zero and one.   For the diffuse case

$$\sin^2 \theta = R_\theta \qquad \phi = 2\pi R_\phi \qquad\qquad (5\text{-}44)$$

From geometrical considerations, with a knowledge of the point of
departure, the direction of departure, and the configuration of the
enclosure, the point of impingement of the energy bundle can be
determined.

Suppose that the surface location on which the bundle is incident
has an absorptance α.   In general, α may depend on the angle of
impingement relative to the surface normal.   A random number R_α is
drawn from a uniformly distributed set between zero and one.   If
$0 \leqslant R_\alpha \leqslant \alpha$, the incident bundle is absorbed; the absorption is tallied
and attention is redirected to the next energy bundle leaving the point
of emission.   On the other hand, if $\alpha < R_\alpha \leqslant 1$, the bundle is reflected.
The angular direction in which the reflected bundle travels is specified
in terms of probability distributions for the local "cone" and polar
angles θ and ϕ.   If the reflecting surface is diffuse, equations (5-44)
continue to apply.   However, in the most general case, the probabilities
for the reflection angles θ and ϕ will depend on the angle under which
the bundle impinges on the surface.

The path of each energy bundle is followed until absorption takes
place.   A very large number of such bundles is employed to represent
adequately the radiation emitted at the preselected surface location.
Once the just-discussed procedure has been completed for one surface

location, a second location is selected, the radiation emitted at that point is studied, and so on.

It is evident that the use of the Monte Carlo method requires access to a digital computer having both high speed and large storage capacity.

REFERENCES

1. J. R. Howell and M. Perlmutter, Monte Carlo solution of thermal transfer through radiant media between gray walls. *J. Heat Transfer*, **C86**, 116–122 (1964).

2. E. R. G. Eckert and E. M. Sparrow, Radiation heat exchange between surfaces with specular reflection. *Intern. J. Heat Mass Transfer*, **3**, 42–54 (1961).

3. E. M. Sparrow, E. R. G. Eckert, and V. K. Jonsson, An enclosure theory for radiative exchange between specularly and diffusely reflecting surfaces. *J. Heat Transfer*, **C84**, 294–300 (1962).

4. R. P. Bobco, Radiation heat transfer in semigray enclosures with specularly and diffusely reflecting surfaces. *J. Heat Transfer*, **C86**, 123–130 (1964).

5. R. V. Dunkle, Radiant interchange in an enclosure with specular surfaces and enclosures with windows or diathermanous walls. *Heat Transfer, Thermodynamics and Education*, Boelter Anniversary Volume, McGraw-Hill, New York, 1964, pp. 133–149.

6. S. H. Lin and E. M. Sparrow, Radiant interchange among curved specularly reflecting surfaces; application to cylindrical and conical cavities. *J. Heat Transfer*, **C87**, 299–307 (1965).

7. M. Perlmutter and R. Siegel, Effect of specularly reflecting gray surfaces on thermal radiation through a tube and from its heated wall. *J. Heat Transfer*, **C85**, 55–62 (1963).

8. E. M. Sparrow and S. H. Lin, Radiation heat transfer at a surface having both specular and diffuse reflectance components. *Intern. J. Heat Mass Transfer*, **8**, 769–779 (1965).

9. A. F. Sarofim and H. C. Hottel, Radiative exchange among non-Lambert surfaces. *J. Heat Transfer*, **C88**, 37–44 (1966).

10. J. T. Bevans and D. K. Edwards, Radiation exchange in an enclosure with directional wall properties. *J. Heat Transfer*, **C87**, 388–396 (1965).

11. J. R. Howell and M. Perlmutter, Monte Carlo solution of radiant heat transfer in a nongray, nonisothermal gas with temperature dependent properties. *Am. Inst. Chem. Engr. J.*, **10**, 562–567 (1964).

12. M. Perlmutter and J. R. Howell, Radiant transfer through a gray gas between concentric cylinders using Monte Carlo. *J. Heat Transfer*, **C86**, 169–179 (1964).

13. J. R. Howell, M. K. Strite, and H. Renkel, Heat transfer analysis of rocket nozzles using very high-temperature propellants. *AIAA J.*, **3**, 669–673 (1965).

14. J. R. Howell, M. K. Strite, and H. Renkel, Analysis of heat-transfer effects in rocket nozzles operating with very high-temperature hydrogen. *NASA TR R-220*, February 1965.

15. J. R. Howell and M. K. Strite, Heat transfer in rocket nozzles using high-temperature hydrogen propellant with real property variations. *J. Spacecraft Rockets*, **3**, 1063–1068 (1966).

16. L. G. Polgar and J. R. Howell, Directional thermal-radiative properties of conical cavities. *NASA TN D-2904*, June 1965.

17. R. C. Corlett, Direct numerical simulation of thermal radiation in vacuum. *J. Heat Transfer*, **88**, 376–382 (1966).

18. M. M. Weiner, J. W. Tindall, and L. M. Candell, Radiative interchange factors by Monte Carlo. *ASME Paper No. 65-WA/HT-51* (1965).

19. J. S. Toor and R. Viskanta, A numerical experiment of radiant heat exchange by the Monte Carlo method. *Intern. J. Heat Mass Transfer*, 883–897 (1968).

20. A. Haji-Sheikh and E. M. Sparrow, Probability distributions and error estimates for Monte Carlo solutions of radiation problems. In *Progress in Heat and Mass Transfer*, vol. 2, Pergamon Press, Oxford, 1969, pp. 1–12.

21. J. R. Howell, Application of Monte Carlo to heat transfer problems. In *Advances in Heat Transfer*, vol. 5, Academic Press, New York, 1968.

PROBLEMS

5-1.   An enclosure consists of N surfaces, all of which are diffuse emitters of radiant energy.   Let there be one or more surfaces that have a specular component of reflectance in accordance with equation (5-1). For a typical surface i, there are exchange factors $E_{A_i-A_1}, E_{A_i-A_2}, \ldots, E_{A_i-A_N}$.   Demonstrate that

$$\sum_{j=1}^{N} E_{A_i-A_j} \geqslant 1$$

(A valid exchange factor conservation rule is examined in Problem 4-10.) Hint: Utilize the angle factor conservation rule stated by equation (3-2).

5-2.   Verify the finding of Problem 5-1 by summing the exchange factors for surface 1 of Fig. 5-1.

5-3.   Consider a three-walled enclosure in the form of an isosceles triangle as shown in the figure at the top of page 168.   The walls of the enclosure are very long in the direction into and out of the plane of the figure.   All of the surfaces are diffuse emitters; in addition, all surfaces are gray.   Surfaces 1 and 3 are diffuse reflectors, while surface 2 has a specular component of reflectance $p_2{}^s$. Draw a diagram showing the images formed by the specular surface. For $\phi \leq 45°$, derive expressions for E_{A1-A1}, E_{A3-A3}, E_{A1-A3}, and E_{A3-A1}. The diffuse angle factor F_{A1-A2} is equal to $1 - \sin \phi$. Verify that $A_1 E_{A1-A3} = A_3 E_{A3-A1}$.

5-4. Discuss the various exchange factors of Problem 5-3 when $\phi = 45°$ and when $\phi > 45°$.

5-5. Consider a rectangular enclosure such as that pictured in Fig. 5-2, but with different radiation properties. Let surfaces 1 and 3 now possess specular components of reflectance, $\rho_1{}^s$ and $\rho_3{}^s$, respectively, while surfaces 2 and 4 are purely diffuse reflectors ($\rho^s = 0$). All surfaces are gray and are diffuse emitters. Derive an expression for the exchange factor E_{A4-A2}. Hint: As a result of the back and forth specular reflections between surfaces 1 and 3, E_{A4-A2} will be represented by infinite series. Further assistance may be obtained by consulting reference 3, cited on page 166.

5-6. Refer to the figure in Problem 3-8. Suppose now that the upper and the lower walls have identical radiation properties $\rho = \rho^s + \rho^d = 1 - \epsilon$. The walls are diffuse emitters. Show that the exchange factor $dE_{dA1-dA2}$ is equal to

$$dE_{dA_1-dA_2} = \sum_{n=0}^{\infty} (\rho^s)^{2n} \Omega(n)$$

where

$$\Omega(n) = \frac{(2n + 1)^2 h^2 dx_2}{2[(x_2 - x_1)^2 + (2n + 1)^2 h^2]^{3/2}}$$

Hint: draw the successive images of dA_1 and note that only the images which lie below the lower wall can irradiate dA_2. Also, make use of the expression for $dF_{dA_1-dA_2}$ given in Problem 3-8.

5-7. For the enclosure pictured in Fig. 5-1, suppose that $\rho_3 = \rho_3{}^s$ (that is, $\rho_3{}^d = 0$). The temperatures of the respective surfaces are prescribed as T_1, T_2, T_3, and T_4. All surfaces are gray. Write the equations for the radiosities B_1, B_2, and B_4 of the diffuse surfaces.

Express the participating exchange factors in terms of appropriate angle factors and of $\rho_3{}^s$. Hint: Utilize equations (5-23), (5-24), and (5-25) to set up the radiosity equations.

5-8. Modify Problem 5-7 so that the specular surface 3 is now prescribed to be adiabatic ($Q_3 = 0$). Correspondingly, T_3 is unknown. Reformulate the radiosity equations for surfaces 1, 2, and 4 so that T_3 no longer appears. Hint: Express $\sigma T_3{}^4$ in terms of B_1, B_2, and B_4 by applying equations (5-29) and (5-30) for $i = 3$.

5-9. Generalize the conditions of Problem 5-7 so that $\rho_3 = \rho_3{}^s + \rho_3{}^d$. Derive a radiosity equation for each surface of the enclosure. Hint: Use equations (5-32) and (5-33).

5-10. Consider an enclosure made up of N surfaces of the type treated in Section 5-3. Show that an exchange factor conservation rule for any surface i can be expressed as

$$\sum_{j=1}^{N} (1 - \rho_j{}^s) E_{A_i - A_j} = 1$$

Hint: For an enclosure in which all surfaces are at the same uniform temperature T, the second law of thermodynamics requires that $Q_i = 0$ for all i. Correspondingly, equation (5-37) gives that $B_i = (1 - \rho_i{}^s)\sigma T_i{}^4$ when $\rho^d \neq 0$. The same expression for B_i also holds when $\rho^d = 0$. With this, the aforementioned conservation rule follows directly from equations (5-32) and (5-33), provided $\rho^d \neq 0$. An identical end result for $\rho^d = 0$ is obtained by applying equation (5-38) with $Q_i = 0$.

CHAPTER 6

APPLICATIONS AND SOLUTIONS
OF THE EQUATIONS OF
RADIANT INTERCHANGE

The equations that govern the exchange of radiant energy among surfaces have been solved for a wide range of technically interesting situations. These problems may be classified into three general categories: (1) situations in which the radiant interchange is analyzed without consideration of other modes of heat transfer, (2) situations in which heat conduction within the participating surfaces interacts with the radiant interchange, and (3) situations in which convection in the gas that occupies the space between the participating surfaces interacts with the radiant interchange. Each of these groupings is, in itself, a specialization of the general case where radiant interchange, wall conduction, and gas convection act simultaneously. The general case, although capable of formulation, has not yet been solved to any significant extent at the time of this writing.

In this chapter a range of specific problems will be presented to illustrate the preceding three categories of solutions. For the case of purely radiative transport, attention is focused on the radiation characteristics of isothermal-walled cavities, Section 6-1. Interaction between radiation and conduction is illustrated by consideration of the radiating fin in Section 6-2. Interest in such fins has been stimulated by applications to space-vehicle thermal control systems and power plants. Finally, in Section 6-3, boundary layer and pipe flow problems are discussed by way of illustrating the interaction between radiation and convection.

It is not the present intent, nor would it be practical, to bring together all existing solutions for radiant interchange among surfaces. Rather, the aim of this chapter is to illustrate the type of results that have been (and can be) obtained by applying the analytical procedures that were described in earlier chapters.

6-1 Purely Radiative Systems: Cavities

A cavity is a partially enclosed space. When radiation from an external source enters a cavity, the ensuing reflections and re-reflections among the cavity walls provide additional opportunities for energy absorption. Thus the energy absorbed within a cavity will generally exceed that which would be absorbed by a plane surface of identical absorptance stretched tightly across the cavity opening. Similarly, the multiple reflections within a heated isothermal-walled cavity act to augment the emissive power of the cavity relative to that of a plane surface of identical temperature and emittance stretched across its opening. The just-described augmentation of the absorptance and emittance is sometimes termed the cavity effect. In practice, the cavity effect is employed in the design of energy sources such as black bodies and in the design of absorbers of solar radiation.

Analysis of the radiant interchange. For purposes of analysis, the bounding walls of the cavity are taken to be isothermal and to be gray, diffuse emitters of radiant energy. All the walls of the cavity have a common emittance ϵ. Radiant energy entering the cavity opening from an external source is assumed to be uniformly and diffusely distributed; such radiation can be characterized by an effective black-body temperature T_∞. In the case of purely radiative transport, the governing equations are linear in T^4. Therefore the cavity problem characterized by a wall temperature T_w and an external environment temperature T_∞ is fully equivalent to the cavity problem characterized by a wall temperature $(T_w{}^4 - T_\infty{}^4)^{1/4}$ and an environment temperature of absolute zero. The latter problem is somewhat simpler to analyze and is therefore the one usually investigated.

When the cavity walls are diffuse reflectors of radiant energy, the equations governing the radiant transport are written by specializing equation (3-40) of Chapter 3 in accordance with the conditions discussed in the foregoing paragraph. From this, there follows

$$B_i(\mathbf{r}_i) = \epsilon\sigma(T_w{}^4 - T_\infty{}^4) + (1 - \epsilon) \sum_{j=1}^{N} \int_{A_j} B_j(\mathbf{r}_j)K(\mathbf{r}_i, \mathbf{r}_j)\, dA_j, \quad 1 \leqslant i \leqslant N$$

$$(6\text{-}1)$$

These equations respectively apply at the walls $1, 2, \ldots, N$ that bound the cavity, but not at the cavity opening. For a fictitious black surface stretched over the opening, it is easily understood that $B = 0$ inasmuch as $\rho = 1 - \epsilon = 0$ and the temperature is zero.

It may be recalled from Chapter 3 that the $\mathbf{r}_i$ and $\mathbf{r}_j$ simply denote coordinates on the surfaces i and j, and K is related to the angle factor

$dF_{dA_i - dA_j}$ in accordance with equation (3-11). For a specific illustration of the procedure for further specializing equation (6-1), readers are referred to pages 96–97 of Chapter 3. References to published literature on cavities where the radiosity equations are formulated will be given later.

Once the radiosity distribution is determined by solving the integral equations (6-1), the corresponding distribution of the local heat flux q_i is readily evaluated from equation (3-41), where T_i^4 is replaced by $(T_w^4 - T_\infty^4)$ and ϵ_i is replaced by ϵ. With this information, the rate Q at which radiant energy streams out of the cavity opening is calculable by integration, that is,

$$Q = \sum_{j=1}^{N} \int_{A_j} q_j \, dA_j \qquad (6-2)$$

Q may be regarded as the effective emissive power of the cavity.

When the reflectance of the cavity walls is represented as a sum of specular and diffuse components, ρ^s and ρ^d, respectively, then the formulation of Chapter 5 is applicable. The governing equations for the radiosity distributions bear a close resemblance to those for pure diffusely reflecting surfaces. In fact, by respectively replacing $(1 - \epsilon)$ and $K(\mathbf{r}_i, \mathbf{r}_j) \, dA_j$ by ρ^d and $dE_{dA_i - dA_j}$ in the foregoing equation (6-1), one arrives at the appropriate governing equations for the specular-diffuse case. The local heat flux is related to the radiosity in accordance with equation (5-43) of Chapter 5, where one now appends a subscript j to q and B. Once q_j has been evaluated, the effective emissive power Q is determined as indicated in equation (6-2).

As is evident from the foregoing, the analysis for cavities whose walls are either pure diffuse reflectors or specular-diffuse reflectors requires that an integral equation system be solved. However, when the walls are pure specular reflectors, this is no longer the case. By direct application of the exchange-factor concept of Chapter 5, the radiant energy streaming out of the opening of a specularly reflecting cavity follows as

$$Q = \epsilon\sigma(T_w^4 - T_\infty^4)A_0 \sum_{j=1}^{N} E_{A_0 - A_j} \qquad (6-3)$$

where A_0 is the area of a fictitious surface stretched across the cavity opening.

In specifying the overall performance of a cavity, it is customary to compare the rate of energy efflux from the cavity opening with that streaming out of a black-walled cavity. The latter quantity, to be denoted by Q_b, is precisely equal to the radiant emission of a black surface stretched across the opening of the black-walled cavity.   An apparent emittance ϵ_a may then be defined as

$$\epsilon_a = \frac{Q}{Q_b} = \frac{Q}{\sigma A_0 (T_w{}^4 - T_\infty{}^4)} \tag{6-4}$$

It can readily be demonstrated that ϵ_a exceeds the actual emittance ϵ of the cavity walls.   For instance, for diffusely reflecting cavity walls, an expression for Q, alternate to equation (6-2), is

$$Q = A_0 \sum_{j=1}^{N} \int_{A_j} B_j(r_j)\, dF_{A_0 - dA_j} \tag{6-5}$$

But, $B_j \geqslant \epsilon\sigma(T_w{}^4 - T_\infty{}^4)$ and $\sum_{j=1}^{N} F_{A_0 - A_j} = 1$; consequently,

$$Q \geqslant A_0 \epsilon\sigma(T_w{}^4 - T_\infty{}^4) \tag{6-6}$$

A corresponding result can be derived for the case in which the walls are partially or wholly specularly reflecting.   The departure of ϵ_a from ϵ is a measure of the cavity effect.   The maximum value of ϵ_a that can be attained is unity.

In general, the number of possible cavity shapes is virtually boundless.   However, there appears to be a group of cavity configurations that frequently recurs, and it is these configurations that will be considered here.   Schematic diagrams of these cavities are presented in Fig. 6-1 along with dimensional nomenclature.   The rectangular groove and the V-groove, diagrams (b) and (d), respectively, are regarded as being very long in the direction into and out of the plane of the figure.

The formulation and the solution of the governing equations for the various cavities may be found in the published literature as follows: circular cylindrical, refs. 1, 2, and 3; conical cavity, refs. 2, 3, and 4; rectangular groove, refs. 5 and 6; V-groove, ref. 7; sphere, refs. 8 and 9. The aforementioned literature citations are believed to represent the most recent investigations for the respective cavity configurations. Earlier contributions are, in turn, listed in the bibliographies appended to the just-cited papers.   These earlier works are of both historical and technical value.

Presentation of results.   The solutions of the governing equations have provided information on both the local and the overall heat transfer characteristics for the various cavities.   The sum total of the available results is rather massive, and only a selected portion can be displayed here.   It was decided to omit the local heat transfer results from this presentation and to refer interested readers to the just-cited

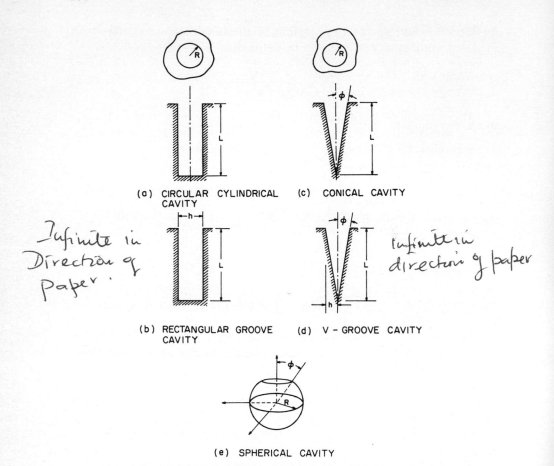

Infinite in Direction of Paper.

Infinite in direction of paper

(a) CIRCULAR CYLINDRICAL CAVITY

(b) RECTANGULAR GROOVE CAVITY

(c) CONICAL CAVITY

(d) V - GROOVE CAVITY

(e) SPHERICAL CAVITY

Fig. 6-1   Various cavity configurations.

literature.   The overall heat transfer results, expressed in terms of the apparent emittance ϵ_a, will now be presented.

Figures 6-2 and 6-3 pertain to the circular cylindrical cavity. The first of these figures shows the apparent emittance as a function of the cavity depth-to-radius ratio.   Results are given for values of surface emittance ranging from 0.1 to 0.9; solid and dashed lines are respectively employed to designate diffusely and specularly reflecting cavity walls.   Inspection of the figure shows that, as expected, $\epsilon_a \geqslant \epsilon$ for all values of the parameters.   The departure of ϵ_a from ϵ increases with increasing cavity depth.   In the case of diffusely reflecting walls, the apparent emittance reaches a limiting value (less than unity) such that further increases in cavity depth have no effect.   These limiting

values are achieved in surprisingly short cavity depths.   For instance,
for a surface emittance of 0.5, a cavity of depth-to-radius ratio $L/R = 3$
behaves, in essence, like a cavity of infinite depth.   For specularly
reflecting walls, the apparent emittance continues to increase toward
unity with increasing cavity depth.

   Further inspection of Fig. 6-2 indicates that the cavity effect is
most pronounced when the emittance of the cavity wall is low.

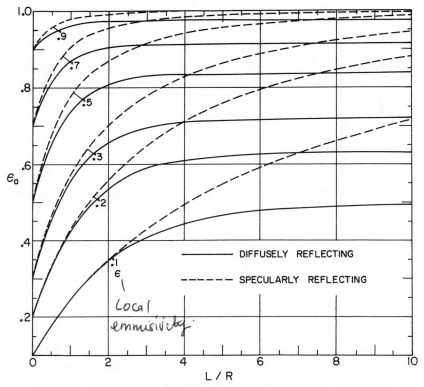

Fig. 6-2   Apparent emittance results for diffusely and specularly
reflecting circular cylindrical cavities.

Furthermore, for a given emittance ϵ, a specularly reflecting cavity wall
gives rise to higher values of ϵ_a than does a diffusely reflecting wall.

   A detailed picture of the way in which ϵ_a is affected by the directional reflectance characteristics of the cavity walls is presented in Fig.
6-3.   These results are for the cylindrical cavity of infinite depth.   The
figure consists of two parts.   The upper portion shows ϵ_a as a function
of the surface emittance ϵ for various parametric values of ρ^s/ρ

$(\rho = \rho^s + \rho^d)$. The lower portion of the figure is a rephrasing of this same information with ρ^s/ρ as independent parameter. The figure shows that the increase of ϵ_a with increasing values of ρ^s/ρ (i.e., with a relative increase in the specular component) is most marked when the surface emittance ϵ is low. Furthermore, for such surfaces, the sharpest increases occur as ρ^s/ρ approaches unity.

The available results for the apparent emittance of rectangular groove cavities are shown in Fig. 6-4. This figure presents information similar to that provided by Fig. 6-2 for the cylindrical cavity. The

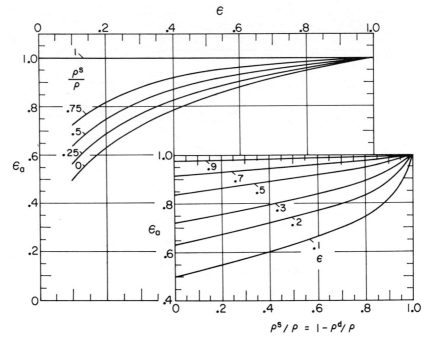

Fig. 6-3   Effect of ρ^s/ρ on the apparent emittance of a deep cylindrical cavity.

general characteristics of the results for the rectangular groove are identical to those already discussed for the cylinder. It is also interesting to note that at corresponding values of L/R and L/h, the values of ϵ_a from Figs. 6-2 and 6-4 do not differ significantly, especially for the case of diffusely reflecting surfaces.

The apparent emittance results for conical cavities are plotted in Fig. 6-5 as a function of the cone-opening angle 2ϕ. The curves are parameterized by the surface emittance ϵ. Furthermore, for each ϵ, results are given for reflectances that are purely specular ($\rho^s/\rho = 1$), purely diffuse ($\rho^s/\rho = 0$), and half specular and half diffuse, ($\rho^s/\rho = 0.5$). In all cases the apparent emittance increases as the cone-opening angle becomes smaller. In the limit as $\phi \to 0°$, the conical cavity behaves like a circular cylindrical cavity of infinite depth. It may be verified that the ordinate intercepts in Fig. 6-5 are consistent with the informa-

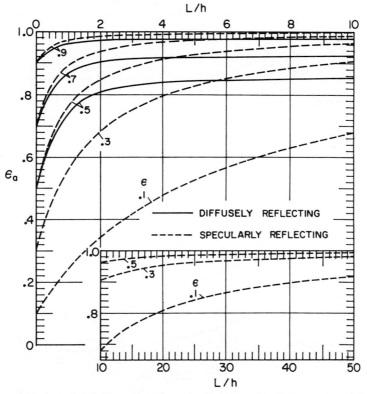

Fig. 6-4   Apparent emittance results for diffusely and specularly reflecting rectangular groove cavities.

tion presented in Fig. 6-3. At the other limit, as the opening angle approaches 180°, the apparent emittance approaches the surface emittance ϵ.

At a fixed value of ϵ, the specularly reflecting cavity radiates most energy and the diffusely reflecting cavity least. The curves corresponding to $\rho^s/\rho = 0.5$ do not necessarily lie midway between the aforementioned extremes, although this behavior is nearly achieved when the surface emittance is high. At lower values of surface emittance, the curves for $\rho^s/\rho = 0.5$ lie closer to those for diffuse reflection when ϕ is small and closer to those for specular reflection when ϕ is large. In general, it appears that the apparent emittance is independent of the directional distribution of the reflectance when $2\phi \geqslant 80°$.

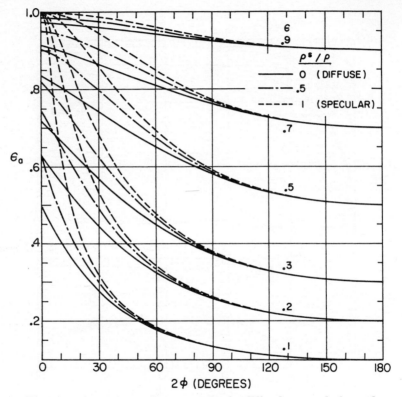

Fig. 6-5   Apparent emittance results for diffusely, specularly, and specularly-diffusely reflecting conical cavities.

The available results for the V-groove, presented in Fig. 6-6, show trends identical to those just discussed for the conical cavity. In fact, for given values of the parameters, the apparent emittance values for the diffuse case are essentially the same for the two cavities. For the specular case, there are small deviations between the results for the two cavities.

Among the basic cavity configurations considered here, only the spherical cavity permits a closed-form solution for the case of diffusely reflecting walls.   In terms of the "cone" angle ϕ defined in Fig. 6-1, the expression for ϵ_a is

$$\epsilon_a = \frac{\epsilon}{1 - 0.5(1 - \epsilon)(1 + \cos\phi)} \tag{6-7}$$

A plot of this information is presented in Fig. 6-7.   For any value of the surface emittance, the apparent emittance approaches unity as the opening angle becomes smaller.   As far as the authors know, results for the specularly reflecting spherical cavity are not now available.

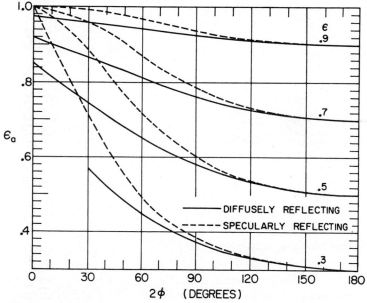

Fig. 6-6   Apparent emittance results for diffusely and specularly reflecting V-groove cavities.

Closely related to the radiant interchange in cavities is the transport of radiant energy through passages, both ends of which are open to radiating environments.   The analytical formulation and the method of solution for the two groups of problems are essentially identical.   Consequently, no further discussion will be given here. Interested readers may wish to consult refs. 3, 10, and 11, which treat the transport of radiant energy through a circular tube, and ref. 12, which gives results for tapered tubes and tapered plane-walled passages.

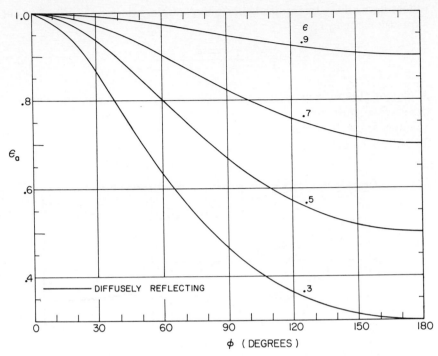

Fig. 6-7 Apparent emittance results for diffusely reflecting spherical cavities.

6-2 Radiative-Conductive Systems: Fins

In atmosphere-free space, radiation is a primary mechanism for transferring waste heat from space vehicles. It is widely contemplated to employ extended surfaces, that is, fins, to augment the area available for radiation. In such a situation, the heat transfer is controlled by the interaction between heat conduction within the fin material and radiation at the fin surface.

The single plate-type fin. A typical radiator configuration consisting of plate-type fins attached to circular tubes is pictured schematically in Fig. 6-8, lower diagram. The tubes contain a flowing fluid from which heat is to be extracted. In the simplest analytical model of this system, it is assumed that the radiant interchange between the fin and the tubes is negligible. Such an assumption is reasonable in cases where the tube radius is much smaller than the spacing between the tubes. In addition, the fin heat conduction in the z direction

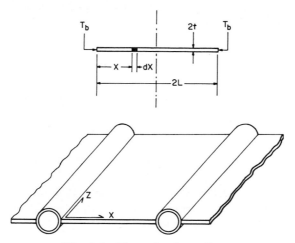

Fig. 6-8   Fin and tube radiator.

(streamwise direction) is usually small compared with the heat conduction in the x direction (ref. 13) and is therefore neglected.

On the basis of these simplifications, the problem can be modeled as shown at the top of Fig. 6-8. If the fin thickness $2t$ is much less than the fin height $2L$, it is reasonable to invoke the thin-fin assumption such that $T = T(x)$ alone.

Under steady state conditions, an energy balance on a typical element of fin requires that the net conductive inflow equals the net radiative outflow. By applying Fourier's law, the net conductive inflow is

$$k(2t) \frac{d^2T}{dx^2} \, dx \tag{6-8}$$

per unit length normal to the page. The net radiation is the difference between the emitted energy and the absorbed portion of the incident energy, that is,

$$[\epsilon \sigma T^4(x) - \alpha H(x)]2dx \tag{6-9}$$

In writing the preceding expression, it is assumed that the radiation incident on the upper and lower surfaces of the fin is the same. Asymmetries in the radiation incident on these surfaces can be taken into account without difficulty. In general, α may not equal ϵ, since the wavelength range of the incident radiation may differ from that of the emitted radiation. After assuming that H is independent of x, one may define an equivalent temperature of space T^* such that

$$\alpha H(x) \equiv \epsilon \sigma T^{*4} \tag{6-10}$$

Upon combining equations (6-8), (6-9), and (6-10) and introducing dimensionless variables, there follows

$$\frac{d^2\theta}{dX^2} = N_c(\theta^4 - \theta^{*4}) \qquad (6\text{-}11)$$

in which

$$\theta = \frac{T}{T_b}, \quad X = \frac{x}{L}, \quad N_c = \frac{\epsilon\sigma T_b{}^3 L^2}{kt} \qquad (6\text{-}12)$$

and T_b is the temperature at the fin base, $x = 0$. The dimensionless grouping N_c is sometimes referred to as the conductance parameter. It is seen to be a ratio of two conductances, $\epsilon\sigma T_b{}^3 L$ for radiation and kt/L for conduction. In practice, N_c values of 1 or 2 are realistic. If thermal symmetry is assumed, that is, $T(0) = T(2L) = T_b$, then the boundary conditions can be stated as

$$\theta(0) = 1 \qquad \left(\frac{d\theta}{dX}\right)_{X=1} = 0 \qquad (6\text{-}13)$$

The nonlinear mathematical system consisting of equations (6-11) and (6-13) is readily solved by numerical means. By employing a forward-integration procedure, such as the Runge-Kutta method, one may determine which of a succession of guessed values of $d\theta/dX$ at $X = 0$ leads to a solution that satisfies the boundary condition $d\theta/dX(1) = 0$. When programmed for a digital computer, such a solution scheme proceeds quite rapidly.

The result of greatest practical interest is the overall rate of heat transfer Q_f from the fin. This information is usually presented in terms of a fin effectiveness η defined as the ratio of the actual fin heat loss to that from an ideal fin having infinite thermal conductivity. The temperature of such an ideal fin would be equal to T_b throughout. Thus

$$\eta = \frac{Q_f}{Q_{\text{ideal}}}, \quad Q_{\text{ideal}} = 2L\epsilon\sigma(T_b{}^4 - T^{*4}), \quad Q_f = -k(2t)\left(\frac{dT}{dx}\right)_{x=0}$$

$$(6\text{-}14)$$

in which both Q_f and Q_{ideal} pertain to that portion of the fin between $x = 0$ and $x = L$ and to a unit length in the z direction.

A presentation of fin-effectiveness results (e.g., from refs. 14, 15) is made in Fig. 6-9. It is seen that the effectiveness decreases quite rapidly with increasing N_c when N_c is small. For $N_c = 1$, the effectiveness has already dropped to about 50 percent. This suggests that there is little motivation for using fins having N_c values much in excess of unity. Indeed, the optimum fin (that which provides the maximum

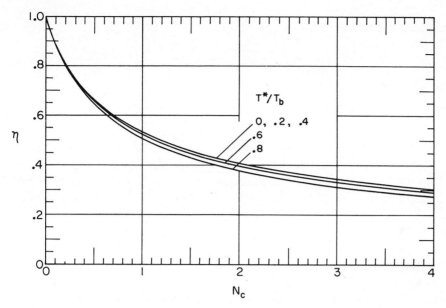

Fig. 6-9   Effectiveness of plate-type radiating fins.

heat transfer for a given fin weight) is characterized by N_c values ranging from 0.87 to 0.6 as θ^* ranges from 0 to 0.9 (ref. 15).

The particularly simple form of the governing equation for the single plate-type fin is due to the absence of radiant interaction between the fin and other portions of the heat rejection system.   Such radiant interaction will now be considered.

Radiant interaction between fin and tube.   When the diameter of the tubes is not small relative to the spacing $2L$, the radiant interchange between the fin and the tubes must be taken into account.   The foregoing analysis may be easily extended to include such interactions when the participating surfaces are *black* and when the tube surfaces are isothermal at temperature T_b (ref. 16).   Indeed, for this case, it is only necessary to append to the right-hand side of equation (6-11) a term that describes the radiant energy emitted by the tubes and arriving at a typical element of fin.

Upon considering Fig. 6-10 (lower sketch), which is an end view of the fin-tube system, one notes that distinct portions of the tube surfaces can be seen from each fin location x.   These areas on the tube surfaces are labeled A_1 and A_2.   It is clear that A_1 and A_2 are functions of x.

If the area A_1 emits an energy quantity $\sigma T_b{}^4 A_1$, then $\sigma T_b{}^4 F_{dA_x - A_1}$

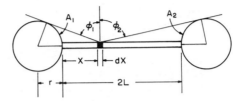

Fig. 6-10   End view of fin-tube radiator.

arrives at the fin element per unit area.   A similar term applies to the
radiation leaving A_2.   When these contributions are introduced into
the energy balance, equation (6-11), there follows

$$\frac{d^2\theta}{dX^2} = N_c[\theta^4 - (H^*/\sigma T_b{}^4) - (F_{dA_x - A_1} + F_{dA_x - A_2})] \quad (6\text{-}15)$$

in which H^* is the radiation from space that is incident at X.

The angle factors are readily evaluated by employing the methods
of Chapter 4.   For this purpose, it is assumed that the extension of the
fin and tube surfaces in the direction into and out of the plane of the
figure is very large.   Equation (4-64) of Chapter 4 may then be applied,
which gives

$$F_{dA_x - A_1} = \tfrac{1}{2}(1 - \sin \phi_1) \quad (6\text{-}16a)$$

$$F_{dA_x - A_2} = \tfrac{1}{2}(1 - \sin \phi_2) \quad (6\text{-}16b)$$

In turn, the sines of the angles ϕ_1 and ϕ_2 are readily evaluated by
trigonometry in terms of the lengths r, L, and x.

The solution of equation (6-15) can be carried out by the same
numerical technique as was described for equation (6-11).   Actual
numerical results have been obtained only for the limiting case in which
$H^* = 0$ (ref. 16).   The fin heat-transfer results Q_f are represented as a
ratio

$$\frac{Q_f}{Q_{f0}} \quad (6\text{-}17)$$

where Q_{f0} is the fin heat-transfer rate when $r/L = 0$.   This information
is presented in Fig. 6-11 as a function of r/L with N_c as curve parameter.

The departure of the curves from unity immediately indicates the reduction in fin heat transfer due to the presence of the black tube surfaces. Practical design considerations suggest that r/L values in excess of 0.4 are quite unlikely. Therefore, for practical values of N_c (~ 1 or 2), the reductions in fin heat loss due to radiant interaction with black tube surfaces is limited to 10 to 15 percent.

If the participating surfaces are nonblack, the radiant interaction introduces a significant degree of complication into the analysis. Whereas the foregoing fin problems were described by differential

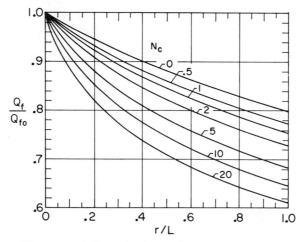

Fig. 6-11   Effect of tube surface on fin heat loss.

equations, the problem now under consideration requires a mathematical description that is integrodifferential. Although solutions are not now available, it is instructive to indicate the governing equations.

The method of analysis described in Section 3-3 will now be applied. The equations characterizing the radiant interchange may be written either in terms of radiosities or heat fluxes by respectively specializing equation (3-40) or equation (3-45). It appears that the latter approach is somewhat more direct. In formulating the problem, it is assumed that all surfaces are gray and that both tubes have the same uniform temperature T_b, the same distribution of incident radiation H_t^*, and the same emittance ϵ_t. Furthermore, for simplicity, it is supposed that symmetrical conditions exist on the top and bottom sides of the system. The subscripts t and f are used to denote tube and fin, respectively.

The heat conduction equation for the fin has a form similar to that for the preceding fin analyses, that is,

$$\frac{d^2\theta}{dX^2} = \frac{N_c}{\epsilon_f}\frac{q_f}{\sigma T_b{}^4} \tag{6-18}$$

in which q_f is the net rate of radiative heat loss per unit area at a fin location x. Additional relationships between the unknowns are obtained by specializing equation (3-45) in accordance with the foregoing assumptions. Cognizance is also taken of the notation of Fig. 6-10, upper sketch. The integral equations that result are

$$\frac{q_f(X)}{\epsilon_f \sigma T_b{}^4} = \theta^4(X) - (H_f^*/\sigma T_b{}^4) - (F_{dA_x - A_1} + F_{dA_x - A_2})$$

$$+ \frac{1-\epsilon_t}{\epsilon_t}\left[\int_{A_1(x)} \frac{q_1(\delta_1)}{\sigma T_b{}^4} K(X, \delta_1)\, dA_1 + \int_{A_2(x)} \frac{q_2(\delta_2)}{\sigma T_b{}^4} K(X, \delta_2)\, dA_2\right]$$

$$\tag{6-19}$$

$$\frac{q_1(\delta_1)}{\epsilon_t \sigma T_b{}^4} = 1 - (H_t^*/\sigma T_b{}^4) - F_{dA_1 - A_2} - \int_{A_f(\delta_1)} \theta^4(X) K(\delta_1, X)\, dA_x$$

$$+ \frac{1-\epsilon_f}{\epsilon_f}\int_{A_f(\delta_1)} \frac{q_f(X)}{\sigma T_b{}^4} K(\delta_1, X)\, dA_x$$

$$+ \frac{1-\epsilon_t}{\epsilon_t}\int_{A_2(\delta_1)} \frac{q_2(\delta_2)}{\sigma T_b{}^4} K(\delta_1, \delta_2)\, dA_2 \tag{6-20}$$

In the integrals that appear in equation (6-19), $A_1(x)$ and $A_2(x)$ denote the surface areas of tubes 1 and 2 that are respectively seen from position x on the fin. Correspondingly, in equation (6-20), $A_f(\delta_1)$ and $A_2(\delta_1)$ are the areas of the fin and of tube 2 that can be seen from position δ_1 on tube 1.

It may be observed that the preceding equations contain four unknown functions $\theta(X)$, $q_f(X)$, $q_1(\delta_1)$, and $q_2(\delta_2)$. However, because of the assumed symmetry of the situation, $q_1(\delta_1) = q_2(\delta_2)$ when $\delta_1 = \delta_2$. Thus there are actually three unknown functions. To complete the specification of the problem, it only remains to state the boundary conditions. These are given by equation (6-13).

Inspection of equations (6-18) to (6-20) reveals that although several simplifying assumptions were incorporated into the formulation, the resulting mathematical problem is quite complex. An analysis of this same problem that uses a two-band approximation (as in Section 3-4) to account for differences between the infrared and solar absorptances adds another pair of integral equations (ref. 17, pp. 439–440).

Radiant interaction between fin surfaces. There are technically interesting radiator configurations in which radiant interchange takes place between neighboring fins of an ensemble. Such a situation is illustrated in the left-hand diagram of Fig. 6-12, which shows a tube to which are attached longitudinal fins. Heat is to be extracted from a fluid that flows through the tube.

As a limiting case of such a configuration, one may consider the situation in which the tube radius is much smaller than the height of the fins. Furthermore, if the thermal conditions and the radiation properties are the same for each fin, it is sufficient to single out for analytical treatment a typical pair of fins as pictured in the right-hand diagram of the figure. Clearly, results for such a model should also apply for corrugated fins attached to a plane wall.

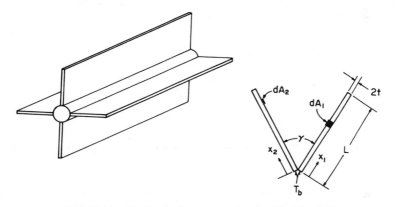

Fig. 6-12 Radiatively interacting longitudinal fins.

The analysis of this problem closely parallels that just described for the case of radiant interaction between fin and tube. An energy balance on a typical element of fin (shown shaded in the figure) yields a differential equation identical to equation (6-18) except that now the subscript f is no longer needed. The dimensionless variables θ, X, and N_c are defined as in equation (6-12). Another relationship between the heat flux q and the temperature θ, valid for gray surfaces, is readily derived by specializing equation (3-45) to the situation pictured in Fig. 6-12. From this, one obtains

$$\frac{q(X_1)}{\epsilon \sigma T_b{}^4} = \theta^4(X_1) - \int_{A_2} \theta^4(X_2) K(X_1, X_2)\, dA_2$$

$$+ \frac{1 - \epsilon}{\epsilon} \int_{A_2} \frac{q(X_2)}{\sigma T_b{}^4} K(X_1, X_2)\, dA_2 \quad (6\text{-}21)$$

In writing equation (6-21), the contribution of radiation arriving from the external environment has been omitted in recognition of the solutions that are available in the literature (ref. 18).

Owing to symmetry, it is evident that $\theta(X_1) = \theta(X_2)$ and $q(X_1) = q(X_2)$ when $X_1 = X_2$. Hence there are two unknowns, $\theta(X)$ and $q(X)$. The required angle factor is readily evaluated when it is assumed that the fin surfaces are very long in the direction into and out of the plane of the figure. Under these circumstances, the angle factor is obtained by differentiation as indicated by equation (4-63) of Chapter 4.

The boundary condition at $X = 0$ is $T = T_b$ or $\theta = 1$. However, the location $x/L = 1$ is not a symmetry line. Thus the convenient boundary condition that $d\theta/dX = 0$ at $X = 1$ is justified on the basis of a negligible heat loss from the fin tip. This condition is commonly used for conducting-convecting fins.

Results for the radiating-fin configuration just described have been obtained in ref. 18 on the basis of numerical solutions of the integro-differential system of governing equations. A typical set of results for the overall heat loss is shown in Fig. 6-13, which pertains to fins with surface emittance $\epsilon = 0.75$. On the ordinate, Q_f represents the rate of heat loss from the two surfaces that form the V-cavity as pictured in Fig. 6-12. The quantity $2\sigma T_b^4 L \sin(\gamma/2)$ is the rate of heat loss from the same configuration, but with black fins of infinite thermal conductivity. The abscissa is the parameter N_c/ϵ.

The figure shows that the fin heat loss decreases with increasing N_c, and it appears that the use of fins characterized by values of N_c much in excess of unity would not be advantageous. Comparable information for fin ensembles having emittances of 1.0 and 0.5 is given in the aforementioned reference. These results show that although the fin heat loss increases with increasing values of the emittance, the increase in Q_f is not as great as the proportionate increase in ϵ.

Consideration has also been given to a radiator configuration similar to that of Fig. 6-12 but with fins of trapezoidal profile rather than of rectangular profile (ref. 19). The radiative participation of the tube was neglected as in the foregoing formulation. A somewhat more complex physical situation occurs when there is radiant interaction between fin and tube as well as between neighboring fins of an ensemble. Such a state-of-affairs occurs in the longitudinal fin-tube configuration (Fig. 6-12) when the tube radius is not small relative to the fin height. Another technically interesting case where such interactions occur is the radiator consisting of a tube and a series of uniformly spaced annular fins (ref. 20). The method of analysis for this problem is

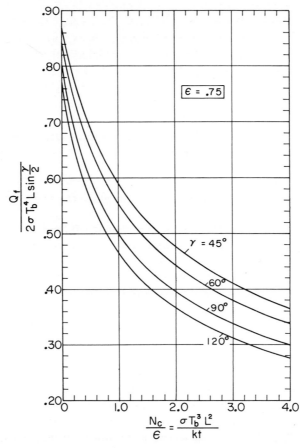

Fig. 6-13   Effectiveness of radiatively interacting longitudinal fins.

similar to that described above, the most difficult aspect in the formulation being the derivation of the angle factors.

    Mention may be made of a class of problems closely related to the
radiating-conducting fin.   These problems involve thin-walled bodies
subject to incident thermal radiation over a portion of their external
surfaces; for example, the solar heating of a space vehicle.   The resulting distribution of temperature along the surface is controlled by the
interaction between thermal radiation at the surface and heat conduction within the thickness of the wall.   Representative studies of
this type of problem may be found in refs. 21, 22, and 23.

    All the problems discussed in this section have involved heat
conduction in solids of small thickness.   Correspondingly, it was

permissible to employ a one-dimensional model where the surface heat transfers are included directly in the energy balance. Such a model is necessarily approximate inasmuch as energy conservation is not satisfied at every point within the solid; rather, it is satisfied only on the average for an element that spans the entire thickness of the solid. If energy conservation is imposed at each and every point within a solid, only conductive contributions enter the energy balance (e.g., Laplace's equation for the temperature). In such an exact formulation, the surface heat transfers (i.e., radiative effects) constitute the boundary conditions on the differential equation of heat conduction.

6-3   Radiative-Convective Systems: Boundary Layers and Ducts

At moderate temperature levels, the transfer of heat by convection in forced-flow systems is usually much greater than that by radiation. However, at higher temperature levels, radiation may begin to play a significant role. In natural convection systems, the radiative and convective contributions may be competitive even at moderate temperatures.

When the temperatures of the participating surfaces are prescribed, the convective and radiative heat transfers can be determined independently of each other. However, for other boundary conditions, there is a dynamic interaction between the two modes of heat transfer. In this section such dynamic interactions will be illustrated for two types of forced-convection flows. The first problem to be studied involves the boundary layer flow over a flat plate, while the second is concerned with flow in a circular tube. In all cases the flowing fluid is a radiatively nonparticipating gas.

Surface temperature of a flat plate.   Consideration is given to a flat plate aligned parallel to a uniform free-stream flow having velocity U_∞ and static temperature T_∞. The flow is in the positive x direction. The flat plate, which has its leading edge at $x = 0$, extends indefinitely to the right along the positive x axis. The plate simultaneously experiences convective heat transfer, radiative exchange with the environment, aerodynamic heating, and internal heating or cooling from sources or sinks. The aim of the analysis is to determine the equilibrium-temperature distribution of the plate surface that results from the interaction of the aforementioned energy transfers. Longitudinal heat conduction within the plate is neglected.

The first step in the analysis is to perform an energy balance at a typical location x on the plate surface. The various ingredients of the energy balance will now be discussed. The radiant energy arriving per unit time and area at the plate surface from various sources in the environment is described by quantities $e_{r1}, e_{r2}, \ldots, e_{rn}$. In general, the absorptance of the surface for each of these radiation quantities may be different. If e_R denotes the rate at which radiant energy is locally absorbed per unit area and $\alpha_1, \alpha_2, \ldots, \alpha_n$ are the absorptances corresponding to $e_{r1}, e_{r2}, \ldots, e_{rn}$, it follows that

$$e_R = \sum_{i=1}^{n} \alpha_i e_{ri} \tag{6-22}$$

The local emission of radiant energy per unit time and unit area is $\epsilon \sigma T_w{}^4(x)$. In addition, owing to a heat source or sink within the plate, there may be an energy flux e_p that must be transferred from the plate to the fluid and the environment. Finally, there is a convective heat flux q_c from the plate to the fluid.

Under steady state conditions, the various energies must be in balance, that is,

$$q_c(x) = e_R(x) - \epsilon \sigma T_w{}^4(x) + e_p(x) \tag{6-23}$$

In general, e_R and e_p may be regarded as known quantities. Although there is no essential difficulty in accounting for the x dependence of e_R and e_p, it is simpler, for present purposes, to take them as constant and to deal only with their sum $e = e_R + e_p$. With this, equation (6-23) becomes

$$q_c(x) = e - \epsilon \sigma T_w{}^4(x) \tag{6-23a}$$

The foregoing constitutes a single relation between the two unknowns: the convective heat flux and the surface temperature. A second relationship stems from convective heat transfer theory. For laminar flow, it can be shown that (ref. 24)

$$T_w(x) - T_{aw} = \frac{C/3}{\mathrm{Re}^{1/2} \, \mathrm{Pr}^{1/3}} \frac{x^{2/3}}{k} \int_0^x \frac{q_c(x_0)}{(x - x_0)^{2/3}} \, dx_0 \tag{6-24}$$

in which T_{aw} is the equilibrium temperature of an adiabatic plate induced by aerodynamic heating alone, that is, in the absence of all radiation effects and internal sources or sinks within the plate. T_{aw} is independent of x and is given by

$$T_{aw} - T_\infty = R \frac{U_\infty{}^2}{2c_p}, \qquad R = \mathrm{Pr}^{1/2} \tag{6-25}$$

In equation (6-24), Re and Pr are the well-known Reynolds and Prandtl numbers, respectively defined as

$$\mathrm{Re} = \frac{U_\infty x}{\nu} \qquad \mathrm{Pr} = \frac{c_p \mu}{k} \tag{6-26}$$

The symbols c_p, k, μ, and ν represent the specific heat, thermal conductivity, dynamic viscosity, and kinematic viscosity of the gas. C is a nearly constant parameter whose value is 2.187 for $Pr = 0.7$ and 2.179 for $Pr = 1$; x_0 is a dummy variable of integration.

After substitution of equation (6-23a) into (6-24) and subsequent integration, there follows after nondimensional rearrangement

$$\theta(X) = 1 + X^{\frac{1}{2}}\left(\frac{e}{\epsilon \sigma T_{aw}^{\ 4}}\right) - \frac{X^{\frac{1}{6}}}{3} \int_0^X \frac{\theta^4(\xi)}{(X - \xi)^{\frac{2}{3}}} \, d\xi \qquad (6\text{-}27)$$

where
$$\theta = \frac{T_w}{T_{aw}} \qquad X = \left(\frac{h_{\text{rad}}}{h_{\text{UHF}}}\right)^2 \sim x \qquad (6\text{-}28a)$$

and
$$h_{\text{rad}} = \epsilon \sigma T_{aw}^{\ 3} \qquad h_{\text{UHF}} = \frac{k}{Cx} \, \text{Re}^{\frac{1}{2}} \, \text{Pr}^{\frac{1}{3}} \qquad (6\text{-}28b)$$

θ is readily identified as a dimensionless temperature and X as a dimensionless x coordinate. Thus equation (6-27) constitutes a nonlinear integral equation for the distribution of the temperature along the plate surface.

It is interesting to illuminate the physical meaning of the X variable. h_{rad} is a general measure of the strength of the surface radiation and has units of a heat transfer coefficient. Inasmuch as T_{aw} is independent of x, so is h_{rad}. On the other hand, h_{UHF} is the local convective heat transfer coefficient for laminar flow over a flat plate with uniform heat flux; clearly, $h_{\text{UHF}} \sim x^{-\frac{1}{2}}$. Therefore $h_{\text{rad}}/h_{\text{UHF}}$ is a measure of the relative strength of the surface radiation and the convection. At the leading edge, $h_{\text{UHF}} \to \infty$ and $h_{\text{rad}}/h_{\text{UHF}} = 0$. With increasing downstream distances, $h_{\text{rad}}/h_{\text{UHF}}$ increases monotonically.

The solution of equation (6-27) was performed numerically in ref. 24. The essential step in preparing the problem for solution was to approximate the integral by a summation. A very small step size was employed to ensure highly accurate results. The details of the computational procedure are available in the reference.

The surface-temperature distributions obtained from the solutions of the governing integral equation are shown as solid lines in Fig. 6-14. The curves are parameterized by values of $e/\epsilon \sigma T_{aw}^{\ 4}$ ranging from 0 to 2. For cases characterized by $e/\epsilon \sigma T_{aw}^{\ 4} < 1$, the wall temperature is less than T_{aw} and decreases with increasing x. The opposite is true when $e/\epsilon \sigma T_{aw}^{\ 4} > 1$. In the former cases, heat flows from the gas to the wall by convection. In the latter, the direction of convective heat flow is reversed.

At the far right of the figure is an array of line segments that are labeled *asymptotes*. These apply to the condition $h_{\text{rad}}/h_{\text{UHF}} \to \infty$.

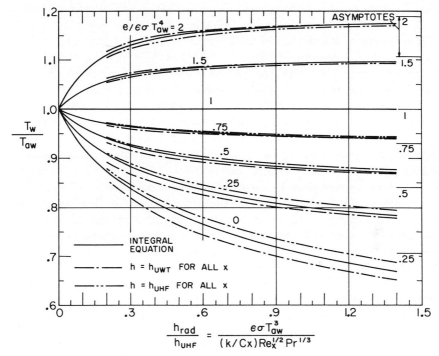

Fig. 6-14   Surface-temperature distribution for laminar flow over a flat plate.

This implies that the convective heat transfer q_c approaches zero. Correspondingly, from the energy balance (6-23a), there follows

$$\frac{T_w}{T_{aw}} = \left(\frac{e}{\epsilon \sigma T_{aw}{}^4}\right)^{1/4} \tag{6-29}$$

It is seen from the figure that the asymptotes are approached more rapidly for cases in which $e/\epsilon \sigma T_{aw}{}^4$ is nearly unity.

An approximate, but much simpler formulation of the foregoing problem can be made if the convective heat transfer coefficient is specified in advance.   To proceed with the approximate analysis, the convective heat flux q_c that appears in the energy balance (6-23a) is replaced by $h^*(T_w - T_{aw})$, where h^* represents a convective heat transfer coefficient.   Then, upon rearrangement, one obtains

$$\frac{T_w(x)}{T_{aw}} = \left(\frac{e}{\epsilon \sigma T_{aw}{}^4}\right)\frac{h_{\text{rad}}}{h^*} - \left[\frac{T_w(x)}{T_{aw}}\right]^4 \frac{h_{\text{rad}}}{h^*} + 1 \tag{6-30}$$

In reality, h^* depends on the specific distribution of the surface temperature or the surface heat flux; therefore, h^* will not be known *a priori*. However, as an approximation, it may be assumed that h^* is insensitive to the details of the thermal boundary condition. This is tantamount to assuming that *local similarity* exists in the thermal boundary layer. In evaluating h^*, it is natural to use the heat transfer coefficients for the two standard cases of uniform wall temperature (UWT) and uniform heat flux (UHF). Solutions of equation (6-30) corresponding to these two cases have been carried out and are respectively plotted in Fig. 6-14 as dash-dot and dash-double dot lines. In general, it is seen that the two approximate solutions bracket the exact solution for the temperature distribution. Indeed, the bracketing curves form rather close bounds on the exact solution.

Surface temperature distributions for turbulent flow over a flat plate, comparable to those of Fig. 6-14, are also presented in ref. 24. For the turbulent case, the results from the approximate formulation, equation (6-30), are essentially indistinguishable from those obtained by solving the governing integral equation.

As a final remark, it may be noted that the essential role of radiation in the problem just discussed is to modify the thermal boundary condition that is imposed on the convective boundary layer.

Surface temperature of a tube. Attention is now turned to a non-participating gas flowing through a circular tube. There is a source of heat within the tube wall, for instance, electrical dissipation or nuclear fission. The outside surface of the tube is insulated. It is evident that the temperature of the gas will increase during its course of flow through the tube. Correspondingly, the temperature of the tube wall increases in the direction of flow. At sufficiently high temperature levels, there may be an appreciable amount of radiant energy transferred from downstream to upstream elements on the tube wall. The effect of this radiative transfer is to provide a wall-temperature distribution that is more uniform than that for the case of purely convective transport.

The mathematical formulation of the problem is discussed with the aid of Fig. 6-15. A transparent gas at temperature T_i enters a tube of diameter D and length L, flowing from left to right. The gas has a mean velocity $\bar{U}$. The ends of the tube open into environments that have effective black-body temperatures T_1 and T_2. The tube wall is a gray, diffuse emitter and reflector of radiant energy. Because of a uniformly distributed heat source within the wall, there is an energy quantity e that must be dissipated from the inner surface of the tube per unit time and area.

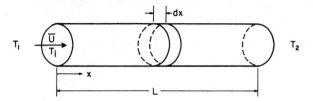

Fig. 6-15   Diagram for analysis of simultaneous radiative and convective transport in a tube.

The analysis begins with an energy balance on a typical ring element on the tube surface as shown in the figure. All quantities entering the energy balance are taken per unit area of the surface. The radiant energy incident on the element is $H(x)$, while the leaving radiation is $B(x)$; consequently, the net radiative loss is $[B(x) - H(x)]$. The convective heat transfer from the element to the gas is $q_c(x)$. Under steady state conditions, the sum of these transports must be equal to e, that is,

$$e = B(x) - H(x) + q_c(x) \qquad (6\text{-}31)$$

The incident flux $H(x)$ may be eliminated from the foregoing equation by employing equation (3-18) of Chapter 3, that is,

$$B(x) = \epsilon \sigma T_w{}^4(x) + (1 - \epsilon)H(x) \qquad (6\text{-}32)$$

where $T_w(x)$ is the local wall temperature, and ρ has been replaced by $(1 - \epsilon)$ in accordance with the graybody condition. Furthermore, the convective heat flux is expressible as

$$q_c(x) = h[T_w(x) - T_g(x)] \qquad (6\text{-}33)$$

in which h is the heat transfer coefficient and $T_g(x)$ is the local bulk temperature of the gas. In general, h may be a function of x, but it is taken as a constant to simplify the problem. For turbulent flows, the assumption of a constant h is not unreasonable provided that the tube is not too short.

Upon eliminating H and q_c from equation (6-31), one obtains

$$e = \frac{\epsilon}{1 - \epsilon}[\sigma T_w{}^4(x) - B(x)] + h[T_w(x) - T_g(x)] \qquad (6\text{-}34)$$

As written, equation (6-34) contains three unknown functions, $T_w(x)$, $T_g(x)$, and $B(x)$. An additional relationship is derived by specializing equation (3-40) to the present problem. In applying this equation, it is helpful to recall that the radiation entering the ends of the tube is assumed black at temperatures T_1 and T_2, respectively. From this, one obtains

$$B(x) = \epsilon\sigma T_w{}^4(x) + (1 - \epsilon)\left[\sigma T_1{}^4 F_{dA_x - A_1} + \sigma T_2{}^4 F_{dA_x - A_2}\right.$$
$$\left. + \int_{\xi=0}^{L} B(\xi)K(x, \xi)\pi D\, d\xi\right] \qquad (6\text{-}35)$$

In this expression, $F_{dA_x - A_1}$ and $F_{dA_x - A_2}$ are the respective angle factors from the ring element at x to the inlet and exit of the tube. In addition, $K(x, \xi)\pi D\, d\xi$ is the angle factor from the ring at x to another ring at ξ.

A third equation connecting the unknowns is derived from an energy balance on the gas passing through a cylindrical control volume of length dx and diameter D. From this, there follows

$$\frac{dT_g}{dx} = \frac{4q_c(x)}{\rho \overline{U} c_p D} \qquad (6\text{-}36)$$

Equations (6-34), (6-35), and (6-36), taken together with the boundary condition $T_g = T_i$ at $x = 0$, constitute a complete specification of the problem. This system has been solved by numerical means in ref. 25. As a prelude to the numerical solution, the integral equation (6-35) was reduced to a second-order differential equation by approximating the kernel K with an exponential function as described in Section 3-5. Then B is eliminated from the problem by double differentiation of equation (6-34). This leaves a pair of coupled nonlinear ordinary differential equations for $T_w(x)$ and $T_g(x)$.

The details of the solution are well-documented in the reference. As noted there, there are *six* independent parameters whose values must be specified for each case for which computations are performed. The presence of so many parameters precludes a concise display of representative results. It is suggested that interested readers consult ref. 25, which contains an extensive graphical presentation.

Extensions of the analysis just described to the entrance region of a circular tube and to an asymmetrically heated parallel-plate channel are reported in refs. 26 and 27, respectively. The former also includes the effects of heat conduction in the tube wall.

REFERENCES

1. E. M. Sparrow, L. U. Albers, and E. R. G. Eckert, Thermal radiation characteristics of cylindrical enclosures. *J. Heat Transfer*, **C84**, 73–81 (1962).

2. S. H. Lin and E. M. Sparrow, Radiant interchange among curved specularly reflecting surfaces—application to cylindrical and conical cavities. *J. Heat Transfer*, **C87**, 299–307 (1965).

3. E. M. Sparrow and S. H. Lin, Radiant heat transfer at a surface having both specular and diffuse reflectance components. *Intern. J. Heat Mass Transfer*, **8**, 769–779 (1965).

4. E. M. Sparrow and V. K. Jonsson, Radiant emission characteristics of diffuse conical cavities. *J. Opt. Soc. Am.*, **53**, 816–821 (1963).

5. E. M. Sparrow and J. L. Gregg, Radiant emission from a parallel-walled groove. *J. Heat Transfer*, **C84**, 270–271 (1962).

6. E. M. Sparrow and V. K. Jonsson, Thermal radiation absorption in rectangular-groove cavities. *J. Appl. Mech.*, **E30**, 237–244 (1963).

7. E. M. Sparrow and S. H. Lin, Absorption of thermal radiation in V-groove cavities. *Intern. J. Heat Mass Transfer*, **5**, 1111–1115 (1962).

8. E. M. Sparrow and V. K. Jonsson, Absorption and emission characteristics of diffuse spherical cavities. *NASA TN D-1289* (1962).

9. E. M. Sparrow and V. K. Jonsson, Absorption and emission characteristics of diffuse spherical enclosures. *J. Heat Transfer*, **C84**, 188–189 (1962).

10. C. M. Usiskin and R. Siegel, Thermal radiation from a cylindrical enclosure with specified wall heat flux. *J. Heat Transfer*, **C82**, 369–374 (1960).

11. M. Perlmutter and R. Siegel, Effect of specularly reflecting gray surface on thermal radiation through a tube and from its heated wall. *J. Heat Transfer*, **C85**, 55–62 (1963).

12. E. M. Sparrow and V. K. Jonsson, The transport of radiant energy through tapered tubes or tapered gaps. *J. Heat Transfer*, **C86**, 132 (1964).

13. E. M. Sparrow, V. K. Jonsson, and W. J. Minkowycz, Heat transfer from fin-tube radiators including longitudinal heat conduction and radiant interchange between longitudinally nonisothermal finite surfaces. *NASA TN D-2077*, December 1963.

14. S. Lieblein, Analysis of temperature distribution and radiant heat transfer along a rectangular fin. *NASA TN D-196* (1959).

15. J. G. Bartas and W. H. Sellers, Radiation fin effectiveness. *J. Heat Transfer*, **C82**, 73–75 (1960).

16. E. M. Sparrow and E. R. G. Eckert, Radiant interaction between fins and base surfaces. *J. Heat Transfer*, **C84**, 12–18 (1962).

17. E. M. Sparrow, Radiation heat transfer between surfaces. *Adv. Heat Transfer*, **2**, 399–449 (1965).

18. E. M. Sparrow, E. R. G. Eckert, and T. F. Irvine, Jr., The effectiveness of radiating fins with mutual irradiation. *J. Aerospace Sci.*, **28**, 763–772 (1961).

19. B. V. Karlekar and B. T. Chao, Mass minimization of radiating trapezoidal fins with negligible base cylinder interaction. *Intern. J. Heat Mass Transfer*, **6**, 33–48 (1963).

20. E. M. Sparrow, G. B. Miller, and V. K. Jonsson, Radiating effectiveness of annular-finned space radiators, including mutual irradiation between radiator elements. *J. Aerospace Sci.*, **29**, 1291–1299 (1962).

21. L. D. Nichols, Surface-temperature distribution on thin-walled bodies subjected to solar radiation in interplanetary space. *NASA TN D-584* (1961).

22. W. E. Olmstead and S. Raynor, Solar heating of a rotating spherical space vehicle. *Intern. J. Heat Mass Transfer*, **5**, 1165–1177 (1962).

23. P. Hrycak, Influence of conduction on spacecraft skin temperatures. *AIAA J.*, **1**, 2619–2621 (1963).

24. E. M. Sparrow and S. H. Lin, Boundary layers with prescribed heat flux-application to simultaneous convection and radiation. *Intern. J. Heat Mass Transfer*, **8**, 437–448 (1965).

25. R. Siegel and M. Perlmutter, Convective and radiant heat transfer for flow of a transparent gas in a tube with a gray wall. *Intern. J. Heat Mass Transfer*, **5**, 639–660 (1962).

26. R. Siegel and E. G. Keshock, Wall temperatures in a tube with forced convection, internal radiation exchange, and axial wall conduction. *NASA TN D-2116* (1964).

27. E. G. Keshock and R. Siegel, Combined radiation and convection in an asymmetrically heated parallel plate flow channel. *J. Heat Transfer*, **C86**, 341–350 (1964).

PROBLEMS

6-1. A deep circular cylindrical cavity may, for purposes of analysis, be treated as a cavity of infinite depth provided that the reflection process is diffuse (see Fig. 6-2). Show that the governing integral equation for the radiosity distribution in such a cavity follows from equation (6-1) as

$$B(x) = \epsilon\sigma(T_w{}^4 - T_\infty{}^4) + (1 - \epsilon)\int_{\xi=0}^{\infty} B(\xi)\, dF_{dA_x - dA_\xi}$$

in which ξ is a dummy variable of integration and the angle factor is given by equation (4-62). The coordinate x is measured as in Fig. 5-5.

6-2. Generalize the radiosity equation derived in Problem 6-1 to take account of specular and diffuse components of reflectance ρ^s and ρ^d, respectively. Hint: See equation (5-42).

6-3. For Problems 6-1 and 6-2, write expressions relating the local heat flux $q(x)$ with the radiosity distribution $B(x)$. Then, express the cavity heat loss Q in terms of $q(x)$.

6-4. Apply the finding of Problem 5-1 to equation (6-3) and demonstrate that $\epsilon_a \geq \epsilon$.

6-5. On the basis of Problems 3-13 and 3-14, an expression for the apparent emittance of a diffusely reflecting V-groove cavity can be written as

$$\epsilon_a = \frac{\epsilon}{\epsilon + (1 - \epsilon) \sin \phi}$$

This result is approximate in that, in its derivation, the variation of the radiosity along the surfaces of the cavity was suppressed. Make numerical comparisons of the ϵ_a values from the foregoing equation with those of Fig. 6-6 and delineate the parameter ranges where the level of agreement is satisfactory.

6-6. Derive the expression for the apparent emittance of a spherical cavity, equation (6-7), by applying the radiosity solution (3-83). Hint: In equation (3-83a), replace T^4 by $T_w{}^4 - T_\infty{}^4$, thereby making $H^* = 0$. Also, $A/4\pi R^2 = \frac{1}{2}(1 + \cos \phi)$.

6-7. A sheet of mild steel has an emittance ϵ of 0.2. It is proposed to increase the radiant emission of the sheet by machining circular cylindrical holes into the surface. The depth of each hole is three times its diameter. The holes are uniformly dispersed and occupy 40 % of the surface area. What is the rate of emission from an area A of the sheet consisting of 40 % holes and 60 % surface? How does this compare with the energy emitted by the same area in the absence of holes? Assume that the holes are diffusely reflecting and isothermal.

6-8. As a modification of Problem 6-7, suppose that the holes are conical instead of cylindrical. The ratio of depth to hole opening diameter is as before. Calculate the rate of emission for an area A as specified in Problem 6-7. How much has the rate of emission decreased owing to the use of conical holes instead of cylindrical holes?

6-9. When a noncondensing fluid such as a gas or a liquid passes through the tubes of a radiator such as that of Fig. 6-8, the fluid temperature decreases in the flow direction (i.e., the z direction). The temperature decrease is a result of radiative heat losses from the fins and tubes. It is usually assumed that the net fin heat conduction in the z direction can be neglected, so that the analysis and results of pages 180–183 can be applied locally, at each z. However, in order to apply the results, it is necessary to know the fin base temperature T_b at each z. Outline a calculation method for determining $T_b(z)$. For computational simplicity, assume that at each z, the fluid temperature and fin base temperature are identical. In addition, neglect the radiative heat loss from the tube. Corresponding to these assumptions, the variation of $T_b(z)$ is given by

Isaac Kuria

$$T_b(z) - T_b(0) = \frac{2}{\dot{m}c_p} \int_0^z Q_f \, dz$$

in which $\dot{m}$ and c_p are, respectively, the flow rate and specific heat of the fluid, $T_b(0)$ is the known value of T_b at $z = 0$, and Q_f is the fin heat transfer rate defined by equation (6-14). Note that Q_f is a function of T_b. Hint: On the basis of equation (6-14) and Fig. 6-9, show that Q_f and T_b can be related. This relationship, when introduced into the foregoing equation involving $T_b(z)$, $T_b(0)$ and Q_f, yields an integral equation for either $Q_f(z)$ or $T_b(z)$.

6-10. For a fin-tube radiator, Fig. 6-10, the r/L ratio is given as 0.2. When $N_c = 1$, how does Q_f compare with Q_{ideal} defined by equation (6-14)? Assume that $T^* = 0$.

6-11. The longitudinal fins of Fig. 6-12 form a V-groove cavity. How are the results of Fig. 6-13 related to those of Fig. 6-6?

6-12. A wire in a vacuum is stretched tightly between two supports. The wire has diameter D and length $2L$ ($D \ll 2L$). The supports are maintained at a temperature T_b, and the equivalent temperature of the environment is T^* (see equation (6-10)). Show that by a redefinition of N_c, the analysis and results of pages 182–183 apply for the wire.

6-13. For the flat plate problem analyzed on pages 180–184, a sufficiently accurate solution for T_w/T_{aw} can be obtained by applying the approximate formulation, equation (6-30), provided that the flow is turbulent. For the turbulent case, $h_{UHF} \approx h_{UWT} = h^* = 0.03(k/x)Re^{4/5}Pr^{3/5}$. Compute curves, appropriate to turbulent flow, of T_w/T_{aw} versus h_{rad}/h^* for $e/\epsilon\sigma T_{aw}^4 = 0$ and 2.

6-14. Reformulate the analysis of pages 194–196 for the case of a black-walled tube.

PART THREE

RADIANT ENERGY TRANSFER THROUGH ABSORBING, EMITTING, AND SCATTERING MEDIA

In the following chapters attention is directed toward radiant energy transfer within media that absorb, emit, and scatter thermal radiation; that is, the medium surrounding a given surface or surfaces actively participates in the overall radiation transfer process. In addition, when the medium either conducts and/or convects thermal energy, a mutual interaction between the separate modes of energy transfer will take place.

Historically, radiant transfer in participating media is an area that has received considerable attention from astrophysicists for quite a number of years (for example, refs. 1 and 2 of Chapter 7). From the point of view of the engineer, however, it is only recently that a formal approach to such problems has been undertaken. In the past, it was usually deemed sufficient to employ the concept of an effective gas emittance and, correspondingly, to treat the participating medium as a radiating surface. While such an approach may be adequate in applications where large factors of safety are employed, recent technological problems such as reentry and plasma physics have necessitated a more rigorous approach to radiation transfer in participating media.

In Chapter 7 the basic equations that govern radiation transfer in absorbing, emitting, and scattering media are derived. Although this development can be accomplished with total disregard of the physical mechanism of radiative transport, it will be extremely beneficial to interpret the results within the framework of photon transfer. Indeed, it will be shown that the transfer of heat by molecular conduction and the transfer of heat by photon radiation are phenomenologically very analogous.

The remaining four chapters of Part Three are concerned with specific heat transfer processes involving radiation-participating media. Chapter 8 deals with the case of radiative equilibrium; that is, problems in which radiation transfer is the sole mode of energy transfer, while Chapters 9 and 10 are concerned with coupled modes of energy transfer. Combined conduction and radiation are treated in Chapter 9, and Chapter 10 is concerned with combined convection and radiation. Chapter 11 deals with the incorporation of spectroscopic properties of infrared active gases into the basic equations of radiative transfer.

CHAPTER 7

BASIC EQUATIONS FOR
ENERGY TRANSFER IN ABSORBING,
EMITTING, AND SCATTERING MEDIA

The fundamental principle governing the temperature field within an absorbing, emitting, and scattering medium is the same as that for a nonparticipating medium; that is, the expression of conservation of energy within the medium. The addition that must be made to the energy equation when dealing with a radiation-participating medium is the inclusion of the radiation flux within the medium, and this is treated in Section 7-1. In Section 7-2 the physical regimes of radiative transport are discussed with primary emphasis on the photon point of view, while Sections 7-3 to 7-7 are concerned specifically with the equation for the radiation flux and its limiting forms. Finally, Section 7-8 describes various mean absorption and scattering coefficients, and Section 7–9 presents a discussion of the differential approximation for radiation transfer. The publications by R. and M. Goulard (ref. 3) and R. Viskanta (refs. 4 and 5) have been utilized extensively throughout this chapter.

7-1   Conservation of Energy

For generality, consider a moving compressible fluid that is a radiation-participating medium. This general case is, of course, directly reducible to stationary solids. The presence of a radiation field within the fluid will alter the classical conservation equations of momentum and energy as the result of three effects:

(1)   The radiation energy flux constitutes an additional mechanism for heat transfer to or from a fluid element.

(2)   A fluid element will contain radiant energy as well as molecular energy.

(3)   The radiation pressure tensor will augment the conventional fluid dynamic pressure tensor.

Even at high temperatures the last two effects are generally insignificant and are assumed to be negligible in the present development. In addition, the fluid will be assumed to be a molecular continuum and to be in local thermodynamic equilibrium. The requirement of local thermodynamic equilibrium is, of course, necessary in order that the temperature be definable throughout the medium.

With the preceding assumptions, the equations expressing conservation of mass and momentum remain unaltered, while conservation of energy may be expressed as* (ref. 6)

$$\rho c_v \frac{DT}{Dt} = -\operatorname{div} \mathbf{q} - p \operatorname{div} \mathbf{U} + \mu \Phi \tag{7-1}$$

where $\mathbf{U}$ is the fluid velocity vector, $\mathbf{q}$ is the heat flux vector that is induced within the fluid as the result of temperature gradients, and Φ is the Rayleigh dissipation function.

It may be recalled that equation (7-1) is nothing more than a statement of the first law of thermodynamics applied to a moving element of fluid. The left-hand side of the equation denotes the time rate of change of internal energy of the element, $\operatorname{div} \mathbf{q}$ is the net heat addition to the element, $p \operatorname{div} \mathbf{U}$ is the reversible compression work done by the element, and $\mu \Phi$ represents the irreversible work done on the element which is dissipated into heat as the result of fluid friction.

As written, equation (7-1) does not differ from the form employed for nonradiating fluids—the sole difference lies in the formulation of the heat flux vector $\mathbf{q}$. For a nonradiating fluid, $\mathbf{q}$ represents thermal conduction within the fluid and is expressed by the Fourier conduction law as

$$\mathbf{q} = -k \operatorname{grad} T \tag{7-2}$$

where k is the thermal conductivity of the fluid. For a fluid that absorbs, emits, and scatters radiation, there will, in addition, be a radiation heat flux present within the fluid, and this vector will be denoted by $\mathbf{q}_R$. Thus

$$\mathbf{q} = -k \operatorname{grad} T + \mathbf{q}_R$$

and equation (7-1) becomes

$$\rho c_v \frac{DT}{Dt} = \operatorname{div} (k \operatorname{grad} T) - p \operatorname{div} \mathbf{U} + \mu \Phi - \operatorname{div} \mathbf{q}_R \tag{7-3}$$

---

* For simplicity, the internal energy of the fluid is considered to be a function solely of temperature. Also, internal heat sources, chemical reactions, etc., can be incorporated into equation (7-1) in the usual manner.

An alternate and generally more useful form of the energy equation is (ref. 6)

$$\rho c_p \frac{DT}{Dt} = \text{div } (k \text{ grad } T) + \beta T \frac{Dp}{Dt} + \mu \Phi - \text{div } \mathbf{q}_R \qquad (7\text{-}4)$$

where β is the coefficient of thermal expansion of the fluid.

It is evident that before equation (7-4) can be applied to any problem involving a radiation-participating medium, a formulation for the radiation flux vector $\mathbf{q}_R$ as a function of temperature must be obtained. First, however, it will be convenient to discuss the various regimes of radiation transfer from an entirely physical point of view.

7-2   Radiation Regimes

One of the most important dimensionless parameters associated with radiation-participating media is the optical thickness of the medium. Consider for the moment a nonscattering medium, and let L be a characteristic physical dimension. If the absorption coefficient κ_λ is independent of temperature, the monochromatic optical thickness of the medium is defined as

$$\tau_{0\lambda} = \kappa_\lambda L$$

Recalling from Section 1-3 that the photon mean free path is $\lambda_p = 1/\kappa_\lambda$, the optical thickness may be rephrased as

$$\tau_{0\lambda} = \frac{L}{1/\kappa_\lambda} = \frac{L}{\lambda_p}$$

and thus $\tau_{0\lambda}$ is the ratio of the characteristic length to the photon mean free path. In molecular transfer problems one encounters the Knudsen number, which is the ratio of the molecular mean free path to a characteristic dimension. Consequently, $\tau_{0\lambda}$ may be regarded as simply a reciprocal photon Knudsen number. As in molecular transfer problems, it is also possible to classify various radiation transfer regimes.

Perhaps the simplest case is that of $\tau_{0\lambda} \gg 1$. Under this condition the photon mean free path is much smaller than the characteristic dimension, and one may think in terms of a photon continuum. In other words, every element of the medium is directly affected only by its neighbors, and as in the case of continuum molecular conduction, the radiation transfer within the medium becomes a diffusion process. More specifically, it will be shown in Section 7-6 that the monochromatic radiation flux is given by

$$\mathbf{q}_{R\lambda} = -\frac{4}{3\kappa_\lambda} \text{ grad } e_{b\lambda} \qquad (7\text{-}5)$$

This is often referred to as the Rosseland approximation or optically thick approximation although "limit" rather than "approximation" is more appropriate, for equation (7-5) is a correct asymptotic limit for large $\tau_{0\lambda}$.

Consider next the opposite extreme of $\tau_{0\lambda} \ll 1$, which is referred to as the optically thin limit. In this situation the photon mean free path is much larger than the characteristic physical dimension. Correspondingly, photons emitted by a given fluid element will travel directly to the bounding surfaces, and any intervening absorption of photons by the fluid will be negligible. Therefore every element of the fluid exchanges radiation directly with the bounding surfaces, such that there is no radiative interaction between various fluid elements. This is sometimes referred to as "negligible self-absorption," meaning that the fluid does not absorb any of its own emitted radiation although it will absorb radiation emitted by the bounding surfaces. One should note that the limit of $\tau_{0\lambda} = 0$ denotes a nonparticipating medium, such that photons travel from surface to surface with no intervening absorption or emission. This, of course, has its parallel in free molecule flow, in which molecules travel from surface to surface with no intervening collisions.

The final, and doubtless most important, regime involves intermediate values of $\tau_{0\lambda}$. In this case radiation exchange takes place between all fluid elements, and, as will be shown in Section 7-4, this results in an integral expression for the radiation flux vector. Consequently, the energy equation, equation (7-4), becomes an integrodifferential equation. Again drawing an analogy to molecular transfer, one can note that the present case of intermediate $\tau_{0\lambda}$ values corresponds to the transition regime in rarefied gasdynamics. In this respect it should be mentioned that the expression for the radiation flux vector applicable to intermediate values of $\tau_{0\lambda}$ can be derived through application of the kinetic theory Boltzmann equation to photon energy transfer (ref. 7). However, in Sections 7-3 and 7-4 a different approach will be employed, one that consists of considering a pencil of rays impinging on an elemental volume and then formulating a radiation energy balance on the volume.

In summary, it has been illustrated that a very definite analogy exists between photon and molecular transport phenomena. However, there are distinct practical differences. As an example, consider carbon dioxide at standard temperature and pressure. The molecular mean free path is on the order of a millionth of an inch, and it would be a rare situation indeed when one could not assume a molecular continuum. On the other hand, the photon mean free path is roughly one inch,* and an extremely large characteristic dimension would be required before one could employ the optically thick (or photon continuum) result.   This is characteristic of the majority of radiation problems, so that the intermediate range of optical thicknesses is by far the most important regime.

Before proceeding, it should again be emphasized that for a transparent or nonparticipating medium $\kappa_\lambda = 0$; consequently div $\mathbf{q}_R = 0$, for there is no absorption or emission within the medium.   On the other hand, when $\kappa_\lambda \to \infty$ the medium is completely opaque to thermal radiation and $\mathbf{q}_R = 0$.   The radiation term in equation (7-4) thus vanishes for the limiting cases of either a transparent or an opaque medium.

7-3 Equations of Transfer

The first stage in arriving at the equation for the radiation flux vector in an absorbing, emitting, and scattering medium is to develop the expression for the intensity of radiation, I_λ, within the medium; the present section is concerned with this development.   From this point on, attention will be directed toward one-dimensional radiation transfer.   Although a more general treatment could be given, it is felt that the one-dimensional case constitutes a reasonable compromise between utility and clarity of development.

Consider a radiating medium bounded by two infinite, parallel, opaque surfaces as illustrated in Fig. 7-1.   The intensity of radiation within the medium, $I_\lambda(y, \theta)$, will be divided into two contributions: the intensity directed in the positive y direction, which will be denoted by $I_\lambda^+(y, \theta)$, and that directed in the negative direction, denoted by $I_\lambda^-(y, \theta)$.   These two contributions to the total intensity will be considered separately.

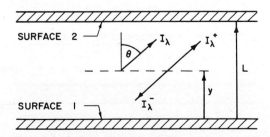

Fig. 7-1   Coordinate system for one-dimensional radiative transfer.

* This value is an average over all wavelengths.

In arriving at equations defining I_λ^+ and I_λ^-, a radiation balance will be invoked on a volume element of the medium. This should not be confused with the overall conservation of energy given by equation (7-4). The radiation energy balance will describe I_λ^+ and I_λ^- in terms of the local temperature within the medium, but only overall conservation of energy yields the actual temperature field. Furthermore, in applying the radiation balance, the medium will be assumed to be stationary. Radiation propagates with the velocity of light, so that it is necessary to account for motion of the medium only under relativistic conditions; that is, when the velocity of the medium is comparable to that of light. A final assumption regards scattering. For simplicity it will be assumed that scattering within the medium is both isotropic and coherent (see Section 1-3).

Figure 7-2 shows how the radiation balance is applied to an elemental volume of the medium. The volume consists of an elemental slab of thickness dy with a monochromatic beam of intensity I_λ^+ impinging on the lower surface. Recall that I_λ^+ denotes the time rate of radiant energy transfer per unit solid angle and per unit area normal to the pencil of rays. As the beam passes through the slab, its magnitude is changed by the amount dI_λ^+. This change is the result of three separate effects.

(1) I_λ^+ is augmented as the result of emission from the elemental volume.

(2)   Both absorption and scattering will result in the attenuation of I_λ^+ as it passes through the slab.

(3) I_λ^+ will be augmented by energy scattered within the slab due to all incident beams of energy.

To evaluate the augmentation of I_λ^+ as the result of emission, we have from equation (1-24) that the monochromatic rate of energy emission per unit volume within the medium is

$$4\kappa_\lambda e_{b\lambda}(y)$$

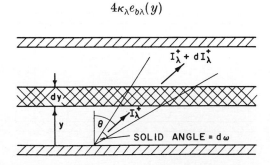

Fig. 7-2   Physical model for radiation balance.

where $e_{b\lambda}(y)$ is a function of position through its dependency on temperature and κ_λ is the monochromatic absorption coefficient. Since the slab volume per unit surface area is dy, and dividing by the total solid angle 4π, the emission of the slab per unit solid angle and per unit surface area is

$$\frac{\kappa_\lambda e_{b\lambda}(y)\, dy}{\pi}$$

Recasting this in terms of a unit area normal to I_λ^+ (i.e., normal to the pencil of rays), I_λ^+ is augmented as the result of emission by the amount

$$\frac{\kappa_\lambda e_{b\lambda}(y)\, dy}{\pi \cos \theta} \tag{7-6}$$

The attenuation of I_λ^+ due both to absorption and scattering is given in terms of the monochromatic extinction coefficient β_λ by equation (1-19) as

$$\beta_\lambda I_\lambda^+ \frac{dy}{\cos \theta} \tag{7-7}$$

where $dy/\cos \theta$ is the path length through the element.

To characterize the augmentation of I_λ^+ due to scattering of all incident beams of energy, let $I_\lambda(y, \theta')$ be the intensity of radiation incident on the slab through the angle θ'. The scattering path length is $dy/\cos \theta'$, and from equation (1-17) the scattered energy for isotropic coherent scattering is expressed in terms of the monochromatic scattering coefficient γ_λ by

$$\gamma_\lambda I_\lambda(y, \theta') \frac{dy}{\cos \theta'}$$

based on a unit area normal to $I_\lambda(y, \theta')$. Converting this to a unit surface area, the scattered energy per unit solid angle and per unit surface area is

$$\gamma_\lambda I_\lambda(y, \theta')\, dy$$

The scattered energy per unit surface area due to *all* incident radiation is obtained by integrating over the entire solid angle and noting that $d\omega = \sin \theta'\, d\theta'\, d\phi$; then

$$\int_{4\pi} \gamma_\lambda I_\lambda(y, \theta')\, dy\, d\omega = 2\pi\gamma_\lambda\, dy \int_0^\pi I_\lambda(y, \theta') \sin \theta'\, d\theta'$$

$$= \gamma_\lambda G_\lambda(y)\, dy \tag{7-8}$$

The function $G_\lambda(y)$ denotes the incident energy per unit area within the medium, and its definition follows from equation (7-8):

$$G_\lambda(y) = 2\pi \int_0^\pi I_\lambda(y, \theta') \sin \theta'\, d\theta' \tag{7-9}$$

Augmentation of I_λ^+ as the result of scattering may thus be written as

$$\frac{\gamma_\lambda G_\lambda(y) \, dy}{4\pi \cos \theta} \tag{7-10}$$

The differential equation for $I_\lambda^+(y, \theta)$ may now be formulated, and since dI_λ^+ is equal to the sum of equations (7-6) and (7-10) minus equation (7-7),

$$\cos \theta \frac{dI_\lambda^+}{dy} + \beta_\lambda I_\lambda^+ = \frac{\kappa_\lambda}{\pi} e_{b\lambda}(y) + \frac{\gamma_\lambda}{4\pi} G_\lambda(y) \tag{7-11}$$

In the same manner, an identical equation is obtained for $I_\lambda^-(y, \theta)$; that is,

$$\cos \theta \frac{dI_\lambda^-}{dy} + \beta_\lambda I_\lambda^- = \frac{\kappa_\lambda}{\pi} e_{b\lambda}(y) + \frac{\gamma_\lambda}{4\pi} G_\lambda(y) \tag{7-12}$$

Equations (7-11) and (7-12) are the *equations of transfer*. It should be noted that the $e_{b\lambda}(y)$ and $G_\lambda(y)$ appearing on the right-hand sides of these equations are unknown functions of y. Planck's Function $e_{b\lambda}$ is a function of y through its dependency on temperature; however, knowledge of the temperature field must await the solution of equation (7-4). The equation for $G_\lambda(y)$ will be discussed in the next section.

It is convenient now to define the quantities

$$\tau_\lambda = \int_0^y \beta_\lambda \, dy, \quad \tau_{0\lambda} = \int_0^L \beta_\lambda \, dy, \quad \mu = \cos \theta$$

The definition of $\tau_{0\lambda}$ is simply an extension of the optical thickness as defined in Section 7-2 to include scattering as well as a temperature-dependent extinction coefficient, while τ_λ is an optical coordinate. Equations (7-11) and (7-12) now become

$$\mu \frac{dI_\lambda^+}{d\tau_\lambda} + I_\lambda^+ = \frac{1}{\pi\beta_\lambda} \left[\kappa_\lambda e_{b\lambda}(\tau_\lambda) + \frac{\gamma_\lambda}{4} G_\lambda(\tau_\lambda) \right] \tag{7-13a}$$

$$\mu \frac{dI_\lambda^-}{d\tau_\lambda} + I_\lambda^- = \frac{1}{\pi\beta_\lambda} \left[\kappa_\lambda e_{b\lambda}(\tau_\lambda) + \frac{\gamma_\lambda}{4} G_\lambda(\tau_\lambda) \right] \tag{7-13b}$$

One may note that I_λ^+ corresponds to positive values of μ, whereas I_λ^- is associated with negative μ values.

The boundary conditions for equations (7-13) will, for the time being, be taken as

$$I_\lambda^+(\tau_\lambda, \mu) = I_\lambda^+(0, \mu), \qquad \tau_\lambda = 0$$

$$I_\lambda^-(\tau_\lambda, \mu) = I_\lambda^-(\tau_{0\lambda}, \mu), \qquad \tau_\lambda = \tau_{0\lambda}$$

The actual evaluation of $I_\lambda^+(0, \mu)$ and $I_\lambda^-(\tau_{0\lambda}, \mu)$ requires specification of the directional emission and reflection characteristics of the surfaces, and this will be treated in later sections. By using either an integrating factor or the method of variation of parameters, the solutions to equations (7-13) are found to be

$$I_\lambda^+(\tau_\lambda, \mu) = I_\lambda^+(0, \mu)e^{-\tau_\lambda/\mu}$$

$$+ \frac{1}{\pi}\int_0^{\tau_\lambda} \frac{1}{\beta_\lambda}\left[\kappa_\lambda e_{b\lambda}(t) + \frac{\gamma_\lambda}{4} G_\lambda(t)\right]e^{-(\tau_\lambda - t)/\mu} \frac{dt}{\mu} \quad (7\text{-}14a)$$

$$I_\lambda^-(\tau_\lambda, \mu) = I_\lambda^-(\tau_{0\lambda}, \mu)e^{(\tau_{0\lambda} - \tau_\lambda)/\mu}$$

$$- \frac{1}{\pi}\int_{\tau_\lambda}^{\tau_{0\lambda}} \frac{1}{\beta_\lambda}\left[\kappa_\lambda e_{b\lambda}(t) + \frac{\gamma_\lambda}{4} G_\lambda(t)\right]e^{-(\tau_\lambda - t)/\mu} \frac{dt}{\mu} \quad (7\text{-}14b)$$

Thus equations (7-14) describe the radiation field in terms of the temperature field within the medium (through Planck's Function $e_{b\lambda}$) and the incident radiation function $G_\lambda(\tau_\lambda)$.

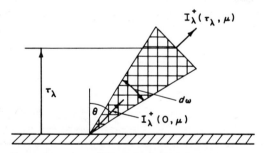

Fig. 7-3   Interpretation of the solution of the equation of transfer.

The physical interpretation of equations (7-14) is quite evident. With reference to Fig. 7-3, the first term in equation (7-14a), $I_\lambda^+(0, \mu)e^{-\tau_\lambda/\mu}$, denotes energy that originated at the lower surface and that has been attenuated as the result of absorption and scattering by the factor $e^{-\tau_\lambda/\mu}$, where τ_λ/μ is the optical path length. The integral in equation (7-14) represents augmentation of I_λ^+ due to energy emitted and scattered in the θ direction from the cross-hatched volume in Fig. 7-3. The presence of the integral expression is due to this augmented energy originating over a finite path length, which, as previously discussed, will cause the energy equation, equation (7-4), to become an integrodifferential equation.

7-4   Radiation Flux Equation

Referring to Fig. 7-4, the monochromatic radiation flux is expressed in terms of the intensity of radiation as

$$q_{R\lambda}(\tau_\lambda) = \int_{4\pi} I_\lambda(\tau_\lambda, \mu) \cos\theta \, d\omega = 2\pi \int_{-1}^{1} I_\lambda(\tau_\lambda, \mu)\mu \, d\mu$$

and recalling that I_λ^+ and I_λ^- correspond to positive and negative values of μ, respectively, then

$$q_{R\lambda}(\tau_\lambda) = 2\pi \int_0^1 I_\lambda^+ \mu \, d\mu - 2\pi \int_0^{-1} I_\lambda^- \mu \, d\mu \qquad (7\text{-}15)$$

Upon substituting equations (7-14) into equation (7-15), the expression for the one-dimensional radiation flux becomes

$$q_{R\lambda}(\tau_\lambda) = 2\pi \int_0^1 I_\lambda^+(0, \mu)e^{-\tau_\lambda/\mu}\mu \, d\mu - 2\pi \int_0^1 I_\lambda^-(\tau_{0\lambda}, -\mu)e^{-(\tau_{0\lambda} - \tau_\lambda)/\mu}\mu \, d\mu$$

$$+ \, 2\int_0^{\tau_\lambda} \left[\frac{\kappa_\lambda}{\beta_\lambda} e_{b\lambda}(t) + \frac{\gamma_\lambda}{4\beta_\lambda} G_\lambda(t)\right] E_2(\tau_\lambda - t) \, dt$$

$$- \, 2\int_{\tau_\lambda}^{\tau_{0\lambda}} \left[\frac{\kappa_\lambda}{\beta_\lambda} e_{b\lambda}(t) + \frac{\gamma_\lambda}{4\beta_\lambda} G_\lambda(t)\right] E_2(t - \tau_\lambda) \, dt \qquad (7\text{-}16)$$

where the $E_n(t)$ are the exponential integral functions defined by

$$E_n(t) = \int_0^1 \mu^{n-2} e^{-t/\mu} \, d\mu$$

A more complete discussion of these functions, together with tabulated values, is given in Appendix B. The total radiation flux is correspondingly expressed as

$$q_R(y) = \int_0^\infty q_{R\lambda}(\tau_\lambda) \, d\lambda$$

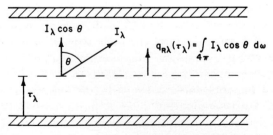

Fig. 7-4   The radiation heat flux.

Consider now the application of equation (7-16) to the energy equation, equation (7-4).   We have in one dimension

$$\text{div } \mathbf{q}_R = \frac{dq_R}{dy} = \int_0^\infty \frac{dq_{R\lambda}}{dy}\, d\lambda$$

or, alternately,

$$\text{div } \mathbf{q}_R = \int_0^\infty \beta_\lambda \frac{dq_{R\lambda}}{d\tau_\lambda}\, d\lambda \qquad (7\text{-}17)$$

and upon differentiating equation (7-16),

$$-\frac{dq_{R\lambda}}{d\tau_\lambda} = 2\pi \int_0^1 I_\lambda^+(0, \mu)e^{-\tau_\lambda/\mu}\, d\mu + 2\pi \int_0^1 I_\lambda^-(\tau_{0\lambda}, -\mu)e^{-(\tau_{0\lambda}-\tau_\lambda)/\mu}\, d\mu$$

$$+ 2 \int_0^{\tau_{0\lambda}} \left[\frac{\kappa_\lambda}{\beta_\lambda} e_{b\lambda}(t) + \frac{\gamma_\lambda}{4\beta_\lambda} G_\lambda(t) \right] E_1(|\tau_\lambda - t|)\, dt$$

$$- 4 \frac{\kappa_\lambda}{\beta_\lambda} e_{b\lambda}(\tau_\lambda) - \frac{\gamma_\lambda}{\beta_\lambda} G_\lambda(\tau_\lambda) \qquad (7\text{-}18)$$

Equations (7-17) and (7-18) thus describe, at least in principle, the radiation term div $\mathbf{q}_R$ appearing in equation (7-4) with restriction to one-dimensional radiation.   It remains to obtain the equation governing the incident radiation function $G_\lambda(\tau_\lambda)$.   From equation (7-9) we may write

$$G_\lambda(\tau_\lambda) = 2\pi \int_0^1 I_\lambda^+(\tau_\lambda, \mu')\, d\mu' - 2\pi \int_0^{-1} I_\lambda^-(\tau_\lambda, \mu')\, d\mu'$$

such that, from equations (7-14), the function $G_\lambda(\tau_\lambda)$ is described by the integral equation

$$G_\lambda(\tau_\lambda) = 2\pi \int_0^1 I_\lambda^+(0, \mu)e^{-\tau_\lambda/\mu}\, d\mu + 2\pi \int_0^1 I_\lambda^-(\tau_{0\lambda}, -\mu)e^{-(\tau_{0\lambda}-\tau_\lambda)/\mu}\, d\mu$$

$$+ 2 \int_0^{\tau_{0\lambda}} \left[\frac{\kappa_\lambda}{\beta_\lambda} e_{b\lambda}(t) + \frac{\gamma_\lambda}{4\beta_\lambda} G_\lambda(t) \right] E_1(|\tau_\lambda - t|)\, dt \qquad (7\text{-}19)$$

To summarize the preceding results, substitution of equations (7-17) and (7-18) into equation (7-4) yields the energy equation for a radiating medium.   Since Planck's Function $e_{b\lambda}$ is a function of temperature, and since $e_{b\lambda}$ appears under the integral sign in equation (7-18), the energy equation is subsequently an integrodifferential equation with temperature as the dependent variable.   Furthermore, $e_{b\lambda}$ is related to temperature in a nonlinear manner; hence the energy equation is additionally nonlinear.   Coupled with this is equation (7-19), and the system of equations describing energy transfer in a radiating medium thus consists of two simultaneous, nonlinear, integrodifferential equations that, with appropriate integration over wavelength, describe $T(y)$ and $G(y)$.

Although this system of equations appears to be quite formidable, it should be borne in mind that the equations are very general in nature. There are a number of specific situations in which the complexity of the governing equations is reduced considerably, and these specific cases will now be discussed.

The gray-medium approximation. Probably the greatest simplification is the gray-medium approximation in which it is assumed that the absorption and the scattering coefficients are both independent of wavelength. Although this is rarely a physically realistic approximation, it at least serves as an initial stepping stone toward nongray analyses.

To apply equations (7-16), (7-18), and (7-19) to a gray medium, one may note that τ_λ is now independent of λ and may be written as τ. The only remaining monochromatic quantities are $I_\lambda^+(0, \mu)$, $I_\lambda^-(\tau_0, \mu)$, $G_\lambda(\tau)$, and $e_{b\lambda}(\tau)$. Upon integrating the three equations over wavelength and noting that

$$\int_0^\infty e_{b\lambda} \, d\lambda = \sigma T^4, \quad \int_0^\infty I_\lambda \, d\lambda = I, \quad \int_0^\infty G_\lambda \, d\lambda = G$$

the relevant radiation equations become, for a gray medium,

$$q_R(\tau) = 2\pi \int_0^1 I^+(0, \mu) e^{-\tau/\mu} \mu \, d\mu - 2\pi \int_0^1 I^-(\tau_0, -\mu) e^{-(\tau_0 - \tau)/\mu} \mu \, d\mu$$

$$+ 2 \int_0^\tau \left[\frac{\kappa\sigma}{\beta} T^4(t) + \frac{\gamma}{4\beta} G(t) \right] E_2(\tau - t) \, dt$$

$$- 2 \int_\tau^{\tau_0} \left[\frac{\kappa\sigma}{\beta} T^4(t) + \frac{\gamma}{4\beta} G(t) \right] E_2(t - \tau) \, dt \qquad (7\text{-}20)$$

$$-\frac{dq_R}{d\tau} = 2\pi \int_0^1 I^+(0, \mu) e^{-\tau/\mu} \, d\mu + 2\pi \int_0^1 I^-(\tau_0, -\mu) e^{-(\tau_0 - \tau)/\mu} \, d\mu$$

$$+ 2 \int_0^{\tau_0} \left[\frac{\kappa\sigma}{\beta} T^4(t) + \frac{\gamma}{4\beta} G(t) \right] E_1(|\tau - t|) \, dt$$

$$- 4 \frac{\kappa\sigma}{\beta} T^4(\tau) - \frac{\gamma}{\beta} G(\tau) \qquad (7\text{-}21)$$

$$G(\tau) = 2\pi \int_0^1 I^+(0, \mu) e^{-\tau/\mu} \, d\mu + 2\pi \int_0^1 I^-(\tau_0, -\mu) e^{-(\tau_0 - \tau)/\mu} \, d\mu$$

$$+ 2 \int_0^{\tau_0} \left[\frac{\kappa\sigma}{\beta} T^4(t) + \frac{\gamma}{4\beta} G(t) \right] E_1(|\tau - t|) \, dt \quad (7\text{-}22)$$

These equations more clearly illustrate the nonlinear and integro-differential structure of the energy equation. Upon substituting equation (7-21) in equation (7-4), temperature appears explicitly under the integral sign and in a nonlinear manner; that is, as T^4.

Radiative equilibrium. When radiation is the predominant mode of energy transfer and the system is in steady state, we have the state of radiative equilibrium; that is, div $\mathbf{q}_R = 0$, or, in one dimension, $dq_R/dy = 0$. Conservation of energy is expressed by an integral equation; thus the governing equations are two simultaneous integral equations.

A much greater simplification ensues for radiative equilibrium of a gray medium. Comparing equations (7-21) and (7-22) for $dq_R/d\tau = 0$, we see that*

$$G(\tau) = 4\frac{\kappa\sigma}{\beta} T^4(\tau) + \frac{\gamma}{\beta} G(\tau) \qquad (7\text{-}23)$$

or
$$G(\tau) = 4\sigma T^4(\tau) \qquad (7\text{-}24)$$

From this result, together with equation (7-21), conservation of energy is expressed by

$$2\sigma T^4(\tau) = \pi \int_0^1 I^+(0, \mu)e^{-\tau/\mu}\, d\mu + \pi \int_0^1 I^-(\tau_0, -\mu)e^{-(\tau_0 - \tau)/\mu}\, d\mu$$
$$+ \sigma \int_0^{\tau_0} T^4(t)E_1(|\tau - t|)\, dt \quad (7\text{-}25)$$

and this single integral equation describes the temperature field within the medium. The absorption and the scattering coefficients no longer appear separately but are present only in the form of the extinction coefficient, $\beta = \kappa + \gamma$, contained in the definitions for τ and τ_0. In addition, equation (7-25) is a linear equation in T^4.

Nonscattering media. Equations (7-16) and (7-18) may be applied directly to media that do not scatter radiation by setting $\gamma_\lambda = 0$, and the incident radiation function $G_\lambda(\tau_\lambda)$ will no longer appear in these equations. Although equation (7-19) with $\gamma_\lambda = 0$ is still a valid equation, $G_\lambda(\tau_\lambda)$ is redundant to the problem and the latter equation is no longer necessary. Nonscattering media are thus described by a single equation, which is the integrodifferential energy equation.

Pure scattering. At the other extreme, consider pure scattering

---

* For a nongray medium $dq_{R\lambda}/dy \neq 0$ even though $dq_R/dy = 0$, for energy absorbed at a given wavelength will be emitted at different wavelengths.

$(\kappa_\lambda = 0)$; that is, the medium neither absorbs nor emits but only scatters radiation.   For this case equations (7-16) and (7-19) reduce to

$$q_{R\lambda}(\tau_\lambda) = 2\pi \int_0^1 I_\lambda^+(0, \mu)e^{-\tau_\lambda/\mu}\mu\, d\mu - 2\pi \int_0^1 I_\lambda^-(\tau_{0\lambda}, -\mu)e^{-(\tau_{0\lambda}-\tau_\lambda)/\mu}\mu\, d\mu$$

$$+ \frac{1}{2}\int_0^{\tau_\lambda} G_\lambda(t)E_2(\tau_\lambda - t)\, dt - \frac{1}{2}\int_{\tau_\lambda}^{\tau_{0\lambda}} G_\lambda(t)E_2(t - \tau_\lambda)\, dt \quad (7\text{-}26)$$

and

$$G_\lambda(\tau_\lambda) = 2\pi \int_0^1 I_\lambda^+(0, \mu)e^{-\tau_\lambda/\mu}\, d\mu + 2\pi \int_0^1 I_\lambda^-(\tau_{0\lambda}, -\mu)e^{-(\tau_{0\lambda}-\tau)/\mu}\, d\mu$$

$$+ \frac{1}{2}\int_0^{\tau_{0\lambda}} G_\lambda(t)E_1(|\tau_\lambda - t|)\, dt \quad (7\text{-}27)$$

Setting $\kappa_\lambda = 0$ in equation (7-18) and comparing the result with equation (7-27), one finds that

$$\frac{dq_{R\lambda}}{d\tau_\lambda} = 0$$

and consequently the energy equation is uncoupled from the radiation transfer process.   This is physically evident, for there can be no absorption or emission by any fluid element.

Similarly, equations (7-26) and (7-27) illustrate that the radiation transfer is independent of the temperature field within the medium. Again the physical interpretation is evident, since the medium only scatters incident radiation.   One should note that for a gray-scattering medium equation (7-24) is no longer applicable, for equation (7-23) simply reduces to the identity $G(\tau) = G(\tau)$, and the function $G_\lambda(\tau_\lambda)$ is defined by equation (7-27) regardless of whether the medium is gray or nongray.*

Diffuse surfaces.   As mentioned previously, evaluation of the surface intensities $I_\lambda^+(0, \mu)$ and $I_\lambda^-(\tau_{0\lambda}, \mu)$ requires specification of the directional emission and reflection characteristics of the surfaces. The case in which the surfaces emit and reflect in a diffuse manner will now be considered, while in Chapter 8 a specific example of a surface that is a diffuse emitter and specular reflector will be treated.

For a diffuse surface the quantities $I_\lambda^+(0, \mu)$ and $I_\lambda^-(\tau_{0\lambda}, \mu)$ are independent of direction; that is, independent of μ.   Furthermore,

* For a gray medium the subscript λ is deleted in equations (7-26) and (7-27).

from equation (1-9) these quantities may be expressed in terms of the surface radiosities $B_{1\lambda}$ and $B_{2\lambda}$ by

$$I_\lambda^+(0, \mu) = \frac{B_{1\lambda}}{\pi} \qquad I_\lambda^-(\tau_{0\lambda}, \mu) = \frac{B_{2\lambda}}{\pi}$$

such that

$$2\pi \int_0^1 I_\lambda^+(0, \mu)e^{-\tau_\lambda/\mu}\mu\, d\mu = 2B_{1\lambda}E_3(\tau_\lambda) \qquad (7\text{-}28a)$$

$$2\pi \int_0^1 I_\lambda^-(\tau_{0\lambda}, -\mu)e^{-(\tau_{0\lambda} - \tau_\lambda)/\mu}\mu\, d\mu = 2B_{2\lambda}E_3(\tau_{0\lambda} - \tau_\lambda) \quad (7\text{-}28b)$$

$$2\pi \int_0^1 I_\lambda^+(0, \mu)e^{-\tau_\lambda/\mu}\, d\mu = 2B_{1\lambda}E_2(\tau_\lambda) \qquad (7\text{-}28c)$$

$$2\pi \int_0^1 I_\lambda^-(\tau_{0\lambda}, -\mu)e^{-(\tau_{0\lambda} - \tau_\lambda)/\mu}\, d\mu = 2B_{2\lambda}E_2(\tau_{0\lambda} - \tau_\lambda) \quad (7\text{-}28d)$$

By making the above substitutions, the preceding radiation equations apply directly to the case in which the bounding surfaces are diffuse. For a gray medium and gray surfaces, the subscript λ is deleted in equations (7-28).

It remains to evaluate $B_{1\lambda}$ and $B_{2\lambda}$; consider initially surface 1. Upon substituting equations (7-28a) and (7-28b) into equation (7-16) and letting $\tau_\lambda = 0$, the monochromatic radiation flux from surface 1 is given by

$$q_{R\lambda}(0) = B_{1\lambda} - 2B_{2\lambda}E_3(\tau_{0\lambda})$$

$$- 2\int_0^{\tau_{0\lambda}} \left[\frac{\kappa_\lambda}{\beta_\lambda} e_{b\lambda}(t) + \frac{\gamma_\lambda}{4\beta_\lambda} G_\lambda(t)\right]E_2(t)\, dt \quad (7\text{-}29)$$

Since the first term, $B_{1\lambda}$, is the radiant energy leaving surface 1, then the remaining terms denote the incident energy, and

$$H_{1\lambda} = 2B_{2\lambda}E_3(\tau_{0\lambda}) + 2\int_0^{\tau_{0\lambda}} \left[\frac{\kappa_\lambda}{\beta_\lambda} e_{b\lambda}(t) + \frac{\gamma_\lambda}{4\beta_\lambda} G_\lambda(t)\right]E_2(t)\, dt \quad (7\text{-}30)$$

Further, the surface radiosity may be expressed as

$$B_{1\lambda} = \epsilon_{1\lambda}e_{b1\lambda} + (1 - \epsilon_{1\lambda})H_{1\lambda}$$

Then

$$B_{1\lambda} = \epsilon_{1\lambda}e_{b1\lambda} + 2(1 - \epsilon_{1\lambda})\left\{B_{2\lambda}E_3(\tau_{0\lambda})\right.$$

$$\left. + \int_0^{\tau_{0\lambda}} \left[\frac{\kappa_\lambda}{\beta_\lambda} e_{b\lambda}(t) + \frac{\gamma_\lambda}{4\beta_\lambda} G_\lambda(t)\right]E_2(t)\, dt\right\} \quad (7\text{-}31)$$

In turn, for surface 2,

$$B_{2\lambda} = \epsilon_{2\lambda} e_{b2\lambda} + 2(1 - \epsilon_{2\lambda})\Big\{ B_{1\lambda} E_3(\tau_{0\lambda})$$

$$+ \int_0^{\tau_{0\lambda}} \Big[\frac{\kappa_\lambda}{\beta_\lambda} e_{b\lambda}(t) + \frac{\gamma_\lambda}{4\beta_\lambda} G_\lambda(t) \Big] E_2(\tau_{0\lambda} - t)\, dt \Big\} \quad (7\text{-}32)$$

Equations (7-31) and (7-32) constitute two simultaneous equations for $B_{1\lambda}$ and $B_{2\lambda}$. For black surfaces, $B_{1\lambda} = e_{b1\lambda}$ and $B_{2\lambda} = e_{b2\lambda}$.

7-5   Optically Thin Limit

Additional simplifications of the radiation equations follow from the physical discussion of radiation regimes presented in Section 7-2. These concern the magnitude of the optical thickness $\tau_{0\lambda}$, which, for a medium that scatters as well as absorbs and emits, is defined in the previous section by

$$\tau_{0\lambda} = \int_0^L \beta_\lambda\, dy$$

In the present discussion attention will be directed toward the optically thin limit, which is defined by the condition $\tau_{0\lambda} \ll 1$.

From Appendix B the exponential integrals $E_2(t)$ and $E_3(t)$ may, for small t, be written as

$$E_2(t) = 1 + O(t) \qquad E_3(t) = \tfrac{1}{2} - t + O(t^2)$$

For simplicity, the present development will be restricted to diffuse surfaces, and upon substituting the above expressions into equations (7-16) and (7-28), the monochromatic radiation flux correct to $O(\tau_{0\lambda})$ becomes

$$q_{R\lambda}(\tau_\lambda) = B_{1\lambda}(1 - 2\tau_\lambda) - B_{2\lambda}(1 - 2\tau_{0\lambda} + 2\tau_\lambda)$$

$$+ 2 \int_0^{\tau_\lambda} \Big[\frac{\kappa_\lambda}{\beta_\lambda} e_{b\lambda}(t) + \frac{\gamma_\lambda}{4\beta_\lambda} G_\lambda(t) \Big] dt$$

$$- 2 \int_{\tau_\lambda}^{\tau_{0\lambda}} \Big[\frac{\kappa_\lambda}{\beta_\lambda} e_{b\lambda}(t) + \frac{\gamma_\lambda}{4\beta_\lambda} G_\lambda(t) \Big] dt \quad (7\text{-}33)$$

Now, providing that $\tau_{0\lambda} \ll 1$, one may additionally delete terms of $O(\tau_{0\lambda})$, and the radiation flux under optically thin conditions reduces to

$$q_{R\lambda} = B_{1\lambda} - B_{2\lambda} \quad (7\text{-}34)$$

Neglecting terms of $O(\tau_{0\lambda})$ in equations (7-31) and (7-32), the surface radiosities are found to be

$$B_{1\lambda} = \frac{\epsilon_{1\lambda}e_{b1\lambda} + (1 - \epsilon_{1\lambda})\epsilon_{2\lambda}e_{b2\lambda}}{1 - (1 - \epsilon_{1\lambda})(1 - \epsilon_{2\lambda})} \qquad (7\text{-}35a)$$

$$B_{2\lambda} = \frac{\epsilon_{2\lambda}e_{b2\lambda} + (1 - \epsilon_{2\lambda})\epsilon_{1\lambda}e_{b1\lambda}}{1 - (1 - \epsilon_{1\lambda})(1 - \epsilon_{2\lambda})} \qquad (7\text{-}35b)$$

It is evident that equations (7-34) and (7-35) correspond to radiation transfer through nonparticipating media. In fact, equations (7-35) can alternately be obtained from equation (3-22). This does *not* imply, however, that the radiation term in the energy equation may be neglected. To evaluate $dq_{R\lambda}/d\tau_\lambda$, differentiation of equation (7-33) yields

$$-\frac{dq_{R\lambda}}{d\tau_\lambda} = 2B_{1\lambda} + 2B_{2\lambda} - \frac{4\kappa_\lambda}{\beta_\lambda}e_{b\lambda}(\tau_\lambda) - \frac{\gamma_\lambda}{\beta_\lambda}G_\lambda(\tau_\lambda) \qquad (7\text{-}36)$$

Again, this expression neglects terms of $O(\tau_{0\lambda})$, which would have appeared as $O(\tau_{0\lambda}{}^2)$ in equation (7-33).

By applying the same limiting procedure to equation (7-19), the incident radiation function reduces to

$$G_\lambda(\tau_\lambda) = 2B_{1\lambda} + 2B_{2\lambda}$$

for the optically thin limit, and the final form of equation (7-36) becomes

$$-\frac{dq_{R\lambda}}{d\tau_\lambda} = \frac{2\kappa_\lambda}{\beta_\lambda}[B_{1\lambda} + B_{2\lambda} - 2e_{b\lambda}(\tau_\lambda)] \qquad (7\text{-}37)$$

An interesting feature of this equation is that it may be recast as

$$-\frac{dq_{R\lambda}}{dy} = 2\kappa_\lambda[B_{1\lambda} + B_{2\lambda} - 2e_{b\lambda}(y)] \qquad (7\text{-}38)$$

and one notes that the radiation transfer to or from a volume element is *independent* of the scattering coefficient. Thus, under optically thin conditions, scattering plays no role in overall conservation of energy.

The first two terms in equation (7-38) represent absorption in an elemental volume owing to energy that originated at the surfaces, while the last term denotes energy emitted by the element. It is obvious that equation (7-38) does not include energy exchange between different elements, and this coincides with the physical discussion presented in Section 7-2.

The mathematical simplifications associated with the optically thin limit are quite evident. Since there is no integral term in equation (7-38), the energy equation becomes a differential rather than an integrodifferential equation. In addition, the simultaneous integral equation for $G_\lambda(\tau_\lambda)$ has been eliminated, and the surface radiosities appearing in equation (7-38) are known *a priori* from equations (7-35).

7-6   Optically Thick Limit

Equation (7-5), which is the optically thick expression for the radiation flux, will now be shown to be a limiting form of the general result given by equation (7-16). To accomplish this, let

$$S_\lambda(\tau_\lambda) = \frac{\kappa_\lambda}{\beta_\lambda}\, e_{b\lambda}(\tau_\lambda) + \frac{\gamma_\lambda}{4\beta_\lambda}\, G_\lambda(\tau_\lambda)$$

and expanding $S(t)$ in a Taylor series about $t = \tau_\lambda$

$$S_\lambda(t) = S_\lambda(\tau_\lambda) + \frac{dS_\lambda}{d\tau_\lambda}(t - \tau_\lambda) + \frac{1}{2}\frac{d^2 S}{d\tau_\lambda^2}(t - \tau_\lambda)^2 + \cdots$$

Substituting this into equation (7-16), the expression for the radiation flux becomes (with $z = \tau_\lambda - t$ and $z' = t - \tau_\lambda$)

$$q_{R\lambda} = 2\pi \int_0^1 I_\lambda^+(0,\mu)e^{-\tau_\lambda/\mu}\mu\, d\mu - 2\pi \int_0^1 I_\lambda^-(\tau_{0\lambda}, -\mu)e^{-(\tau_{0\lambda}-\tau_\lambda)/\mu}\mu\, d\mu$$

$$+ 2S(\tau_\lambda)\left[\int_0^{\tau_\lambda} E_2(z)\, dz - \int_0^{\tau_{0\lambda}-\tau_\lambda} E_2(z')\, dz'\right]$$

$$- 2\frac{dS}{d\tau_\lambda}\left[\int_0^{\tau_\lambda} zE_2(z)\, dz + \int_0^{\tau_{0\lambda}-\tau_\lambda} z'E_2(z')\, dz'\right]$$

$$+ \frac{d^2 S}{d\tau_\lambda^2}\left[\int_0^{\tau_\lambda} z^2 E_2(z)\, dz - \int_0^{\tau_{0\lambda}-\tau_\lambda} z'^2 E_2(z')\, dz'\right] + \cdots \qquad (7\text{-}39)$$

To apply this equation to the optically thick limit ($\tau_{0\lambda} \to \infty$), we will, for the time being, restrict ourselves to regions removed from the boundaries, so that $\tau_\lambda \to \infty$ and $\tau_{0\lambda} - \tau_\lambda \to \infty$. Equation (7-39) then reduces to

$$q_{R\lambda} = -4\frac{dS_\lambda}{d\tau_\lambda}\int_0^\infty zE_2(z)\, dz$$

Furthermore,

$$\int_0^\infty zE_2(z)\, dz = \frac{1}{3}$$

and

$$q_{R\lambda} = -\frac{4}{3}\frac{d}{d\tau_\lambda}\left[\frac{\kappa_\lambda}{\beta_\lambda}e_{b\lambda}(\tau_\lambda) + \frac{\gamma_\lambda}{4\beta_\lambda}G_\lambda(\tau_\lambda)\right] \qquad (7\text{-}40)$$

Proceeding in the same manner with equation (7-19), one obtains

$$G_\lambda(\tau_\lambda) = 4S_\lambda(\tau_\lambda) \int_0^\infty E_1(z)\, dz = 4S_\lambda(\tau_\lambda)$$

or
$$G_\lambda(\tau_\lambda) = 4e_{b\lambda}(\tau_\lambda) \qquad (7\text{-}41)$$

Employing this in equation (7-40),

$$q_{R\lambda} = -\frac{4}{3}\frac{de_{b\lambda}}{d\tau_\lambda} = -\frac{4}{3\beta_\lambda}\frac{de_{b\lambda}}{dy} \qquad (7\text{-}42)$$

which is the one-dimensional form of equation (7-5) (with κ_λ replaced by β_λ to include scattering). Once again we have a situation in which the absorption and the scattering coefficients enter only through the extinction coefficient β_λ.

Let us return now to the restriction concerning regions removed from the boundaries; that is, $\tau_\lambda \to \infty$ and $\tau_{0\lambda} - \tau_\lambda \to \infty$. At first glance this would appear to preclude the possibility of employing equation (7-42) to evaluate the radiation heat transfer to or from a surface. Such, however, is not the case. From a photon point of view, the conditions $\tau_\lambda \to \infty$ and $\tau_{0\lambda} - \tau_\lambda \to \infty$ correspond to the requirement that one consider regions at least a photon mean free path from the surface, for beyond a mean free path no photons emitted by the surface will be present, and thus the presence of the surface cannot be detected. Equation (7-42) thus applies to within a mean free path of the boundary. However, if the radiation flux expressed by equation (7-42) were, in turn, inapplicable at the boundary, this would necessitate a discontinuity in radiant flux within a mean free path from the surface, and such a discontinuity is not physically possible. Equation (7-42) may thus be employed at the boundaries of the medium as well as everywhere within the medium provided, of course, that optically thick conditions prevail,* and also provided that equation (7-42) is used in the energy equation to evaluate the temperature distribution throughout the *entire* medium.

A direct parallel to the preceding argument exists in molecular transfer. The Fourier conduction law, equation (7-2), can be derived

---

* This may be shown formally by writing the equation of transfer, equation (7-13a), as

$$\frac{\mu}{\tau_{0\lambda}}\frac{dI_\lambda^+}{d\xi} + I_\lambda^+ = \frac{1}{\pi}S(\tau_\lambda)$$

where $\xi = \tau_\lambda/\tau_{0\lambda} = O(1)$. For $\tau_{0\lambda} \gg 1$, the highest derivative in the above equation is multiplied by a small parameter, and the problem may be treated as a singular perturbation problem (see Sections 9-3 and 10-4). From this it can be shown that $q_{R\lambda}$ is constant across a region of thickness $\tau_\lambda = O(1)$. However, if conduction transfer is included, only the *total* heat flux (conduction plus radiation) will be constant across this region (ref. 22).

for a molecular gas from kinetic theory, and this derivation neglects molecular collisions with the bounding walls. The derivation thus strictly applies only for regions at least a mean free path from the surfaces, but by the preceding arguments applied to molecular transfer, equation (7-2) is indeed applicable at the boundaries as is well known.

One should further note from equation (7-42) that radiation transfer within the medium or from a bounding surface is *independent* of the surface emittance ϵ_λ. This is not surprising when analogy is again drawn to molecular transfer. Here the analogous surface property is the accommodation coefficient, which is a measure of the ability of the surface to exchange energy with colliding molecules. Under the condition of a molecular continuum, conduction heat transfer is no longer dependent on the accommodation coefficient. Equation (7-42) also illustrates that under optically thick conditions the directional properties of the surface become redundant, just as the manner in which a surface reflects molecules plays no role in continuum conduction heat transfer.

A final comment regarding equation (7-42) concerns application to the case of pure scattering. Recall that for pure scattering the temperature within the medium becomes redundant with respect to radiation transfer, and the sole dependent variable is $G_\lambda(\tau_\lambda)$. Thus equation (7-42) applies to pure scattering provided $e_{b\lambda}$ is replaced by $G_\lambda/4$ through use of equation (7-41).

7-7 Radiation Slip

An additional radiation regime is that of radiation slip, which differs in a specific manner from the previous cases of the optically thin and optically thick limits. To illustrate this difference, consider combined conduction and radiation transfer in a one-dimensional medium. Equation (7-4) becomes

$$\frac{d}{dy}\left(k\,\frac{dT}{dy}\right) - \frac{dq_R}{dy} = 0$$

This equation requires two boundary conditions as the result of the presence of conduction heat transfer; these are the continuity of temperature conditions

$$T = T_1 \quad y = 0 \qquad \text{and} \qquad T = T_2 \quad y = L$$

Consider next the absence of conduction. Then $dq_R/dy = 0$, and with the exception of the optically thick limit there will result either an integral equation or, for the optically thin limit, an algebraic equation.

Neither equation requires boundary conditions; hence the subsequent result is that the temperature of the medium adjacent to a surface will differ from the surface temperature. This is physically evident from the fact that the temperature of the medium adjacent to a surface is directly affected not only by the surface but also by all other volume elements and surfaces. The radiation slip method is a means of accounting for these temperature jumps and thus applies only in the absence of conduction.

The following development is somewhat of a modification of Deissler's more general three-dimensional treatment (ref. 8). Basically the radiation slip approach represents an extension of the preceding optically thick limit. The fundamental assumption is that the first departure from optically thick radiation takes place at the boundaries of the medium. Equation (7-42) is assumed to apply everywhere within the medium (including $\tau_\lambda = 0$ and $\tau_\lambda = \tau_{0\lambda}$), while we wish to obtain information for the quantities $e_\lambda(0) - e_{1\lambda}$ and $e_{2\lambda} - e_\lambda(\tau_{0\lambda})$, which correspond to the temperature jumps at the boundaries.

Attention will again be directed toward diffuse surfaces. The radiation transfer from the lower surface may be expressed as

$$q_{R\lambda}(0) = \epsilon_{1\lambda} e_{b1\lambda} - \epsilon_{1\lambda} H_{1\lambda} \qquad (7\text{-}43)$$

where $H_{1\lambda}$ is given by equation (7-30). Since equation (7-42) is to be employed within the medium, we may let $\tau_{0\lambda} \to \infty$ while

$$q_{R\lambda}(0) = -\frac{4}{3}\left(\frac{de_{b\lambda}}{d\tau_\lambda}\right)_{\tau_\lambda = 0} \qquad (7\text{-}44)$$

Combining this with equations (7-30) and (7-43),

$$-\frac{4}{3\epsilon_{1\lambda}}\left(\frac{de_{b\lambda}}{d\tau_\lambda}\right)_{\tau_\lambda = 0} = e_{b1\lambda} - 2\int_0^\infty \left[\frac{\kappa_\lambda}{\beta_\lambda} e_{b\lambda}(t) + \frac{\gamma_\lambda}{4\beta_\lambda} G_\lambda(t)\right] E_2(t)\, dt$$

or, since $G_\lambda(\tau_\lambda) = 4e_{b\lambda}(\tau_\lambda)$ from equation (7-41),

$$-\frac{4}{3\epsilon_{1\lambda}}\left(\frac{de_{b\lambda}}{d\tau_\lambda}\right)_{\tau_\lambda = 0} = e_{b1\lambda} - 2\int_0^\infty e_{b\lambda}(t) E_2(t)\, dt \qquad (7\text{-}45)$$

If we now expand $e_{b\lambda}(\tau_\lambda)$ in a Taylor series about $\tau_\lambda = 0$, then*

$$e_{b\lambda}(\tau_\lambda) = e_{b\lambda}(0) + \left(\frac{de_{b\lambda}}{d\tau_\lambda}\right)_{\tau_\lambda = 0} \tau_\lambda + \frac{1}{2}\left(\frac{d^2 e_{b\lambda}}{d\tau_\lambda^2}\right)_{\tau_\lambda = 0} \tau_\lambda^2 + \cdots$$

---

* The function $e_{b\lambda}(\tau_\lambda)$ is analytic at $\tau_\lambda = 0$ when the radiation flux is described by equation (7-42), which is presently the case. In general, however, $e_{b\lambda}(\tau_\lambda)$ is nonanalytic at the boundaries.

and equation (7-45) yields

$$-\frac{4}{3\epsilon_{1\lambda}}\left(\frac{de_{b\lambda}}{d\tau_\lambda}\right)_{\tau_\lambda=0} = e_{b1\lambda} - 2e_{b\lambda}(0)\int_0^\infty E_2(t)\,dt$$

$$-2\left(\frac{de_{b\lambda}}{d\tau_\lambda}\right)_{\tau_\lambda=0}\int_0^\infty tE_2(t)\,dt - \left(\frac{d^2e_{b\lambda}}{d\tau_\lambda{}^2}\right)_{\tau_\lambda=0}\int_0^\infty t^2E_2(t)\,dt + \cdots \qquad (7\text{-}46)$$

The integrals appearing above have the values

$$\int_0^\infty E_2(t)\,dt = \frac{1}{2}, \quad \int_0^\infty tE_2(t)\,dt = \frac{1}{3}, \quad \int_0^\infty t^2E_2(t)\,dt = \frac{1}{2}$$

and equation (7-46) may be rephrased as

$$e_{b\lambda}(0) - e_{b1\lambda} = \frac{4}{3}\left(\frac{1}{\epsilon_{1\lambda}} - \frac{1}{2}\right)\left(\frac{de_{b\lambda}}{d\tau_\lambda}\right)_{\tau_\lambda=0} - \frac{1}{2}\left(\frac{d^2e_{b\lambda}}{d\tau_\lambda{}^2}\right)_{\tau_\lambda=0} + \cdots \qquad (7\text{-}47)$$

On purely an order-of-magnitude basis, this temperature jump condition may be written

$$\frac{e_{b\lambda}(0) - e_{b1\lambda}}{e_{b2\lambda} - e_{b1\lambda}} = O\left(\frac{1}{\tau_{0\lambda}}\right) + O\left(\frac{1}{\tau_{0\lambda}{}^2}\right) + \cdots$$

and the above illustrates that under optically thick conditions ($\tau_{0\lambda} \to \infty$) continuity of temperature is achieved at the surface. This is compatible with the point of view of a photon continuum and is analogous to the requirement of temperature continuity in a molecular continuum.

The radiation slip regime, in turn, represents an extension of the optically thick limit through the introduction of a temperature jump at the surface as defined by equation (7-47). As pointed out by Probstein (ref. 9), however, "it is *only* for small mean free paths that slip or temperature jump concepts have any theoretical justification," since the optically thick equation (7-42) has been employed in the derivation of equation (7-47). Hence there is no theoretical basis for including the second-order term in equation (7-47), which is $O(1/\tau_{0\lambda}{}^2)$, and the radiation slip condition will be taken as

$$e_{b\lambda}(0) - e_{b1\lambda} = \frac{4}{3}\left(\frac{1}{\epsilon_{1\lambda}} - \frac{1}{2}\right)\left(\frac{de_{b\lambda}}{d\tau_\lambda}\right)_{\tau_\lambda=0} \qquad (7\text{-}48)$$

For the upper wall we have

$$e_{b2\lambda} - e_{b\lambda}(\tau_{0\lambda}) = \frac{4}{3}\left(\frac{1}{\epsilon_{2\lambda}} - \frac{1}{2}\right)\left(\frac{de_{b\lambda}}{d\tau_\lambda}\right)_{\tau_\lambda=\tau_{0\lambda}} \qquad (7\text{-}49)$$

The analogy of equations (7-48) and (7-49) with the expressions for velocity and temperature jumps in a rarefied molecular gas (ref. 10) is quite evident.

One should note, of course, that for pure scattering the concept of a temperature jump has no physical significance. However, by combining equation (7-41) with equations (7-48) and (7-49), the jump condition may be rephrased in terms of the incident radiation function G_λ. At the lower surface, for example, one obtains

$$G_\lambda(0) - 4e_{b1\lambda} = \frac{4}{3}\left(\frac{1}{\epsilon_{1\lambda}} - \frac{1}{2}\right)\left(\frac{dG_\lambda}{d\tau_\lambda}\right)_{\tau_\lambda = 0}$$

and this has a proper physical interpretation for the limit of pure scattering.

7-8 Total Absorption and Scattering Coefficients

In Section 1-3 total absorption and scattering coefficients were defined as

$$\kappa = \frac{\displaystyle\int_0^\infty \kappa_\lambda \int_{4\pi} I_\lambda \, d\omega \, d\lambda}{\displaystyle\int_{4\pi} I \, d\omega} \tag{1-30}$$

$$\gamma = \frac{\displaystyle\int_0^\infty \gamma_\lambda \int_{4\pi} I_\lambda \, d\omega \, d\lambda}{\displaystyle\int_{4\pi} I \, d\omega} \tag{1-31}$$

and from the definition of G_λ given by equation (7-9), these now become

$$\kappa = \frac{\displaystyle\int_0^\infty \kappa_\lambda G_\lambda \, d\lambda}{G} \tag{7-50}$$

$$\gamma = \frac{\displaystyle\int_0^\infty \gamma_\lambda G_\lambda \, d\lambda}{G} \tag{7-51}$$

As pointed out in Section 1-3, neither κ nor γ are properties of the medium, since the incident radiation function G_λ depends in general on all regions of the medium as well as on the bounding surfaces. On the other hand, total emission may be described in terms of the Planck mean absorption coefficient, which is defined by equation (1-29) as

$$\kappa_P = \frac{\displaystyle\int_0^\infty \kappa_\lambda e_{b\lambda} \, d\lambda}{e_b} \tag{1-29}$$

and this quantity *is* a property of the medium.

Unfortunately there is at present no general procedure for handling the nongray radiation equations, although a limited number of approximate techniques have been employed for nonscattering media. One of these is the band approximation in which it is assumed that the medium is gray within distinct wavelength regions. For example, a gas such as carbon dioxide has three absorption-emission bands that are important within the infrared (see Section 1-3), and by the band approximation a constant mean absorption coefficient is considered within each of the bands, while outside the bands the gas is transparent. Although this example is concerned with vibration-rotation bands, the method is equally applicable to electronic-line absorption and emission.

Bevans and Dunkle (ref. 11) and Edwards (ref. 12) discuss the band approximation in detail, while Greif (ref. 13) and Nichols (ref. 14) have employed it in specific solutions involving the governing radiation equations. An alternate approach to nongray media has been employed by Howell and Perlmutter (ref. 15). Instead of the governing radiation equations, the Monte Carlo method, which is a statistical sampling technique applied to photon transfer, was employed.

Although the preceding discussion illustrates that radiation transfer in nongray media is not readily formulated in terms of total coefficients, simplifications result when attention is directed toward the specific cases of the optically thin and optically thick limits as well as to radiation slip.

Optically thin limit. Upon integrating equation (7-38) over all wavelengths, we may express the total radiation term dq_R/dy for the optically thin limit as

$$-\frac{dq_R}{dy} = 2\kappa_m(T,\,T_1)B_1 + 2\kappa_m(T,\,T_2)B_2 - 4\kappa_P(T)\sigma T^4(y) \quad (7\text{-}52)$$

where $\kappa_m(T,\,T_1)$ and $\kappa_m(T,\,T_2)$ are modified Planck coefficients defined by Scala and Sampson (ref. 16) as

$$\kappa_m(T,\,T_1) = \frac{\displaystyle\int_0^\infty \kappa_\lambda(T)B_{1\lambda}\,d\lambda}{B_1} \quad (7\text{-}53a)$$

$$\kappa_m(T,\,T_2) = \frac{\displaystyle\int_0^\infty \kappa_\lambda(T)B_{2\lambda}\,d\lambda}{B_2} \quad (7\text{-}53b)$$

The monochromatic radiosities are given by equations (7-35), and the modified Planck coefficients depend not only on the local state of the medium but also on the temperatures and the emission characteristics of the surfaces.

A particularly simple result is obtained for black surfaces under the assumption that the monochromatic absorption coefficient is independent of temperature; that is, $\kappa_\lambda \neq \kappa_\lambda(T)$. Equations (7-53) now become

$$\kappa_m(T, T_1) = \frac{\int_0^\infty \kappa_\lambda e_{b\lambda}(T_1)\,d\lambda}{e_b(T_1)} = \kappa_P(T_1)$$

$$\kappa_m(T, T_2) = \frac{\int_0^\infty \kappa_\lambda e_{b\lambda}(T_2)\,d\lambda}{e_b(T_2)} = \kappa_P(T_2)$$

such that the modified coefficients reduce directly to the Planck mean absorption coefficient. Equation (7-52) is, in turn,

$$-\frac{dq_R}{dy} = 2\sigma[\kappa_P(T_1)T_1^4 + \kappa_P(T_2)T_2^4 - 2\kappa_P(T)T^4(y)] \quad (7\text{-}54)$$

whereas, for a gray medium ($\kappa = \kappa_P$), equation (7-38) yields

$$-\frac{dq_R}{dy} = 2\sigma\kappa_P(T)[T_1^4 + T_2^4 - 2T^4(y)] \quad (7\text{-}55)$$

Comparison of these results illustrates that, subject to the above assumptions, the difference between nongray and gray radiation is characterized by the temperature dependence of the Planck mean absorption coefficient. It should be evident from equation (1-29) that $\kappa_\lambda \neq \kappa_\lambda(T)$ does *not* imply that κ_P is also independent of temperature.

Optically thick limit. For optically thick radiation, the total radiation heat flux is obtained from equation (7-42) as

$$q_R = -\int_0^\infty \frac{4}{3\beta_\lambda} \frac{de_{b\lambda}}{dy}\,d\lambda$$

or
$$q_R = -\frac{4}{3\beta_R} \frac{de_b}{dy} \quad (7\text{-}56)$$

where β_R is the Rosseland mean extinction coefficient defined by

$$\frac{1}{\beta_R} = \int_0^\infty \frac{1}{\beta_\lambda} \frac{de_{b\lambda}}{de_b}\,d\lambda \quad (7\text{-}57)$$

and it follows that β_R is a property of the medium. For nonscattering media ($\gamma_\lambda = 0$) this reduces to the Rosseland mean absorption coefficient

$$\frac{1}{\kappa_R} = \int_0^\infty \frac{1}{\kappa_\lambda} \frac{de_{b\lambda}}{de_b}\,d\lambda \quad (7\text{-}58)$$

If the medium is gray, $\kappa_R = \kappa_P = \kappa$; otherwise $\kappa_P > \kappa_R$. Stewart and Pyatt (ref. 17) have found that κ_P can be one to two orders of magnitude greater than κ_R for high-temperature nitrogen.

Although the optically thick limit certainly possesses an attractive analytical formulation, it must be borne in mind that for many radiating media absorption and emission take place in distinct wavelength bands. Thus, even though the medium is optically thick within these bands, it may be transparent elsewhere, and one must account for such "windows" when integrating over wavelength to obtain equations (7-57) and (7-58).

To illustrate, consider a nonscattering medium that has N absorption-emission bands, and let each band have the width $\Delta\lambda_i$. Outside these bands the medium is assumed to be transparent. Equation (7-42) will now be integrated over the bands, and let $\bar{q}_R$ denote a radiation heat flux defined by*

$$\bar{q}_R = \sum_{i=1}^{N} \int_{\Delta\lambda_i} q_{R\lambda}\, d\lambda$$

Then

$$\bar{q}_R = -\frac{4}{3} \sum_{i=1}^{N} \int_{\Delta\lambda_i} \frac{1}{\beta_\lambda} \frac{de_{b\lambda}}{dy}\, d\lambda$$

or

$$\bar{q}_R = -\frac{4}{3\kappa_R} \frac{de_b}{dy} \tag{7-59}$$

which is identical with equation (7-56) except that now

$$\frac{1}{\kappa_R} = \sum_{i=1}^{N} \frac{1}{\kappa_{Ri}} \tag{7-60}$$

with

$$\frac{1}{\kappa_{Ri}} = \int_{\Delta\lambda_i} \frac{1}{\kappa_\lambda} \frac{de_{b\lambda}}{de}\, d\lambda \tag{7-61}$$

It is evident that κ_{Ri} is a Rosseland mean absorption coefficient associated with each specific band, while κ_R accounts for all bands. In fact, this is simply an application of the previously discussed band approximation to the optically thick regime. This method was originally applied by Howell and Perlmutter (ref. 15) to high-temperature hydrogen assuming a single absorption-emission band.

---

* In addition to $\bar{q}_R$, there will be a contribution to the radiation heat flux due to energy transfer in the transparent wave length regions. For example, if the walls are black,

$$q_R = \bar{q}_R + (e_{b1} - e_{b2}) - \sum_{i=1.}^{N} \int_{\Delta\lambda_i} (e_{b1\lambda} - e_{b2\lambda})\, d\lambda$$

The integrals appearing in the summation can be evaluated from Table 1-2.

Radiation slip.   If attention is restricted to gray, diffuse surfaces, the jump condition given by equation (7-48) becomes

$$e_{b\lambda}(0) - e_{b1\lambda} = \frac{4}{3\beta_\lambda} \left(\frac{1}{\epsilon_1} - \frac{1}{2}\right)\left(\frac{de_{b\lambda}}{dy}\right)_{y=0} \qquad (7\text{-}62)$$

and integrating over all wavelengths,

$$T^4(0) - T_1{}^4 = \frac{4}{3\beta_R} \left(\frac{1}{\epsilon_1} - \frac{1}{2}\right)\left(\frac{dT^4}{dy}\right)_{y=0} \qquad (7\text{-}63)$$

Consequently, as shown by Deissler (ref. 8), the Rosseland mean extinction coefficient is additionally applicable to the radiation slip regime.

Again, modification must be made for transparent wavelength regions, and this has also been discussed in ref. 15.   Returning to the previous example of a nonscattering medium having N absorption-emission bands, integration of equation (7-62) over each band yields

$$\sum_{i=1}^{N} \left[\int_{\Delta\lambda_i} e_{b\lambda}(0)\, d\lambda - \int_{\Delta\lambda_i} e_{b1\lambda}\, d\lambda\right] = \frac{4}{3\kappa_R} \left(\frac{1}{\epsilon_1} - \frac{1}{2}\right)\left(\frac{de_b}{dy}\right)_{y=0} \qquad (7\text{-}64)$$

where κ_R is given by equation (7-60).   With the surface temperature known, the integrals $\int_{\Delta\lambda_i} e_{b1\lambda}\, d\lambda$ may be evaluated from Table 1-2. This table may also be employed to obtain the relationship between the integrals $\int_{\Delta\lambda_i} e_{b\lambda}(0)\, d\lambda$ and the jump temperature $T(0)$.

Values of κ_P from emittance data.   Although a considerable amount of data has been obtained concerning the emittance of gases, very little information is available regarding mean absorption coefficients.   Following R. and M. Goulard (ref. 3), such emittance data may be converted to values for the Planck mean absorption coefficient. With restriction to nonscattering gases, the monochromatic energy emitted by one surface of a one-dimensional gas slab is

$$2 \int_0^{\tau_{0\lambda}} e_{b\lambda}(t) E_2(t)\, dt \qquad (7\text{-}65)$$

Now, by the definition of the hemispherical monochromatic gas emittance $\epsilon_{g\lambda}$, the energy emitted by one surface of an *isothermal* gas slab is written as (ref. 10)

$$\epsilon_{g\lambda} e_{b\lambda}(T)$$

where $\epsilon_{g\lambda}$ depends on both the state of the gas and on the slab thickness

L.   Letting T be a constant in equation (7-65) and equating it to the above expression, $\epsilon_{g\lambda}$ and κ_λ are related by

$$\epsilon_{g\lambda} = 2 \int_0^{\tau_{0\lambda}} E_2(t)\, dt = 1 - 2E_3(\kappa_\lambda L) \qquad (7\text{-}66)$$

where $\tau_{0\lambda} = \kappa_\lambda L$ under isothermal conditions.

Now, if equation (7-66) were employed to evaluate the absorption coefficient from emittance data, it would be found that κ_λ is dependent on the slab thickness L, whereas the absorption coefficient is actually a gas property.   At this point it should be emphasized that the concept of a gas emittance refers to an isothermal gas, and a gas in radiative equilibrium can approach isothermal conditions only in the limit as $\tau_{0\lambda} \to 0$.   Thus, since

$$E_3(t) = \tfrac{1}{2} - t + O(t^2)$$

the relation between κ_λ and $\epsilon_{g\lambda}$ is found from equation (7-66) to be

$$\kappa_\lambda = \frac{1}{2}\left(\frac{\epsilon_{g\lambda}}{L}\right)_{L\to 0} \qquad (7\text{-}67)$$

Gas-emittance data are conventionally expressed in terms of the total emittance ϵ_g, defined as

$$\epsilon_g = \frac{\displaystyle\int_0^\infty \epsilon_{g\lambda} e_{b\lambda}\, d\lambda}{e_b}$$

Upon multiplying equation (7-67) by $e_{b\lambda}$ and integrating over all wavelengths, the resulting total absorption coefficient is the Planck mean coefficient; that is,

$$\kappa_P = \frac{1}{2}\left(\frac{\epsilon_g}{L}\right)_{L\to 0} \qquad (7\text{-}68)$$

such that values for κ_P can be obtained through extrapolation of the quantity ϵ_g/L.   This procedure has been applied to water vapor, carbon dioxide, and ammonia; the results are shown in Fig. 7-5.   The emittance data were taken from Figs. 4-13, 4-15, and 4-23 of Hottel's chapter in McAdams (ref. 18).   In performing such calculations, L must be replaced by $L_e/2$ in equation (7-68), where $L_e(= 2L)$ is the appropriate mean beam length (ref. 18).

Also shown in Fig. 7-5 are values of κ_P for water vapor, carbon dioxide, and carbon monoxide, which have been calculated by Tien and Abu-Romia (ref. 19).   These values were obtained from equation (1-29) together with spectral absorption coefficient data (see, for

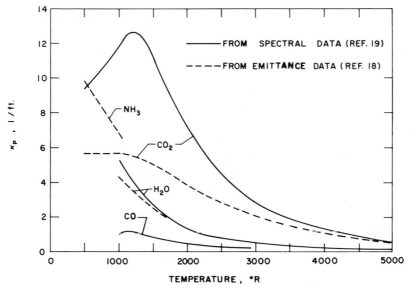

Fig. 7-5   Planck mean absorption coefficients at one atmosphere pressure.

example, Section 1-3). One may note a considerable discrepancy between the two methods of calculation for CO_2, and Tien and Abu-Romia's results should be regarded as the most reliable. This discrepancy is evidently the result of inaccurate values for the gas emittance, ϵ_g, under optically thin conditions.

7-9 Differential Approximation

In this final section of Chapter 7, a brief discussion will be given to the differential approximation for radiation transfer. For a more complete treatment, the reader is referred to Vincenti and Kruger (ref. 20). There are a number of ways in which the general radiation flux equation (7-16) may be converted to an approximate differential equation describing $q_{R\lambda}(\tau_\lambda)$. One of these involves satisfying certain moments of the equations of transfer, while another proceeds by expanding the intensity I in a series of spherical harmonics.

By either method, one obtains the same differential equation expressing $q_{R\lambda}(\tau_\lambda)$. For nonscattering media and one-dimensional radiation, this differential equation is

$$\frac{d^2 q_{R\lambda}}{d\tau_\lambda^2} - 3q_{R\lambda} = 4\frac{de_{b\lambda}}{d\tau_\lambda} \qquad (7\text{-}69)$$

The differential approximation thus involves replacing the general expression for the radiation flux, equation (7-16), by equation (7-69). The equations describing energy transfer within the medium then consist of two simultaneous differential equations—that is, equation (7-69) with appropriate integration over wave length together with conservation of energy. In contrast, use of equation (7-16) results in an integrodifferential equation for nonscattering media.

It remains to formulate boundary conditions on $q_{R\lambda}$ as required by equation (7-69). The surface boundary condition has been given by Cheng (ref. 21; see also ref. 20) for black surfaces, and this may be extended to nonblack diffuse surfaces through replacement of the black-body surface emission by the surface radiosity. For example, at surface 1 this gives

$$2q_{R\lambda}(0) - \left(\frac{dq_{R\lambda}}{d\tau_\lambda}\right)_{\tau_\lambda = 0} = 4B_{1\lambda} - 4e_{b\lambda}(0) \tag{7-70}$$

The radiosity $B_{1\lambda}$ may now be eliminated by noting that

$$B_{1\lambda} = \epsilon_{1\lambda}e_{b1\lambda} + \rho_\lambda H_{1\lambda}$$

or

$$B_{1\lambda} = \epsilon_{1\lambda}e_{b1\lambda} + (1 - \epsilon_{1\lambda})[B_{1\lambda} - q_{R\lambda}(0)]$$

A similar procedure applies to surface 2, and the final boundary conditions for equation (7-69) are

$$\left(\frac{1}{\epsilon_{1\lambda}} - \frac{1}{2}\right) q_{R\lambda}(0) - \frac{1}{4}\left(\frac{dq_{R\lambda}}{d\tau_\lambda}\right)_{\tau_\lambda = 0} = e_{b1\lambda} - e_{b\lambda}(0) \tag{7-71a}$$

$$\left(\frac{1}{\epsilon_{2\lambda}} - \frac{1}{2}\right) q_{R\lambda}(\tau_{0\lambda}) + \frac{1}{4}\left(\frac{dq_{R\lambda}}{d\tau_\lambda}\right)_{\tau_\lambda = \tau_{0\lambda}} = e_{b\lambda}(\tau_{0\lambda}) - e_{b2\lambda} \tag{7-71b}$$

Note that the right sides of these expressions differ from zero only if molecular conduction is neglected.

One may observe that, relative to $q_{R\lambda}$, the second derivative in equation (7-69) is of $O(1/\tau_{0\lambda}^2)$. Furthermore, by reasoning similar to that following equation (7-47), the first derivatives in equations (7-71) are also of $O(1/\tau_{0\lambda}^2)$. With these terms of $O(1/\tau_{0\lambda}^2)$ deleted, the present differential approximation reduces directly to the radiation slip method given by equations (7-44), (7-48), and (7-49).* Thus, while radiation slip constitutes a first-order departure from optically thick

---

* Note that for radiative equilibrium of a gray medium there is no difference between the radiation slip method and the differential approximation, since

$$dq_R/d\tau = d^2q_R/d\tau^2 = 0.$$

radiation, the differential approximation apparently represents a second-order approximation. Indeed, this is implied by the previously mentioned spherical harmonic expansion.

On the other hand, consider the optically thin limit. For $\tau_{0\lambda} \ll 1$, equation (7-69) becomes

$$\frac{d^2 q_{R\lambda}}{d\tau_\lambda{}^2} = 4 \frac{de_{b\lambda}}{d\tau_\lambda}$$

If one performs a single integration and evaluates the constant of integration from equations (7-70) and (7-34), the optically thin form of equation (7-69) is thus

$$-\frac{dq_{R\lambda}}{dy} = 2\kappa_\lambda[B_{1\lambda} + B_{2\lambda} - 2e_{b\lambda}(y)]$$

This is precisely the optically thin expression given by equation (7-38), indicating that the differential approximation may actually be a useful approximation for all values of optical thickness.

A particular advantage of the differential approximation is that it can be formulated in three dimensions. However, this extension to problems other than the one-dimensional plane layer is open to some question, and such extensions should be made with caution.

Although direct application of the differential approximation will not be made in the following chapters, the approximation is actually comparable to the exponential kernel approximation that is employed in Sections 8-1, 9-3, and 10-3. For example, if the approximate expressions for $E_2(t)$ and $E_3(t)$ given by equations (8-6) and (8-7) are substituted into equations (7-16), (7-28a), and (7-28b), and restriction is made to diffuse surfaces, the integrals in equation (7-16) may be eliminated by differentiating twice. This results in a differential equation of the same form as equation (7-69). The only difference is that the coefficient of the second derivative is 4/3 instead of unity.

REFERENCES

1. V. Kourganoff, *Basic Methods in Transfer Problems*, Dover Publications, New York, 1963.

2. S. Chandrasekhar, *Radiative Transfer*, Dover Publications, New York, 1960.

3. R. and M. Goulard, One-dimensional energy transfer in radiant media. *Intern. J. Heat Mass Transfer*, **1**, 81–91 (1960).

4. R. Viskanta, Heat transfer in thermal radiation absorbing and scattering media. *AEC Research and Development Report ANL-6170*, May 1960.

5. R. Viskanta, Radiation transfer and interaction of convection with radiation heat transfer. In *Advances in Heat Transfer*, vol. 3, Academic Press, New York, 1966.

6. W. M. Rohsenow and H. Choi, *Heat, Mass and Momentum Transfer*, Prentice-Hall, Englewood Cliffs, N.J., 1961.

7. D. H. Sampson, Radiative contributions to energy and momentum transport in a gas. General Electric Co., M.S.D. *TIS R63SD14*, March 1963.

8. R. G. Deissler, Diffusion approximation for thermal radiation in gases with jump boundary conditions. *J. Heat Transfer*, **C86**, 240–246 (1964).

9. R. F. Probstein, Author's Reply. *AIAA J.*, **2**, 976 (1964).

10. E. R. G. Eckert and R. M. Drake, Jr., *Heat and Mass Transfer*, McGraw-Hill, New York, 1959.

11. J. T. Bevans and R. V. Dunkle, Radiant interchange within an enclosure. *J. Heat Transfer*, **C82**, 1–19 (1960).

12. D. K. Edwards, Radiation interchange in a nongray enclosure containing an isothermal carbon-dioxide-nitrogen gas mixture. *J. Heat Transfer*, **C84**, 1–11 (1962).

13. R. Greif, Energy transfer by radiation and conduction with variable gas properties. *Intern. J. Heat Mass Transfer*, **7**, 891–900 (1964).

14. L. D. Nichols, Temperature profile in the entrance region of an annular passage considering the effects of turbulent convection and radiation. *Intern. J. Heat Mass Transfer*, **8**, 589–608 (1965).

15. J. R. Howell and M. Perlmutter, Monte Carlo solution of radiant heat transfer in a nongrey, nonisothermal gas with temperature dependent properties. *Amer. Inst. Chem. Engr. J.*, **10**, 562–567 (1964).

16. S. M. Scala and D. H. Sampson, Heat transfer in hypersonic flow with radiation and chemical reaction. General Electric Co., M.S.D. *TIS R63SD46*, March 1963.

17. J. C. Stewart and K. D. Pyatt, Jr., Theoretical study of optical properties. Research Directorate, *Report AFSWC TR 61–71*, vols. I, II, III, September 1961.

18. W. H. McAdams, *Heat Transmission*, McGraw-Hill, New York, 1964.

19. C. L. Tien and M. M. Abu-Romia, Department of Mechanical Engineering, University of California, Berkeley, personal communication.

20. W. G. Vincenti and C. H. Kruger, Jr., *Introduction to Physical Gas Dynamics*, Wiley, New York, 1965.

21. P. Cheng, Study of the Flow of a Radiating Gas by a Differential Approximation. Ph.D. thesis in engineering, Stanford University, 1965.

22. J. L. Novotny, M. D. Kelleher, and P. Schimmel, Radiation Interaction in Free-Convection Stagnation Flow, presented at 5th U.S. National Congress of Applied Mechanics, University of Minnesota, June 1966.

PROBLEMS

7-1. For an isothermal system, show that equation (7-18) yields the result $dq_{R\lambda}/d\tau_\lambda = 0$, as would be expected physically. Hint: Recall that for an isothermal enclosure the intensity of radiation from the bounding surfaces corresponds to black-body radiation.

7-2. Consider the optically thin limit with both plates at the same temperature ($T_2 = T_1$). Show that for this case $G_\lambda(\tau_\lambda) = 4e_{b1\lambda}$.

7-3. From the footnote on page 209, it follows that for $\tau_{0\lambda} \to \infty$, then $I_\lambda{}^+ = S(\tau_\lambda)/\pi$, as well as $I_\lambda{}^- = S(\tau_\lambda)/\pi$, for regions removed from the boundaries. Show how this information, utilizing equations (7-13) and (7-15), directly yields the optically thick equation (7-42).

7-4. Consider the application of equations (7-53) to a case in which both surfaces are black. Assume further that the absorbing-emitting medium is carbon monoxide with only the fundamental band being considered. Following the same procedure as in Problem 1-11, show that

$$\kappa_m(T,T_i) \stackrel{\sim}{=} \kappa_p(T_i)\frac{T_i}{T}$$

where $i = 1, 2$.

7-5. Employing the exponential kernel approximation, as given by equations (8-6) and (8-7), show that equation (7-16) with restriction to diffuse surfaces and a nonscattering medium may be recast as a differential equation which is of the same form as equation (7-69), except that the coefficient of the second derivative is $4/3$ instead of unity.

7-6. Following the same procedure as in the previous problem, derive boundary conditions for $q_{R\lambda}$ which are of the same form as equations (7-71), except that the coefficient of $1/4$ is replaced by $1/3$.   Hint: Note that equation (7-31) may be rephrased as

$$B_{1\lambda} = \epsilon_{1\lambda}e_{b1\lambda} + 2(1 - \epsilon_{1\lambda})[B_{1\lambda} - q_{R\lambda}(0)]$$

7-7. Show that by employing $a = 1$ and $b = (3)^{1/2}$ in the equation preceding equation (8-6), use of the exponential kernal approximation yields exactly equation (7-69). On the other hand, show that the corresponding boundary conditions differ from equations (7-71).

7-8. Consider the conditions of Problem 7-5, except let the medium scatter as well as absorb and emit radiation. Show that the resulting differential equation is again of the same form as equation (7-69), with the coefficient of the second derivative being $(4 - \omega_0)/(3 - \omega_0)$, where $\omega_0 = \gamma_\lambda/\beta_\lambda$. Assume that ω_0 is independent of temperature. Hint: Upon combining equations (7-18) and (7-19), note that

$$G_\lambda(\tau_\lambda) = 4e_{b\lambda}(\tau_\lambda) - \frac{\beta_\lambda}{\gamma_\lambda} \frac{dq_{R\lambda}}{d\tau_\lambda}$$

7-9. Derive the appropriate boundary conditions on $q_{R\lambda}$ for the previous problem.

CHAPTER 8

RADIATIVE EQUILIBRIUM

Attention will now be directed toward solutions of the governing radiation equations, and the present chapter is concerned with radiation transfer under the condition of radiative equilibrium. Consider two infinite parallel surfaces at temperatures T_1 and T_2 as illustrated in Fig. 8-1. A radiation-participating medium is contained between the surfaces, and as discussed in Section 7-4 the energy equation, equation (7-4), reduces to

$$\frac{dq_R}{dy} = 0$$

for one-dimensional radiative equilibrium. It follows, of course, that the radiation heat flux q_R is constant throughout the medium. The case of radiative equilibrium is much less complicated than problems involving simultaneous conduction and/or convection heat transfer.

The present chapter will devote particular attention to methods of handling the radiation equations. In Section 8-1 the assumption of a gray medium is employed, and several methods for solving the resulting

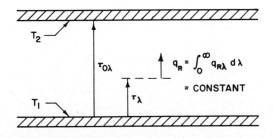

Fig. 8-1   One-dimensional radiative equilibrium.

integral equation are discussed and compared. Section 8-2 is concerned with bounding surfaces that reflect in a specular rather than a diffuse manner, while procedures for treating nongray media are presented in Section 8-3.

8-1   Gray Medium

The first case to be considered is that of a gray, absorbing, emitting, and scattering medium; it will additionally be assumed that the bounding surfaces are gray and diffuse.   With these assumptions, the radiation flux is expressed by equations (7-20), (7-24), and (7-28) as

$$q_R = 2B_1 E_3(\tau) - 2B_2 E_3(\tau_0 - \tau) + 2\sigma \int_0^\tau T^4(t) E_2(\tau - t)\, dt$$

$$- 2\sigma \int_\tau^{\tau_0} T^4(t) E_2(t - \tau)\, dt \quad (8\text{-}1)$$

where it is again recalled that q_R is a constant for radiative equilibrium. Upon differentiating equation (8-1), one also obtains

$$2\sigma T^4(\tau) = B_1 E_2(\tau) + B_2 E_2(\tau_0 - \tau) + \sigma \int_0^{\tau_0} T^4(t) E_1(|\tau - t|)\, dt \quad (8\text{-}2)$$

which corresponds to equation (7-25) for the case of diffuse surfaces.
    As previously discussed, the absorption and the scattering coefficients are present only in the form of the extinction coefficient $\beta = \kappa + \gamma$, and β appears only in τ.   Furthermore, no restriction has been made in equations (8-1) and (8-2) regarding the temperature dependence of β, and the equations apply for a variable extinction coefficient $\beta(T)$. This will be discussed in more detail later.
    Equation (8-2) is conventionally regarded as the equation expressing conservation of energy for one-dimensional radiative equilibrium, and this is a linear integral equation describing $T^4(\tau)$.   However, since q_R is a constant (although unknown), equation (8-1) may alternately be employed to determine $T^4(\tau)$.   These two equations are, of course, not independent, for equation (8-2) is merely the derivative of equation (8-1).   The choice as to which equation to use depends on the specific solution method employed; equation (8-2) is generally more useful.
    Before proceeding, it will be convenient to recast equations (8-1)

and (8-2) in dimensionless form, and, following Viskanta and Grosh (ref. 1), dimensionless quantities will be defined as*

$$\phi(\tau) = \frac{\sigma T^4(\tau) - B_2}{B_1 - B_2} \qquad Q = \frac{q_R}{B_1 - B_2}$$

With these, equations (8-1) and (8-2) reduce, respectively, to

$$Q = 2E_3(\tau) + 2\int_0^\tau \phi(t)E_2(\tau - t)\,dt - 2\int_\tau^{\tau_0} \phi(t)E_2(t - \tau)\,dt \quad (8\text{-}3)$$

$$2\phi(\tau) = E_2(\tau) + \int_0^{\tau_0} \phi(t)E_1(|\tau - t|)\,dt \qquad\qquad (8\text{-}4)$$

In arriving at equations (8-3) and (8-4) a very definite simplification has resulted: this is the fact that the radiosities B_1 and B_2 are no longer present in the equations. It is readily evident that the dimensionless heat flux Q depends on a single parameter, the optical thickness τ_0. To obtain the actual radiation heat transfer from the dimensionless quantity Q, one need know only the radiosity difference $(B_1 - B_2)$.

Although B_1 and B_2 can be obtained from equations (7-31) and (7-32), a much simpler procedure may be used for the present case of radiative equilibrium. Since $q_{R1} = -q_{R2} = q_R$, then from equation (3-19) we may write

$$q_R = \frac{\epsilon_1}{1 - \epsilon_1}(\sigma T_1{}^4 - B_1)$$

$$= -\frac{\epsilon_2}{1 - \epsilon_2}(\sigma T_2{}^4 - B_2)$$

After some algebraic manipulation these equations yield

$$\frac{B_1 - B_2}{\sigma(T_1{}^4 - T_2{}^4)} = \frac{1}{1 + [(1/\epsilon_2) + (1/\epsilon_1) - 2]Q} \qquad (8\text{-}5)$$

Thus, once Q is known as a function of τ_0, equation (8-5) defines the radiosity difference $(B_1 - B_2)$.

In summary, the function $\phi(\tau)$ is described by *either* equation (8-3) or (8-4), while equation (8-3) additionally yields the dimensionless heat flux Q. As pointed out by Meghreblian (ref. 2), $\phi(\tau)$ possesses the property of antisymmetry about $\tau/\tau_0 = \frac{1}{2}$. A number of methods of solving either equation (8-3) or (8-4) are available in the literature, and several of these will now be discussed.

---

* In ref. 1, $\phi(\tau)$ is actually defined as $\dfrac{\sigma T^4(\tau) - B_1}{B_2 - B_1}$.

Exponential kernel approximation. A commonly employed approximation in radiative transfer problems involves replacing the exponential integral $E_n(t)$ by an exponential function. This procedure is analogous to that described in Section 3-5. For example, if we consider equation (8-3) to be the governing integral equation for the present problem, then the kernel function $E_2(t)$ will be approximated by

$$E_2(t) \simeq ae^{-bt}$$

A certain degree of arbitrariness is involved in evaluating the constants a and b, and following Lick (ref. 3) we will require that the areas and first moments of both the exponential and the exponential integral be equal over the range $t = 0$ to $t = \infty$. The result is

$$E_2(t) \simeq \tfrac{3}{4}e^{-3t/2} \tag{8-6}$$

The function $E_3(t)$ also appears in equation (8-3), and the exponential approximation for this follows directly from

$$E_3(t) = -\int E_2(t)\, dt \simeq \tfrac{1}{2}e^{-3t/2} \tag{8-7}$$

A comparison of $E_2(t)$ and $E_3(t)$ with their respective approximations is illustrated in Fig. 8-2.

By substituting equations (8-6) and (8-7) into equation (8-3), there results the approximate integral equation

$$Q = e^{-3\tau/2} + \tfrac{3}{2}e^{-3\tau/2}\int_0^\tau \phi(t)e^{3t/2}\, dt - \tfrac{3}{2}e^{3\tau/2}\int_\tau^{\tau_0} \phi(t)e^{-3t/2}\, dt \tag{8-8}$$

By differentiating this equation twice, the integrals are found to repeat themselves and may be eliminated, with the result that the integral equation is reduced to the differential equation

$$\frac{d\phi}{d\tau} = -\tfrac{3}{4}Q$$

Thus $$\phi = C - \tfrac{3}{4}Q\tau$$

where C is a constant of integration and Q is an additional unknown. Both constants may be evaluated through substitution back into equation (8-8), and the final results for $\phi(\tau)$ are

$$\phi(\tau) = \left(1 - \frac{Q}{2}\right) - \frac{3Q}{4}\,\tau \tag{8-9}$$

$$Q = \frac{1}{1 + (3\tau_0/4)} \tag{8-10}$$

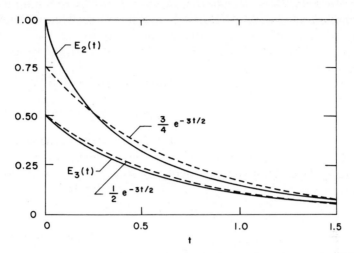

Fig. 8-2 Comparison of exponential integral and exponential functions.

For optically thick conditions ($\tau_0 \to \infty$), equations (8-5) and (8-10) yield*

$$\frac{q_R}{\sigma(T_1{}^4 - T_2{}^4)} = \frac{4}{3\tau_0} \qquad (8\text{-}11)$$

which illustrates that under optically thick conditions the heat flux becomes independent of the surface emittances as discussed in Section 7-6. Furthermore, it is easily shown that equation (8-11) is exactly the result obtained from the Rosseland equation, equation (7-42). The exponential kernel approximation therefore reduces to the correct asymptotic limit for large values of the optical thickness.

Undetermined parameters. Viskanta and Grosh (ref. 1) have employed the method of undetermined parameters to obtain a solution of equation (8-4). This method consists of assuming a solution of the form

$$\phi(\tau) = \sum_{i=0}^{N} D_i \tau^i \qquad (8\text{-}12)$$

The parameters D_i are then determined through substituting equation (8-12) into equation (8-4) and satisfying the integral equation at N locations. The numerical results of Viskanta and Grosh for $N = 5$ will be discussed later together with values for Q, which are obtained from equation (8-3) once $\phi(\tau)$ is known.

* Note from equation (8-5) that for optically thick radiation $B_1 - B_2 = \sigma(T_1{}^4 - T_2{}^4)$ regardless of the surface emittance values.

Iterative solution.   A common means of solving integral equations is through iteration,* and the iterative form of equation (8-4) is

$$2\phi_{i+1}(\tau) = E_2(\tau) + \int_0^{\tau_0} \phi_i(\tau)E_1(|\tau - t|)\,dt \qquad (8\text{-}13)$$

An initial approximation $\phi_0(\tau)$ is chosen, and repeated iterations are then obtained from equation (8-13).

As an example, consider the choice of equation (8-9) for $\phi_0(\tau)$, and

$$\phi_0(\tau) = \left[\frac{1}{(4/3\tau_0)+1}\right]\left(\frac{2}{3\tau_0}+1-\frac{\tau}{\tau_0}\right) \qquad (8\text{-}14)$$

Substituting this into equation (8-13), we obtain

$$\phi_1(\tau) = \left[\frac{1}{1+(3\tau_0/4)}\right]\left[\frac{1}{2}+\frac{3}{4}(\tau_0 - \tau) + \frac{1}{4}E_2(\tau)\right.$$
$$\left. - \frac{1}{4}E_2(\tau_0 - \tau) - \frac{3}{8}E_3(\tau) + \frac{3}{8}E_3(\tau_0 - \tau)\right] \qquad (8\text{-}15)$$

as the first iteration.   The dimensionless heat flux Q corresponding to $\phi_1(\tau)$ may, in turn, be evaluated by substituting equation (8-15) into equation (8-3) and, since Q is a constant, letting $\tau = 0$.   This gives

$$Q = 1 - 2\int_0^{\tau_0} \phi_1(t)E_2(t)\,dt \qquad (8\text{-}16)$$

and results for Q, which will be discussed later, have been obtained by Cess and Sotak (ref. 4) through numerical integration.

A detailed discussion of the iterative method, including a proof of convergence, is given by Heaslet and Fuller (ref. 5).   They additionally employ a cubic expression for $\phi_0(\tau)$.

Numerical methods.   A discussion of numerical methods of solving integral equations is given in Section (3-5) and will not be repeated here.   Usiskin and Sparrow (ref. 6) have obtained a numerical solution of equation (8-4) employing the successive iteration method, and they give results for optical thicknesses up to $\tau_0 = 2$.

Solution in terms of tabulated functions.   Recently Heaslet and Warming (ref. 7) have illustrated that $\phi(\tau)$ and Q may be expressed in terms of functions that have previously been tabulated to a considerable degree of accuracy, and in this sense their results may be considered as exact.   Because the analysis is quite lengthy, only pertinent results will be presented here; the reader is referred to ref. 7 for more detail.

---

* This method applies only to integral equations for which the dependent variable appears outside as well as inside the integral.   For example, it would not apply to equation (8-3).

Consider a function $\Phi(\tau)$ defined by the integral equation*

$$2\Phi(\tau) = E_1(\tau) + \int_0^{\tau_0} \Phi(t)E_1(|\tau - t|)\, dt \qquad (8\text{-}17)$$

The Chandrasekhar X, Y functions are defined in terms of $\Phi(\tau)$ by

$$X(\mu, \tau_0) = 1 + \int_0^{\tau_0} \Phi(t)e^{-t/\mu}\, dt \qquad (8\text{-}18a)$$

$$Y(\mu, \tau_0) = e^{-\tau_0/\mu} + \int_0^{\tau_0} \Phi(\tau_0 - t)e^{-t/\mu}\, dt \qquad (8\text{-}18b)$$

and these functions have been calculated by Chandrasekhar, Elbert, and Franklin (refs. 8 and 9). Further quantities of interest are the nth moments of X and Y defined as

$$\alpha_n(\tau_0) = \int_0^1 X(\mu, \tau_0)\mu^n\, d\mu \qquad (8\text{-}19a)$$

$$\beta_n(\tau_0) = \int_0^1 Y(\mu, \tau_0)\mu^n\, d\mu \qquad (8\text{-}19b)$$

and it is easily verified (ref. 7) that

$$\alpha_0(\tau_0) + \beta_0(\tau_0) = 2$$

The temperature function $\phi(\tau)$ has been shown by Heaslet and Warming to be

$$\phi(\tau) = \tfrac{1}{2} + \tfrac{1}{2}\beta_0(\tau_0)\left[\int_0^{\tau_0 - \tau} \Phi(t)\, dt - \int_0^\tau \Phi(t)\, dt\right] \qquad (8\text{-}20)$$

and their results are presented in Fig. 8-3. The limiting curves for $\tau_0 = 0$ and $\tau_0 = \infty$ correspond to the optically thin and optically thick limits, respectively. In addition, the wall values are

$$\phi(0) = \frac{\alpha_0(\tau_0)}{2} \qquad \phi(\tau_0) = \frac{\beta_0(\tau_0)}{2} \qquad (8\text{-}21)$$

while the dimensionless heat flux Q is given by

$$Q = \beta_0(\tau_0)[\alpha_1(\tau_0) + \beta_1(\tau_0)] \qquad (8\text{-}22)$$

Table 8-1 gives values of α_0, β_0, $\alpha_1 + \beta_1$, and Q taken directly from ref. 7, and for $\tau_0 > 3$ the asymptotic expressions are applicable. The present results serve the important function of a standard against which approximate methods may be compared, and such comparisons will be made later.

* Note that this equation differs from equation (8-4) in that the nonhomogeneous term is $E_1(\tau)$ rather than $E_2(\tau)$. The function $\Phi(\tau)$ has been studied extensively by the astrophysicists.

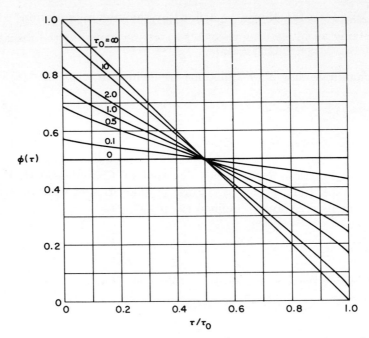

Fig. 8-3   The temperature function $\phi(\tau)$.   From Heaslet and Warming (ref. 7).

Table 8-1   Radiation Functions from Heaslet and Warming (ref. 7)

τ_0	α_0	β_0	$\alpha_1 + \beta_1$	Q
0.1	1.1419	0.8581	1.0672	0.9157
0.2	1.2228	0.7772	1.0926	0.8491
0.3	1.2838	0.7162	1.1080	0.7934
0.4	1.3331	0.6669	1.1185	0.7458
0.5	1.3746	0.6254	1.1259	0.7040
0.6	1.4103	0.5897	1.1316	0.6672
0.8	1.4692	0.5308	1.1392	0.6046
1.0	1.5163	0.4837	1.1440	0.5532
1.5	1.6024	0.3976	1.1501	0.4572
2.0	1.6615	0.3385	1.1525	0.3900
2.5	1.7051	0.2949	1.1538	0.3401
3.0	1.7386	0.2614	1.1542	0.3016
$\tau_0 \gg 1$	$2 - \dfrac{2/\sqrt{3}}{\tau_0 + \gamma}$	$\dfrac{2/\sqrt{3}}{\tau_0 + \gamma}$	$\dfrac{2}{\sqrt{3}}$	$\dfrac{4/3}{\tau_0 + \gamma}$

Note: $\gamma = 1.42089$.

Radiation slip. Unlike the previous methods, which are concerned with solving an integral equation, the starting point for the radiation slip solution is the Rosseland equation

$$q_R = -\frac{4\sigma}{3}\frac{dT^4}{d\tau}$$

and since q_R is constant,

$$q_R = -\frac{4\sigma}{3\tau_0}[T^4(\tau_0) - T^4(0)] \tag{8-23}$$

Now, following Deissler (ref. 10), the temperature jumps are given by equations (7-48) and (7-49) as

$$T^4(0) - T_1{}^4 = -\frac{1}{\sigma}\left(\frac{1}{\epsilon_1} - \frac{1}{2}\right)q_R$$

$$T^4(\tau_0) - T_2{}^4 = \frac{1}{\sigma}\left(\frac{1}{\epsilon_2} - \frac{1}{2}\right)q_R$$

and combining these with equation (8-23), the radiation heat transfer is expressed by

$$\frac{q_R}{\sigma(T_1{}^4 - T_2{}^4)} = \frac{1}{(3\tau_0/4) + (1/\epsilon_1) + (1/\epsilon_2) - 1} \tag{8-24}$$

For $\tau_0 \to \infty$ this reduces to the optically thick result given by equation (8-11) as should be expected. In addition, equation (8-24) can also be obtained through a combination of equation (8-5) with equation (8-10), which is the exponential kernel solution; consequently the radiation slip and the exponential kernel solutions give identical results for the present problem.

Results and discussion. To facilitate a comparison of the preceding methods, Table 8-2 summarizes the various solutions that have been

Table 8-2 Summary of Solutions for Radiative Equilibrium

Reference	Method
Lick (ref. 3)	Exponential kernel approximation
Viskanta and Grosh (ref. 1)	Undetermined parameters, $N = 5$
Cess and Sotak (ref. 4)	Single iteration of exponential kernel solution
Usiskin and Sparrow (ref. 6)	Numerical solution by successive iterations
Heaslet and Warming (ref. 7)	Exact
Deissler (ref. 10)	Radiation slip

described, together with the appropriate references.   Consider first the temperature profile within the medium, which is represented as a dimensionless T^4 profile by $\phi(\tau)$.   Figure 8-4 gives a comparison of the exponential kernel and exact solutions, and the maximum discrepancy is seen to occur at the walls and for small τ_0 values.   For all practical purposes the $\phi(\tau)$ results of refs. 1, 4, and 6 are comparable with the exact results of ref. 7.   The discrepancies are, at most, of the order of 1 or 2 percent.

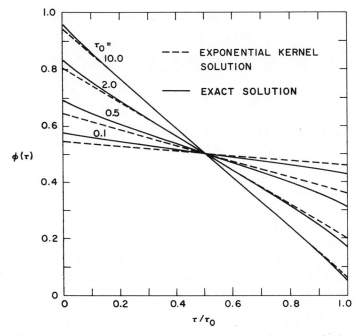

Fig. 8-4   Comparison of exponential kernel and exact solutions.

Comparison of the dimensionless radiation flux Q as given by the exponential kernel and exact solutions is illustrated in Fig. 8-5, where the exponential kernel result is given by equation (8-10).   The agreement is quite satisfactory, with the maximum difference being 3.5 percent.   To further illustrate the close agreement of the other solutions with the exact results, Table 8-3 gives an additional comparison for Q. The discrepancies are indeed quite small, with the maximum departure from exact results occurring at large τ_0.

Recall now that the radiation slip solution is identical to the exponential kernel solution, and the present comparisons would tend to

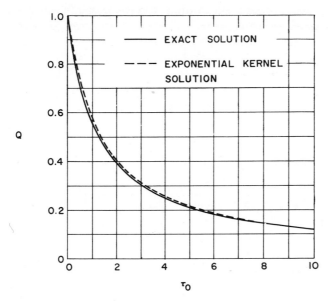

Fig. 8-5   Comparison of exponential kernel and exact solutions.

indicate that the radiation slip method is a reasonable approximation over all values of optical thickness.   In actuality, the present problem does *not* serve as a representative example; the apparent applicability of the radiation slip solution can be traced back to the nearly linear $\phi(\tau)$ profiles.   As will be illustrated in Chapter 10, and as previously discussed in Section 7-7, the radiation slip method is generally applicable only for moderately large values of optical thickness.

As a final point, recall that nowhere in the preceding analyses has the extinction coefficient been assumed to be independent of temperature.   The results as presented are valid for any variation of the extinction coefficient with temperature, and a given variation $\beta(T)$ need be utilized only in the evaluation of the optical thickness τ_0.   If the extinction coefficient is independent of temperature, then $\tau_0 = \beta L$;

Table 8-3   Comparison of Values of Q

τ_0	Heaslet and Warming (ref. 7)	Viskanta and Grosh (refs. 1 and 11)	Cess and Sotak (ref. 4)
0.1	0.9157	0.916	0.916
1.0	0.5532	0.553	0.554
10	0.1167	0.109	0.111

whereas if β is temperature dependent, the relationship between L and τ_0 may be expressed as

$$L = \int_0^{\tau_0} \frac{d\tau}{\beta(\tau)} \tag{8-25}$$

The conversion of $\beta(T)$ to $\beta(\tau)$ is accomplished through use of the known $\phi(\tau)$ profiles, which yield $T(\tau)$.

8-2   Specularly Reflecting Surfaces

In this section we will consider the same problem that has been previously treated with the one exception that specular rather than diffuse surface reflection is assumed.   The restriction to diffuse surface emission will be retained.   From Cess and Sotak (ref. 4) the general problem of two specularly reflecting surfaces involves the function

$$\int_0^1 \frac{e^{-t/\mu}}{1 - \rho_1\rho_2 e^{-2\tau_0/\mu}} \mu^{n-2} \, d\mu$$

This reduces to the exponential integral $E_n(t)$ only when one of the two surfaces is black.   For this reason attention will be directed to the problem of one black surface and one specularly reflecting surface. The physical model is illustrated in Fig. 8-6.   The lower surface is a diffuse emitter and specular reflector, while the upper surface is black.

It should be noted that, in the absence of a radiation-participating medium, the net heat transfer between plates will be independent of the assumption that the lower surface is either a diffuse or specular reflector (for example, see ref. 12).   Thus the present physical model is one that isolates the effect of surface reflection characteristics on radiation transfer within the medium.

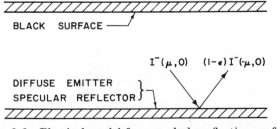

Fig. 8-6   Physical model for specularly reflecting surface.

Turning now to the equation describing the radiation flux q_R, and noting that

$$2\pi \int_0^1 I^-(\tau_0, -\mu)e^{-(\tau_0 - \tau)/\mu}\mu\,d\mu = 2\sigma T_2{}^4 E_3(\tau_0 - \tau)$$

equations (7-20) and (7-24) yield

$$q_R = 2\pi \int_0^1 I^+(0, \mu)e^{-\tau/\mu}\mu\,d\mu - 2\sigma T_2{}^4 E_3(\tau_0 - \tau)$$

$$+ 2\sigma \int_0^\tau T^4(t)E_2(\tau - t)\,dt - 2\sigma \int_\tau^{\tau_0} T^4(t)E_2(t - \tau)\,dt \quad (8\text{-}26)$$

With reference to Fig. 8-6, the surface intensity $I^+(0, \mu)$ for specular reflection becomes

$$I^+(0, \mu) = \frac{\epsilon\sigma T_1{}^4}{\pi} + (1 - \epsilon)I^-(0, -\mu) \quad (8\text{-}27)$$

where the first term represents the diffusely emitted energy and the second term is the contribution due to specular reflection. The emittance of the lower surface is denoted by ϵ. Furthermore, from equation (7-14b)

$$I^-(0, -\mu) = \frac{\sigma T_2{}^4}{\pi} e^{-\tau_0/\mu} + \frac{\sigma}{\pi} \int_0^{\tau_0} T^4(t)e^{-t/\mu} \frac{dt}{\mu} \quad (8\text{-}28)$$

and by combining equations (8-26), (8-27), and (8-28), the radiation flux becomes

$$q_R = 2\epsilon\sigma T_1{}^4 E_3(\tau) - 2\sigma T_2{}^4 E_3(\tau_0 - \tau) + 2\sigma \int_0^\tau T^4(t)E_2(\tau - t)\,dt$$

$$- 2\sigma \int_\tau^{\tau_0} T^4(t)E_2(t - \tau)\,dt + 2\sigma(1 - \epsilon)\left[T_2{}^4 E_3(\tau + \tau_0) \right.$$

$$\left. + \int_0^{\tau_0} T^4(t)E_2(t + \tau)\,dt \right] \quad (8\text{-}29)$$

For comparative purposes it will be convenient to express equation (8-29) in terms of the same dimensionless quantities previously considered; that is,

$$\phi(\tau) = \frac{\sigma T^4(\tau) - B_2}{B_1 - B_2} \qquad Q = \frac{q_R}{B_1 - B_2}$$

Consequently, it is first necessary to obtain an expression for the radiosity B_1, and since*

$$B_1 = 2\pi \int_0^1 I^+(0, \mu)\mu \, d\mu$$

then, from equations (8-27) and (8-28),

$$B_1 = \epsilon\sigma T_1{}^4 + 2\sigma(1 - \epsilon)\left[T_2{}^4 E_3(\tau_0) + \int_0^{\tau_0} T^4(t)E_2(t) \, dt \right] \quad (8\text{-}30)$$

With reference to equation (7-31), this relation is seen to be identical to that for diffuse reflection. The actual values of B_1 will differ for specular or diffuse reflection, however, due to differences in the $T^4(\tau)$ profiles.

One may now express equation (8-29) in the dimensionless form

$$Q = 2E_3(\tau) + 2\int_0^\tau \phi(t)E_2(\tau - t) \, dt - 2\int_\tau^{\tau_0} \phi(t)E_2(t - \tau) \, dt$$

$$+ 2(1 - \epsilon)\int_0^{\tau_0} \phi(t)[E_2(t + \tau) - 2E_3(\tau)E_2(t)] \, dt \quad (8\text{-}31)$$

and upon differentiating

$$2\phi(\tau) = E_2(\tau) + \int_0^{\tau_0} \phi(t)E_1(|\tau - t|) \, dt$$

$$+ (1 - \epsilon)\int_0^{\tau_0} \phi(t)[E_1(t + \tau) - 2E_2(\tau)E_2(t)] \, dt \quad (8\text{-}32)$$

Equations (8-31) and (8-32) constitute the specular reflection counterparts of equations (8-3) and (8-4), and comparison illustrates that for specular reflection the surface emittance ϵ is introduced as a parameter. It may also be shown from equation (8-32) that $\phi(\tau)$ is not antisymmetric as it is for diffuse reflection.

A particularly meaningful approach to the present problem is the exponential kernel approximation. Upon substituting equations (8-6) and (8-7) into equation (8-31), it is seen that the last integral vanishes; this is the term that makes equation (8-31) differ from the diffuse equation (8-3). Hence, within the framework of the exponential kernel approximation, diffuse reflection and specular reflection yield identical results, and these are given by equations (8-9) and (8-10).

A second approximation may be obtained by the iterative method applied to equation (8-32). Identical to the procedure described in the previous section, this method consists of employing equation (8-14) for

---

* Conventionally, the radiosity is used only for diffuse surfaces. However, it is convenient to employ it here in order to facilitate comparison with diffuse results.

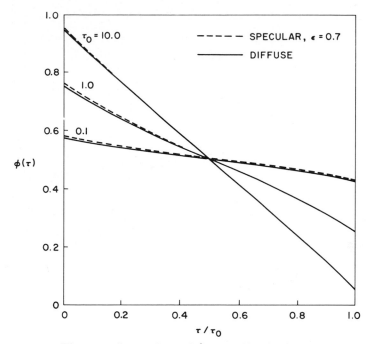

Fig. 8-7   Comparison of $\phi(\tau)$ profiles for $\epsilon = 0.7$.

$\phi_0(\tau)$. The results, taken from ref. 4, are illustrated in Figs. 8-7 and 8-8 for $\epsilon = 0.7$ and 0.3, respectively. The comparable diffuse results as given by equation (8-15) are also shown.

It may be observed that the difference between the specular and the diffuse curves is quite small. It is slightly greater for $\epsilon = 0.3$ than for $\epsilon = 0.7$, since reflection plays a greater role at the lower emittance value. One may also note that the difference decreases as τ_0 becomes large. The reason is that, for large τ_0, radiation transfer within the medium approaches the optically thick limit, thus becoming independent of the reflection characteristics of the bounding surfaces.

A comparison of net heat transfer is illustrated in Table 8-4 (from

Table 8-4   Comparison of Values of Q

		Specular	
τ_0	Diffuse	$\epsilon = 0.7$	$\epsilon = 0.3$
0.1	0.916	0.915	0.914
1	0.554	0.553	0.549
10	0.111	0.109	0.107

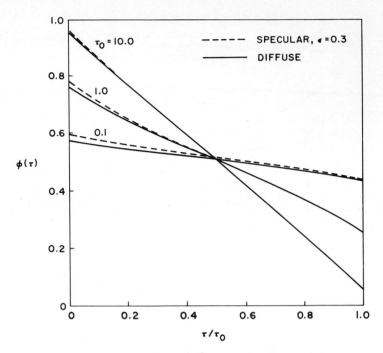

Fig. 8-8   Comparison of $\phi(\tau)$ profiles for $\epsilon = 0.3$.

ref. 4).   As would be expected from the close agreement of the $\phi(\tau)$ profiles, there is no appreciable difference between the specular and the diffuse reflection results for Q.

REFERENCES

1. R. Viskanta and R. J. Grosh, Heat transfer in a thermal radiation absorbing and scattering medium.   International Heat Transfer Conference, Boulder, Colorado, 1961, p. 820.

2. R. V. Meghreblian, An approximate analytic solution for radiation exchange between two flat surfaces separated by an absorbing gas.   *Intern. J. Heat Mass Transfer*, 5, 1051–1052 (1962).

3. W. Lick, Energy transfer by radiation and conduction.   Proceedings of the 1963 Heat Transfer and Fluid Mechanics Institute, Stanford University Press, Palo Alto, Calif., 1963, pp. 14–26.

4. R. D. Cess and A. E. Sotak, Radiation heat transfer in an absorbing medium bounded by a specular reflector.   *ZAMP*, 15, 642–647 (1964).

5. M. A. Heaslet and F. B. Fuller, Approximate predictions of the transport of

thermal radiation through an absorbing plane layer. *NASA TN D-2515,* November 1964.

6. C. M. Usiskin and E. M. Sparrow, Thermal radiation between parallel plates separated by an absorbing-emitting nonisothermal gas. *Intern. J. Heat Mass Transfer,* **1,** 28–36 (1960).

7. M. A. Heaslet and R. F. Warming, Radiative transport and wall temperature slip in an absorbing planar medium. *Intern. J Heat Mass Transfer,* **8,** 979–994 (1965).

8. S. Chandrasekhar, D. Elbert, and A. Franklin, The *X*- and *Y*-functions for isotropic scattering, I. *Astrophys. J.,* **115,** 244–268 (1952).

9. S. Chandrasekhar, D. Elbert, and A. Franklin, The *X*- and *Y*-functions for isotropic scattering, II. *Astrophys. J.,* **115,** 269–278 (1952).

10. R. G. Deissler, Diffusion approximation for thermal radiation in gases with jump boundary conditions. *J. Heat Transfer,* **C86,** 240–246 (1964).

11. R. Viskanta and R. J. Grosh, Effect of surface emissivity on heat transfer by simultaneous conduction and radiation. *Intern. J. Heat Mass Transfer,* **5,** 729–734 (1962).

12. E. R. G. Eckert and R. M. Drake, Jr., *Heat and Mass Transfer,* McGraw-Hill, New York, 1959, pp. 403–406.

PROBLEMS

8-1. Employing the exponential kernel approximation as given in Chapter 10 by equations (10-30), obtain the corresponding expression for Q for radiative equilibrium in a gray medium bounded by parallel gray plates. Compare this result with both equation (8-10) and the exact values given in Table 8-1.

8-2. Employ the method of undetermined parameters to solve equation (8-4) by assuming a solution of the form

$$\phi(\tau) = \tfrac{1}{2}(1 - D_1\tau_0) + D_1\tau$$

This satisfies the antisymmetry condition that $\phi = \tfrac{1}{2}$ for $\tau = \tau_0/2$. Evaluate the constant D_1 by requiring that equation (8-4) be satisfied at $\tau = 0$. Obtain the corresponding expression for Q, and compare this with the exact values given in Table 8-1.

8-3. Consider a gray nonscattering medium bounded by two parallel black plates having the same temperature T_1. Within the medium there is a uniform heat source per unit volume Q. Employing the exponential kernel approximation given by equations (8-6) and (8-7), show that the temperature profile within the medium is given by

$$\frac{\sigma T^4(\tau) - \sigma T_1{}^4}{Q/\kappa} = \frac{\tau_0}{4} + \frac{1}{3} + \frac{3}{8}(\tau_0\tau - \tau^2)$$

8-4. For Problem 8-3 obtain the optically thin and optically thick solutions for the temperature profile. Compare these with the limiting forms of the exponential kernel solution given above, and explain the discrepancy which exists in the optically thin limit.

8-5. Consider again Problem 8-3, except let the bounding surfaces be gray and diffuse, with the two surfaces having the same emissivity ϵ. Show that the temperature profile given in Problem 8-3 applies to the present situation providing $\sigma T_1{}^4$ is replaced by B_1, where B_1 is the radiosity for both of the surfaces. Show further that the equation defining B_1 is

$$\frac{B_1 - \sigma T_1{}^4}{Q/\kappa} = \frac{\tau_0}{2}\left(\frac{1}{\epsilon} - 1\right)$$

8-6. Solve Problem 8-5 using both the differential approximation method (Section 7-9) and the radiation slip method (Section 7-7). Compare these solutions with the preceding solution.

8-7. Resolve Problem 8-3 for the case in which the medium scatters as well as absorbs and emits radiation. Hint: See Problem 7-8

CHAPTER 9

COMBINED CONDUCTION AND RADIATION

The situation in which conduction heat transfer takes place simultaneously with radiation transfer will now be considered; that is, the medium conducts as well as absorbs, emits, and scatters thermal energy. The neglect of convection implies that, if the medium is in a fluid state, internal motions of the medium be absent. Steady state conditions are additionally assumed. As discussed in Chapter 7, the total heat flux vector within the medium will consist of the sum of the conduction and radiation contributions and is given by

$$\mathbf{q} = -k \operatorname{grad} T + \mathbf{q}_R \qquad (9\text{-}1)$$

Furthermore, in the absence of convection

$$\operatorname{div} \mathbf{q} = 0$$

then

$$\operatorname{div} (k \operatorname{grad} T) = \operatorname{div} \mathbf{q}_R \qquad (9\text{-}2)$$

and this simply constitutes the appropriate form of the energy equation, equation (7-4), for the problem at hand.

In the present chapter attention will primarily be directed toward the mutual interaction of the two separate modes of energy transfer; consequently, a very simple physical model will be considered. Specifically, it will be assumed that the medium is gray, that the absorption coefficient, the scattering coefficient, and thermal conductivity of the medium are independent of temperature, and that the bounding surfaces are gray and diffuse. In addition, restriction is made to one-dimensional energy transfer through a medium bounded by two infinite parallel surfaces.

Section 9-1 is concerned with the general formulation (within the confines of the foregoing assumptions) of the equations describing combined conduction and radiation transfer through media that absorb, emit, and scatter thermal radiation, while Section 9-2 deals with solutions for optically thin, optically thick, and purely scattering media. In Section 9-3 the exponential kernel approximation is employed, and the results of this method are discussed in Section 9-4 together with available numerical solutions.

9-1 General Formulation

The physical model and the coordinate system for the present problem are illustrated in Fig. 9-1. With the previously discussed assumptions, the equation expressing conservation of energy, equation (9-2), reduces to

$$k \frac{d^2 T}{dy^2} = \frac{dq_R}{dy} \tag{9-3}$$

This is equivalent to

$$q = -k \frac{dT}{dy} + q_R = \text{constant} \tag{9-4}$$

as is physically obvious.

Now, for a gray, constant-property medium with gray and diffuse bounding surfaces, the applicable radiation equations follow from equations (7-20), (7-21), (7-22), and (7-28):

$$q_R(\tau) = 2B_1 E_3(\tau) - 2B_2 E_3(\tau_0 - \tau)$$

$$+ \frac{2\kappa}{\beta} \int_0^\tau \left[\sigma T^4(t) + \frac{\gamma}{4\kappa} G(t) \right] E_2(\tau - t)\, dt$$

$$- \frac{2\kappa}{\beta} \int_\tau^{\tau_0} \left[\sigma T^4(t) + \frac{\gamma}{4\kappa} G(t) \right] E_2(t - \tau)\, dt \tag{9-5}$$

$$-\frac{dq_R}{d\tau} = 2B_1 E_2(\tau) + 2B_2 E_2(\tau_0 - \tau)$$

$$+ \frac{2\kappa}{\beta} \int_0^{\tau_0} \left[\sigma T^4(t) + \frac{\gamma}{4\kappa} G(t) \right] E_1(|\tau - t|)\, dt$$

$$- 4 \frac{\kappa\sigma}{\beta} T^4(\tau) - \frac{\gamma}{\beta} G(\tau) \tag{9-6}$$

$$G(\tau) = 2B_1 E_2(\tau) + 2B_2 E_2(\tau_0 - \tau)$$

$$+ \frac{2\kappa}{\beta} \int_0^{\tau_0} \left[\sigma T^4(t) + \frac{\gamma}{4\kappa} G(t) \right] E_1(|\tau - t|)\, dt \tag{9-7}$$

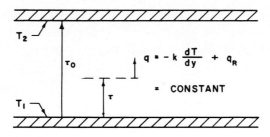

Fig. 9-1   One-dimensional combined conduction and radiation.

while equations (7-31) and (7-32), applied to a gray medium, describe the radiosities B_1 and B_2.

Recalling that $\tau = \beta y$ and $\tau_0 = \beta L$, equations (9-3), (9-6), and (9-7) may be combined to yield the result

$$\frac{d^2 T}{d\tau^2} = \frac{(1 - \gamma/\beta)}{k\beta}[4\sigma T^4(\tau) - G(\tau)] \tag{9-8}$$

and equations (9-7) and (9-8) constitute the two equations describing $T(\tau)$ and $G(\tau)$.   Before proceeding, it will be convenient to recast these equations in dimensionless form, and following Viskanta (ref. 1) dimensionless quantities will be defined as*

$$\theta(\tau) = \frac{T(\tau)}{T_1}, \qquad \theta_2 = \frac{T_2}{T_1}, \qquad \eta(\tau) = \frac{G(\tau)}{\sigma T_1{}^4}$$

$$N = \frac{k\beta}{4\sigma T_1{}^3}, \qquad \omega_0 = \frac{\gamma}{\beta}, \qquad X = \frac{B}{\sigma T_1{}^4}$$

With the above definitions, equations (9-8) and (9-7) reduce, respectively, to

$$N\frac{d^2\theta}{d\tau^2} = (1 - \omega_0)\left[\theta^4(\tau) - \frac{1}{4}\eta(\tau)\right] \tag{9-9}$$

$$\eta(\tau) = 2X_1 E_2(\tau) + 2X_2 E_2(\tau_0 - \tau)$$

$$+ 2\int_0^{\tau_0}\left[(1 - \omega_0)\theta^4(t) + \frac{\omega_0}{4}\eta(t)\right]E_1(|\tau - t|)\, dt \tag{9-10}$$

* The definition of N employed here is the reciprocal of that used in ref. 1.

while from the radiosity equations, equations (7-31) and (7-32), we have

$$X_1 = \epsilon_1 + 2(1 - \epsilon_1)\Big\{ X_2 E_3(\tau_0)$$

$$+ \int_0^{\tau_0} \Big[(1 - \omega_0)\theta^4(t) + \frac{\omega_0}{4}\eta(t) \Big] E_2(t)\, dt \Big\} \qquad (9\text{-}11)$$

$$X_2 = \epsilon_2 \theta_2{}^4 + 2(1 - \epsilon_2)\Big\{ X_1 E_3(\tau_0)$$

$$+ \int_0^{\tau_0} \Big[(1 - \omega_0)\theta^4(t) + \frac{\omega_0}{4}\eta(t) \Big] E_2(\tau_0 - t)\, dt \Big\} \qquad (9\text{-}12)$$

The preceding system of four equations describes the dimensionless quantities $\theta(\tau)$, $\eta(\tau)$, X_1, and X_2. One may observe from equation (9-9) that, in addition, two boundary conditions are required. Since the presence of conduction heat transfer necessitates continuity of temperature at the surfaces, these boundary conditions are

$$\theta(0) = 1 \qquad \theta(\tau_0) = \theta_2 \qquad\qquad (9\text{-}13)$$

Numerical solutions of equations (9-9) through (9-13) have been obtained by Viskanta (ref. 1), and the corresponding results will be discussed in Section 9-4.

Consider now the six parameters entering into the above system of equations: τ_0, ϵ_1, ϵ_2, θ_2, N, and ω_0. Of these, only the last three require amplification. The temperature ratio $\theta_2 = T_2/T_1$, which did not arise in the analysis of Section 8-1 for radiative equilibrium of a gray medium, occurs, for the present equations are nonlinear. The relative role of heat transfer by conduction to that by radiation is indicated by the parameter $N = k\beta/4\sigma T_1{}^3$. For $N = \infty$, heat transfer *within* the medium is only by conduction. This is readily verified from equation (9-9), which reduces to the energy equation for purely conduction heat transfer when $N = \infty$. The opposite extreme of $N = 0$ corresponds to the case in which heat transfer is solely due to radiation. Under this condition, the left side of equation (9-9), which is the term representing conduction heat transfer, vanishes; the resulting equations can be reduced to those of radiative equilibrium given in Section 8-1.

The final parameter $\omega_0 = \gamma/\beta$ is called the *albedo of scattering* and represents the fraction of attenuated energy that is the result of scattering. The limit $\omega_0 = 0$ denotes a nonscattering medium, and for this case equations (9-9) and (9-10) reduce to the single equation

$$N \frac{d^2\theta}{d\tau^2} = \theta^4(\tau) - \frac{\chi_1}{2} E_2(\tau) - \frac{\chi_2}{2} E_2(\tau_0 - \tau)$$

$$- \frac{1}{2} \int_0^{\tau_0} \theta^4(t) E_1(|\tau - t|) \, dt \quad (9\text{-}14)$$

As discussed in Section 7-4, the dimensionless incident radiation function $\eta(\tau)$ becomes redundant for nonscattering media, and for $\omega_0 = 0$ equation (9-14), together with equations (9-11), (9-12), and (9-13), completely describes the combined conduction-radiation problem. The case of pure scattering ($\omega_0 = 1$) will be treated in the following section.

9-2 Limiting Solutions

From the preceding section it is evident that the problem of combined conduction and radiation is a rather difficult one, particularly because of the large number of governing parameters. Consequently, before proceeding with a discussion of either numerical solutions of the governing equations or approximate methods of solution, it will be advantageous to describe three limiting cases that possess very simple solutions. These consist of the optically thin and optically thick radiation regimes and the case of a medium that scatters but does not absorb nor emit thermal radiation (pure scattering).

Optically thin limit. For the case of an optically thin medium, the radiation heat transfer is found from equations (7-34) and (7-35) to be

$$q_R = \frac{\sigma(T_1^4 - T_2^4)}{(1/\epsilon_1) + (1/\epsilon_2) - 1} \quad (9\text{-}15)$$

Again, this is the result for radiation transfer through a nonparticipating medium. Since both q and q_R are independent of position, equation (9-4) may be integrated directly to give

$$q = \frac{k(T_1 - T_2)}{L} + q_R \quad (9\text{-}16)$$

This result simply states that the total heat flux is the sum of the separate and independent conduction and radiation contributions. Furthermore, the total flux q is independent of the albedo of scattering, for equation (9-16) does not depend on the radiative properties of the medium.

Perhaps a word of caution should be injected at this point. The simple manner in which equation (9-16) has been obtained is *not* representative of all optically thin problems but is a direct consequence of the fact that q is constant. When q is not constant, such as in problems involving convection heat transfer, one must deal with the nonlinear energy equation which, for optically thin conditions, is a differential equation.

Optically thick limit. The radiation flux for a gray gas under optically thick conditions follows from equation (7-42):

$$q_R = -\frac{4}{3\beta}\frac{d(\sigma T^4)}{dy} = -\frac{16\sigma T^3}{3\beta}\frac{dT}{dy}$$

Combining this with equation (9-4),

$$q = -\left(k + \frac{16\sigma T^3}{3\beta}\right)\frac{dT}{dy}$$

By direct integration one obtains

$$q = \frac{k}{L}(T_1 - T_2) + \frac{4\sigma}{3\beta L}(T_1^4 - T_2^4) \qquad (9\text{-}17)$$

or, in dimensionless form,

$$\frac{q}{\sigma T_1^4}\tau_0 = 4N(1 - \theta_2) + \frac{4}{3}(1 - \theta_2^4) \qquad (9\text{-}18)$$

It is easily observed from either of these equations that the total heat flux is represented as the sum of heat transfer by pure conduction and heat transfer by pure radiation. This is the same type of result previously obtained for optically thin radiation. Hence, in either the optically thin or optically thick limits, the total heat transfer corresponds to the superposition of the two separate modes of heat transfer as if each were occurring independently of the other.

Equation (9-18) further illustrates that neither the albedo of scattering ω_0 nor the surface emittances ϵ_1 and ϵ_2 appear as parameters in the present limiting case. As discussed in Section 7-6, this is a result of the fact that the scattering coefficient appears only through the extinction coefficient β for optically thick radiation, and that optically thick radiation is independent of the surface emittances. It should again be noted that the albedo of scattering is also absent under optically thin conditions.

Pure scattering. From Section 7-4 it may be recalled that for pure scattering ($\omega_0 = 1$) the energy equation is uncoupled from the radiation transfer process, and, in turn, the radiation transfer is independent of the temperature field within the medium. Thus once again we have a situation where the total heat transfer may be evaluated through superposition of the two separate contributions. The conduction heat transfer is

$$q_C = \frac{k(T_1 - T_2)}{L} \tag{9-19}$$

while, with $\tau_0 = \gamma L$ for pure scattering, the radiation transfer is expressed by

$$q_R = (B_1 - B_2)Q(\tau_0) \tag{9-20}$$

where $Q(\tau_0)$ is the radiative equilibrium solution given in Table 8-1 or Fig. 8-5. The radiosity difference $B_1 - B_2$ is again obtained from equation (8-5).

9-3 Exponential Kernel Solution

As in the case of radiative equilibrium, the exponential kernel approximation offers a convenient means of attacking the present problem of combined conduction and radiation. This procedure has been employed by Lick (ref. 2) for nonscattering media, and his analysis will be described in this section.

In accordance with Section 8-1, the exponential kernel approximation does not deal directly with the energy equation; instead it is concerned with the total heat flux equation, which is the first integral of the energy equation. Upon combining equations (9-4) and (9-5), setting $\omega_0 = 0$ (for scattering is not now being considered), and putting the result in dimensionless form, the total heat flux is given by

$$\frac{q}{\sigma T_1^{\,4}} = -4N \frac{d\theta}{d\tau} + 2X_1 E_3(\tau) - 2X_2 E_3(\tau_0 - \tau)$$

$$+ 2 \int_0^\tau \theta^4(t) E_2(\tau - t)\, dt - 2 \int_\tau^{\tau_0} \theta^4(t) E_2(t - \tau)\, dt \tag{9-21}$$

Now, by employing the exponential kernel approximations for $E_2(t)$ and $E_3(t)$ given by equations (8-6) and (8-7), we have the approximate total flux equation

$$\psi = -4N \frac{d\theta}{d\tau} + X_1 e^{-\frac{3}{2}\tau} - X_2 e^{-\frac{3}{2}\tau_0} e^{\frac{3}{2}\tau} + \frac{3}{2} e^{-\frac{3}{2}\tau} \int_0^\tau \theta^4(t) e^{\frac{3}{2}t}\, dt$$

$$- \frac{3}{2} e^{\frac{3}{2}\tau} \int_\tau^{\tau_0} \theta^4(t) e^{-\frac{3}{2}t}\, dt \tag{9-22}$$

where

$$\psi = \frac{q}{\sigma T_1^{\,4}} = \text{constant}$$

The integrals in equation (9-22) may be eliminated by differentiating twice. The resulting differential equation is

$$4N \frac{d^3\theta}{d\tau^3} - 9N \frac{d\theta}{d\tau} - 3 \frac{d\theta^4}{d\tau} = \frac{9}{4} \psi$$

and this may readily be integrated once to yield

$$4N \frac{d^2\theta}{d\tau^2} - 9N\theta - 3\theta^4 = \frac{9}{4} \psi\tau - \alpha \qquad (9\text{-}23)$$

where the integration constant α and the constant ψ are as yet unknown.

In order to determine α and ψ, equation (9-23) is evaluated at the boundaries, taking cognizance of the boundary conditions given by equation (9-13), with the result that

$$\alpha = 3 + 9N - 4N\left(\frac{d^2\theta}{d\tau^2}\right)_1 \qquad (9\text{-}24)$$

$$\psi\tau_0 = \frac{4}{3}(1 - \theta_2{}^4) + 4N(1 - \theta_2) - \frac{16}{9} N\left[\left(\frac{d^2\theta}{d\tau^2}\right)_1 - \left(\frac{d^2\theta}{d\tau^2}\right)_2\right] \qquad (9\text{-}25)$$

It now remains to evaluate the two second derivatives $(d^2\theta/d\tau^2)_1$ and $(d^2\theta/d\tau^2)_2$, which may be accomplished by evaluating equation (9-22) together with its first derivative at each boundary, and eliminating the integrals from the two sets of equations. Without going into the algebraic details, the result is

$$N\left(\frac{d^2\theta}{d\tau^2}\right)_1 = -\frac{3}{4}X_1 + \frac{3}{4} + \frac{3}{8}\psi + \frac{3}{2} N\left(\frac{d\theta}{d\tau}\right)_1 \qquad (9\text{-}26)$$

$$N\left(\frac{d^2\theta}{d\tau^2}\right)_2 = -\frac{3}{4}X_2 + \frac{3}{4}\theta_2{}^4 - \frac{3}{8}\psi - \frac{3}{2} N\left(\frac{d\theta}{d\tau}\right)_2 \qquad (9\text{-}27)$$

From these expressions, together with equations (9-24) and (9-25), the constants ψ and α are found to be

$$\psi = \left(\frac{4}{\tau_0 + 4/3}\right)\left\{N\left[1 - \theta_2 - \frac{2}{3}\left(\frac{d\theta}{d\tau}\right)_1 - \frac{2}{3}\left(\frac{d\theta}{d\tau}\right)_2\right] + \frac{1}{3}(X_1 - X_2)\right\}$$
$$(9\text{-}28)$$

$$\alpha = 9N + 3X_1 - 6N\left(\frac{d\theta}{d\tau}\right)_1 - \frac{3}{2}\psi \qquad (9\text{-}29)$$

One may easily show that, for $N = 0$, equation (9-28) reduces to

$$\frac{q}{B_1 - B_2} = \frac{1}{1 + 3\tau_0/4}$$

which, as should be expected, is the exponential kernel solution for radiative equilibrium given by equation (8-10). Conversely, for $N \to \infty$, equation (9-28) reduces to that for purely conduction heat transfer.*

It now remains to evaluate the dimensionless radiosities X_1 and X_2. Upon letting $\omega_0 = 0$ in equations (9-11) and (9-12) and making the exponential kernel substitution, the radiosity equations applicable to the present problem become

$$X_1 = \epsilon_1 + (1 - \epsilon_1)\left\{X_2 e^{-\frac{3}{2}\tau_0} + \frac{3}{2}\int_0^{\tau_0} \theta^4(t)e^{-\frac{3}{2}t}\,dt\right\}$$

$$X_2 = \epsilon_2 \theta_2^4 + (1 - \epsilon_2)\left\{X_1 e^{-\frac{3}{2}\tau_0} + \frac{3}{2}e^{-\frac{3}{2}\tau_0}\int_0^{\tau_0} \theta^4(t)e^{\frac{3}{2}t}\,dt\right\}$$

The integrals appearing in these equations may be eliminated through use of equation (9-22) evaluated at each boundary, giving the final results

$$X_1 = 1 - \left(\frac{1 - \epsilon_1}{\epsilon_1}\right)\left\{\psi + 4N\left(\frac{d\theta}{d\tau}\right)_1\right\} \tag{9-30}$$

$$X_2 = \theta_2^4 + \left(\frac{1 - \epsilon_2}{\epsilon_2}\right)\left\{\psi + 4N\left(\frac{d\theta}{d\tau}\right)_2\right\} \tag{9-31}$$

The formulation of the governing equations for the exponential kernel approximation is now complete. Equation (9-23) and equations (9-28) through (9-31) describe the dimensionless quantities $\theta(\tau)$, ψ, X_1, and X_2. The boundary conditions for equation (9-23) are again given by equation (9-13).

Unfortunately, one may observe that the exponential kernel approximation has not afforded any reduction in the number of governing parameters, and, as in the general formulation of Section 9-1, these parameters are τ_0, ϵ_1, ϵ_2, θ_2, and N. The albedo of scattering is absent, for we are now concerned with the special case $\omega_0 = 0$. No general solution of equation (9-23) is obtainable, and any direct solution of the present equations would require recourse to numerical methods.

As an alternate to numerical solutions, Lick (ref. 2) has employed a combination of three approximate solutions that appear to have sufficient overlap in their ranges of applicability so as to yield adequate predictions of total heat transfer for the entire range of governing parameters. Lick's three approximate methods of attack are (1) the

---

* To accomplish this, note from equation (9-21) that

$$(d\theta/d\tau)_1 = (d\theta/d\tau)_2 = -qL/kT_1\tau_0$$

for $N \to \infty$.

optically thick model, (2) a truncated power series, and (3) a singular perturbation solution.

Consider first the optically thick solution. Since this corresponds to $\tau_0 \to \infty$, terms of order $1/\tau_0$ may be deleted. In addition, noting that $d\theta/d\tau$ and ψ are proportional to $1/\tau_0$ for large τ_0, equations (9-28), (9-30), and (9-31) reduce to the single expression, which is correct to $O(1/\tau_0)$,

$$\psi = \frac{4N}{\tau_0} (1 - \theta_2) + \frac{4}{3\tau_0} (1 - \theta_2{}^4) \tag{9-32}$$

Equation (9-32) is identical to (9-18) and illustrates the fact that the exponential kernel solution reduces to the correct optically thick limit. This was previously found to be the case in Section 8-1 for radiative equilibrium.

Turning next to the truncated power series, and letting

$$\xi = \frac{\tau}{\tau_0}$$

the dimensionless temperature $\theta(\xi)$ was expressed by Lick as

$$\theta(\xi) = a_0 + a_1\xi + a_2\xi^2 + a_3\xi^3 \tag{9-33}$$

Upon substituting this into equation (9-23), the a_n's can be evaluated. Lick found that the first two terms represent the linear temperature profile corresponding to pure conduction, while the additional terms are proportional to the dimensionless group

$$\frac{\tau_0{}^2}{N} = \frac{4\kappa\sigma T_1{}^3 L^2}{k}$$

At this point it is advantageous to digress briefly and discuss in more detail the relative role of conduction versus radiation. The parameter N may be written as

$$N = \frac{kT_1/L}{4\sigma T_1{}^4/\kappa L}$$

which clearly represents the ratio of heat transfer by conduction to that by *optically thick* radiation. On the other hand, the group $N/\tau_0{}^2$ is expressible as

$$\frac{N}{\tau_0{}^2} = \frac{kT_1/L}{4\kappa\sigma T_1{}^4 L}$$

Again the numerator corresponds to conduction heat transfer, whereas $4\kappa\sigma T_1{}^4$ denotes energy emission per unit volume, and therefore the denominator corresponds to the energy flux emitted by the one-dimensional medium. Since this does not account for self-absorption, the denominator characterizes *optically thin* radiation. Consequently, both N and N/τ_0^2 are measures of the relative role of conduction versus radiation: the former pertaining to optically thick radiation, while the latter refers to optically thin conditions.

Let us now return to the discussion of equation (9-33). Recall that the first two terms represent pure conduction, whereas the remaining terms are proportional to τ_0^2/N. It is evident that the truncated power series is applicable when conduction is the predominant mode of energy transfer; that is, for $N/\tau_0^2 \gg 1$.* With the temperature profile described by equation (9-33), the dimensionless heat flux ψ is found from equations (9-28) through (9-31).

The third and final solution obtained by Lick consists of a boundary layer (or singular perturbation) solution of equation (9-23). Again letting $\xi = \tau/\tau_0 = y/L$, while

$$\delta = \frac{N}{\tau_0^2}$$

equation (9-23) becomes

$$4\delta \frac{d^2\theta}{d\xi^2} - 9\delta\tau_0^2\theta - 3\theta^4 = \frac{9}{4}\psi\tau_0\xi - \alpha \qquad (9\text{-}34)$$

Consider now the limiting case of $\delta \ll 1$, and additionally assume $\tau_0^2 \leqslant O(1)$; that is, $N \ll 1$. A good approximation to the solution of equation (9-34) is obtained when terms with the coefficient δ are deleted, and this yields

$$\theta^4 = \tfrac{1}{3}\alpha^{(0)} - \tfrac{3}{4}\psi^{(0)}\tau_0\xi \qquad (9\text{-}35)$$

where $\alpha^{(0)}$ and $\psi^{(0)}$ denote α and ψ with $N = 0$. It is obvious that equation (9-35) cannot satisfy the boundary conditions given by equation (9-13); thus it is not a good approximation in the vicinity of the walls. The implication is that equation (9-35) is a valid approximation within the interior of the medium, while an additional solution is required near the boundaries. This solution must include conduction, and the regions near the walls may be regarded as "conduction boundary layers." Problems of this type are called singular perturbation problems. A further discussion of the present solution is given by Mueller and Malmuth (ref. 3).

---

* Actually equation (9-33) is an approximation as far as including terms of order τ_0^2/N is concerned. Tien and Abu-Romia (ref. 8) have indicated that the complete first-order solution is a sixth-degree polynomial.

Considering the application of equation (9-34) to the boundary layer adjacent to surface 2, and letting

$$\bar{\xi} = \frac{\xi}{\sqrt{\delta}}$$

then $$4\frac{d^2\theta}{d\bar{\xi}^2} - 9\delta\tau_0{}^2\theta - 3\theta^4 = \frac{9}{4}\psi\tau_0\xi - \alpha \qquad (9\text{-}36)$$

The second derivative is now of the same order of magnitude as the largest term in the rest of the equation, and since

$$\bar{\xi} = \frac{1}{\sqrt{\delta}}\frac{y}{L}$$

it is clear that the second derivative is of importance within a boundary layer whose thickness is of the order $L\sqrt{\delta} \ll L$. The second term on the left side of equation (9-36) may still be deleted, while $\xi \simeq 1$ within the boundary layer. The boundary layer equation then becomes

$$4\frac{d^2\theta}{d\bar{\xi}^2} - 3\theta^4 = \frac{9}{4}\psi^{(0)}\tau_0 - \alpha^{(0)} \qquad (9\text{-}37)$$

The constants α and ψ have been replaced by $\alpha^{(0)}$ and $\psi^{(0)}$, for the boundary layer equation must properly match the "free-stream" solution given by equation (9-35). From equations (9-28) through (9-31),

$$\psi^{(0)} = \frac{(1 - \theta_2{}^4)}{(3\tau_0/4) + (1/\epsilon_1) + (1/\epsilon_2) - 1} \qquad (9\text{-}38)$$

$$\alpha^{(0)} = 3 - 3\left(\frac{1}{\epsilon_1} - \frac{1}{2}\right)\psi^{(0)} \qquad (9\text{-}39)$$

Upon letting

$$f_2{}^4 = \tfrac{1}{3}\alpha^{(0)} - \tfrac{3}{4}\psi^{(0)}\tau_0$$

equation (9-37) reduces to

$$4\frac{d^2\theta}{d\bar{\xi}^2} - 3\theta^4 = -3f_2{}^4$$

A first integral of this equation is

$$2\left(\frac{d\theta}{d\bar{\xi}}\right)^2 = \frac{3}{5}(\theta^5 - f_2{}^5) - 3f_2{}^4(\theta - f_2)$$

where the constant of integration has been evaluated by matching the free stream and boundary layer solutions (ref. 3). Applying the preceding result at the surface gives

$$\left(\frac{d\theta}{d\tau}\right)_2 = \sqrt{\frac{3}{2N}} \left[\frac{1}{5}(\theta_2{}^5 - f_2{}^5) - f_2{}^4(\theta_2 - f_2)\right]^{\frac{1}{2}} \qquad (9\text{-}40)$$

By following an identical procedure for the boundary layer adjacent to surface 1, there results

$$\left(\frac{d\theta}{d\tau}\right)_1 = \sqrt{\frac{3}{2N}} \left[\frac{1}{5}(1 - f_1{}^5) - f_1{}^4(1 - f_1)\right]^{\frac{1}{2}} \qquad (9\text{-}41)$$

where
$$f_1{}^4 = \tfrac{1}{3}\alpha^{(0)} \qquad (9\text{-}42)$$

The total heat flux (i.e., the sum of conduction and radiation) can now be calculated directly from equations (9-28), (9-40), and (9-41).

In summary, Lick's three methods of solution are as follows:

(1)   Optically thick solution for $\tau_0 \gg 1$.

(2)   Truncated power series for $N/\tau_0{}^2 \gg 1$.

(3)   Singular perturbation solution for $N \ll 1$.

From these three methods Lick has presented total heat transfer results for the conditions $\epsilon_1 = \epsilon_2 = 1$, $\theta_2 = 0.1$ and various values of τ_0 and N. These results will be illustrated in the following section.

9-4   Results and Discussion

Nonscattering media. Figure 9-2 illustrates the total heat flux results obtained by Lick (ref. 2) through combination of his three approximate analyses. Also shown as discrete points are numerical solutions that have been obtained both by Lick and by Viskanta and Grosh (ref. 4). The dashed lines represent the asymptotic limits of pure radiation ($N = 0$) and pure conduction ($N = \infty$), and the figure thus illustrates the transition from radiation transfer to conduction transfer with increasing N.

A very interesting fact is that results virtually identical to those of Fig. 9-2 have been obtained by Probstein (ref. 5) through simple superposition. In other words, the radiation transfer is evaluated as though there were no conduction taking place, while conduction is, in turn, calculated as if there were no radiation transfer occurring. It may be recalled from Section 9-2 that such a superposition yields valid total heat flux results in the optically thin and the optically thick limits.

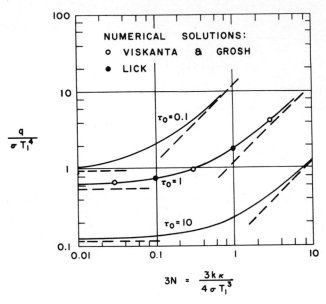

Fig. 9-2   Heat transfer results for $\epsilon_1 = \epsilon_2 = 1$ and $\theta_2 = 0.1$.
From Lick (ref. 2).

However, there is no theoretical justification for its use at intermediate values of optical thickness.

Further illustrations of the applicability of the superposition approximation for black-bounding surfaces are given by Einstein (ref. 6) and Cess (ref. 7).   However, in ref. 7 it is additionally shown that very substantial errors may result when low surface emittances are considered.   Hence the superposition approximation appears to be applicable only to high surface emittances.

Scattering media.   Results of numerical solutions of the radiation equations for combined conduction-radiation through scattering media have been presented by Viskanta (ref. 1), and a portion of his tabulated results are plotted in Figs. 9-3 to 9-5 as a function of the albedo of scattering.   It may be recalled from Section 9-2 that the albedo of scattering is not present as a parameter in either the optically thin or the optically thick limits, while Figs. 9-3 to 9-5 are concerned with the intermediate optical thickness $\tau_0 = 1$.

From Fig. 9-3 it may be seen that the total heat transfer is a fairly weak function of the albedo of scattering for $\epsilon_1 = \epsilon_2 = 1$.   The effect is more pronounced for $N = 0.1$ than for $N = 1$.   The reason is that the relative importance of radiation to conduction diminishes with

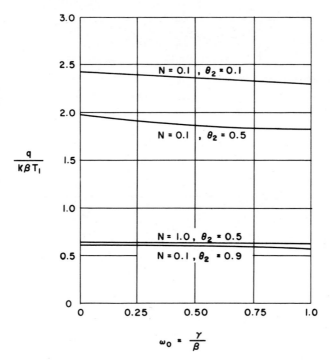

Fig. 9-3 Effect of albedo on heat transfer for $\tau_0 = 1$ and $\epsilon_1 = \epsilon_2 = 1$. From Viskanta (ref. 1).

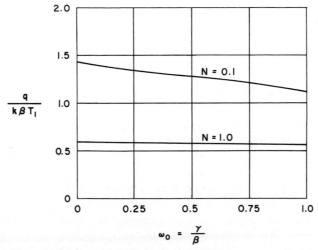

Fig. 9-4 Effect of albedo on heat transfer for $\tau_0 = 1$, $\epsilon_1 = \epsilon_2 = 0.5$, and $\theta_2 = 0.5$. From Viskanta (ref. 1).

increasing N. On the other hand, it may be recalled that the albedo of
scattering does not appear as a parameter for radiative equilibrium
($N = 0$), and consequently the maximum influence of ω_0 on total heat
transfer must occur for intermediate values of N. Unfortunately, the
results of Fig. 9-3 are not sufficient to illustrate this effect. Figures 9-4
and 9-5 show results for $\epsilon_1 = \epsilon_2 = 0.5$ and 0.1, respectively, and it may
be seen that the influence of ω_0 increases with decreasing surface
emittance.

A further interpretation of Figs. 9-3 to 9-5 concerning nonscattering
media may readily be made. In Section 9-2 it was shown that, for pure

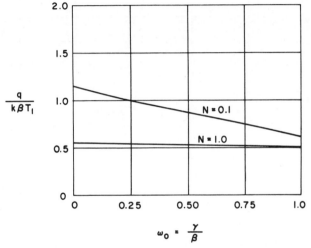

Fig. 9-5   Effect of albedo on heat transfer for $\tau_0 = 1$, $\epsilon_1 = \epsilon_2 = 0.1$,
and $\theta_2 = 0.5$. From Viskanta (ref. 1).

scattering ($\omega_0 = 1$), the total heat flux is *exactly* described by the superposition of solutions for pure conduction and pure radiation. Furthermore, since the pure radiation solution is independent of ω_0 (as is the
pure conduction solution), the results for $\omega_0 = 1$ are numerically
identical with those of the aforementioned superposition approximation
for nonscattering media. An indication of the applicability of the
superposition approximation is thus obtained from Figs. 9-3 to 9-5 by
comparing results for $\omega_0 = 0$ (numerical solution for nonscattering)
with the corresponding $\omega_0 = 1$ values (which coincide with the superposition approximation for nonscattering). As previously discussed, it
is seen that the applicability of the superposition approximation
decreases with decreasing surface emittance.

REFERENCES

1. R. Viskanta, Heat transfer by conduction and radiation in absorbing and scattering materials. *J. Heat Transfer*, **C87**, 143–150 (1965).

2. W. Lick, Energy transfer by radiation and conduction. Proceedings of the 1963 Heat Transfer and Fluid Mechanics Institute, Stanford University Press, Palo Alto, Calif., 1963, pp. 14–26.

3. H. F. Mueller and N. D. Malmuth, Temperature distributions in radiating heat shields by the method of singular perturbations. *Intern. J. Heat Mass Transfer*, **8**, 915–920 (1965).

4. R. Viskanta and R. J. Grosh, Effect of surface emissivity on heat transfer by simultaneous conduction and radiation. *Intern. J. Heat Mass Transfer*, **5**, 729–734 (1962).

5. R. F. Probstein, Radiation slip. *AIAA J.*, **1**, 1202–1204 (1963).

6. T. H. Einstein, Radiant heat transfer with flow and conduction. *NASA TR R-154* (1963).

7. R. D. Cess, The interaction of thermal radiation with conduction and convection heat transfer. In *Advances in Heat Transfer*, vol. 1, Academic Press, New York, 1964.

8. C. L. Tien and M. M. Abu-Romia, Perturbation solutions in the differential analysis of radiation interactions with conduction and convection. *AIAA J.*, **4**, 732–733 (1966).

PROBLEMS

9-1.   For combined conduction and radiation between parallel black plates, derive the differential equation which describes the temperature profile within the medium for the optically thin limit. Show that N/τ_0^2 is the appropriate parameter describing the relative role of conduction versus radiation; i.e., show that the magnitude of the conduction term in the energy equation is of the order N/τ_0^2 relative to the other terms.

9-2.   Repeat Problem 9-1 for the optically thick limit, and illustrate that in this case N is the parameter which characterizes conduction versus radiation.

9-3.   Obtain a system of equations similar in form to equation (9-23) and equations (9-28) through (9-31) by use of the differential approximation (Section 7-9).

9-4.   Evaluate the a_n's appearing in equation (9-33), and obtain the resulting expression for the dimensionless heat flux ψ.

9-5.   Show that for $N \to \infty$, equation (9-28) reduces to that for purely conduction heat transfer.

CHAPTER 10

COMBINED CONVECTION AND RADIATION

The present chapter is concerned with convection heat transfer involving radiating media, and in general this will involve three combined modes of energy transfer: conduction, convection, and radiation. It should be clear from the results of the preceding chapter that problems involving several modes of energy transfer can result in a multitude of parameters; hence, for clarity, several assumptions will initially be imposed. As in the previous case of combined conduction and radiation, a gray medium is assumed, although limited results concerning nongray gases will be presented. In addition, only black-bounding surfaces and nonscattering media are treated.

Section 10-1 deals with the boundary layer form of the energy equation applicable to flow of a radiating media. In Section 10-2 attention is directed toward flow through ducts, while Sections 10-3 and 10-4 cover external flows.

10-1 The Boundary Layer Energy Equation

Most convective heat transfer analyses involving nonradiating media employ the boundary layer form of the energy equation, and such will be the case here. Disregarding for the moment any simplifications of the radiation term appearing in equation (7-4), and assuming steady two-dimensional flow of a constant-property fluid, the boundary layer form of equation (7-4) is (ref. 1)

$$u\frac{\partial T}{\partial x} + v\frac{\partial T}{\partial y} = \alpha\frac{\partial^2 T}{\partial y^2} + \frac{\beta T u}{\rho c_p}\frac{dp}{dx} + \frac{\nu}{c_p}\left(\frac{\partial u}{\partial y}\right)^2 - \frac{1}{\rho c_p}\operatorname{div}\mathbf{q}_R \quad (10\text{-}1)$$

where u and v denote x and y components of velocity, respectively, with x the streamwise coordinate and y the transverse coordinate, and where α is the thermal diffusivity of the fluid. A further simplification ensues if the magnitude of the Eckert number is much less than one, where the Eckert number is defined as

$$\mathrm{Ec} = \frac{u_\infty^2}{c_p \, \Delta T}$$

with u_∞ and ΔT denoting a representative velocity and temperature difference, respectively. In this case the second and the third terms on the right side of equation (10-1) may be neglected.

One of the assumptions involved in arriving at equation (10-1) is that the net conduction heat transfer in the x direction is negligible compared with the net conduction in the y direction, and this simplification is justified providing the condition

$$\mathrm{Pe} = \frac{u_\infty L}{\alpha} \gg 1$$

is satisfied, where Pe denotes the Peclet number and L represents a characteristic dimension.

A similar criterion for the neglect of radiation in the x direction will now be determined. Perhaps the easiest means of accomplishing this is to investigate under what conditions radiation in the x direction is negligible compared to convection, since convection in the x direction will, under most conditions, be a predominant term in the energy equation. The requirement is thus that

$$u \frac{\partial T}{\partial x} \gg \frac{1}{\rho c_p} \frac{\partial q_{Rx}}{\partial x} \qquad (10\text{-}2)$$

The magnitude of $u \, \partial T / \partial x$ is estimated in the usual manner:

$$u \frac{\partial T}{\partial x} \sim u_\infty \frac{(T_w - T_\infty)}{L}$$

In predicting the magnitude of $\partial q_{Rx} / \partial x$, use is made of the fact that the optically thick limit will overestimate radiation heat transfer when applied to conditions that are not optically thick.* Thus in the present situation a conservative criterion would, at most, result through use of the optically thick limit. The order of magnitude of $\partial q_{Rx} / \partial x$ will consequently be estimated from

$$q_{Rx} = -\frac{4}{3\kappa} \frac{\partial e_b}{\partial x} = -\frac{16\sigma T^3}{3\kappa} \frac{\partial T}{\partial x}$$

---

* This may readily be observed, for example, from equation (3-13) of ref. 2.

which corresponds to the case where scattering is not considered ($\beta = \kappa$). Thus

$$\frac{\partial q_{Rx}}{\partial x} \sim \frac{16\sigma T^3}{3\kappa} \frac{(T_w - T_\infty)}{L^2}$$

and from equation (10-2), the condition for radiation transfer in the x direction to be negligible is

$$\frac{u_\infty L}{16\sigma T^3/3\kappa\rho c_p} \gg 1 \qquad (10\text{-}3)$$

The quantity $16\sigma T^3/3\kappa$ can be interpreted as a "radiation conductivity," and the dimensionless group appearing above may, in turn, be considered as a "radiation Peclet number."

With radiation neglected in the x direction, and again letting q_R denote the radiation flux in the y direction, equation (10-1) reduces to

$$u \frac{\partial T}{\partial x} + v \frac{\partial T}{\partial y} = \alpha \frac{\partial^2 T}{\partial y^2} + \frac{\beta T u}{\rho c_p} \frac{dp}{dx} + \frac{\nu}{c_p}\left(\frac{\partial u}{\partial y}\right)^2 - \frac{1}{\rho c_p} \frac{\partial q_R}{\partial y} \quad (10\text{-}4)$$

This may be further simplified to

$$u \frac{\partial T}{\partial x} + v \frac{\partial T}{\partial y} = \alpha \frac{\partial^2 T}{\partial y^2} - \frac{1}{\rho c_p} \frac{\partial q_R}{\partial y} \qquad (10\text{-}5)$$

when the condition Ec $\ll 1$ is satisfied.

10-2 Flow through Ducts

The most complete analyses pertaining to flow of a radiating fluid through heated or cooled ducts are those of Viskanta for slug flow (ref. 3) and laminar flow (ref. 4). Both treatments are concerned with flow through an infinite parallel-plate channel. Viskanta's laminar flow results will now be described.

Consider a parallel-plate channel as illustrated in Fig. 10-1. The

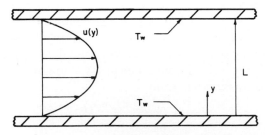

Fig. 10-1 Coordinate system for fully developed laminar heat transfer in a parallel-plate channel.

plates are assumed to be infinite in extent, black, and isothermal with both plates maintained at the same uniform temperature T_w. It is additionally assumed that both the velocity and the temperature profiles are fully developed and that the Eckert number is small (Ec $\ll 1$). Equation (10-5) then becomes

$$u \frac{\partial T}{\partial x} = \alpha \frac{\partial^2 T}{\partial y^2} - \frac{1}{\rho c_p} \frac{\partial q_R}{\partial y} \tag{10-6}$$

while the velocity profile is given by the well-known parabolic result

$$u = 6u_m \left(\frac{y}{L} - \frac{y^2}{L^2} \right) \tag{10-7}$$

with u_m denoting the mean fluid velocity. Furthermore, for a non-scattering gray fluid and black-plate surfaces, we have from equations (7-21) and (7-28)

$$-\frac{dq_R}{d\tau} = 2\sigma T_w{}^4 E_2(\tau) + 2\sigma T_w{}^4 E_2(\tau_0 - \tau)$$

$$+ 2\sigma \int_0^{\tau_0} T^4(t) E_1(|\tau - t|)\, dt - 4\sigma T^4(\tau) \tag{10-8}$$

Now, as a consequence of the assumption of a fully developed temperature profile, the axial temperature gradient $\partial T/\partial x$ may be replaced by (ref. 5)

$$\frac{\partial T}{\partial x} = \left(\frac{T_w - T}{T_w - T_b} \right) \frac{2q_w}{\rho c_p u_m L} \tag{10-9}$$

where T_b is the local bulk-fluid temperature, while q_w represents the total wall-heat flux and is expressed as

$$q_w = -k \left(\frac{\partial T}{\partial y} \right)_{y=0} + 2\sigma T_w{}^4 [1 - E_3(\tau_0)] - 2\sigma \int_0^{\tau_0} T^4(t) E_2(t)\, dt \tag{10-10}$$

Dimensionless quantities will now be defined as

$$N = \frac{k\kappa}{4\sigma T_w{}^3}, \quad \psi = \frac{q_w}{\sigma T_w{}^4}, \quad \theta = \frac{T}{T_w}, \quad \theta_b = \frac{T_b}{T_w}$$

Upon combining equations (10-6) through (10-9), the dimensionless form of the energy equation applicable to the present problem becomes

$$2N \frac{d^2\theta}{d\tau^2} - \frac{6\psi}{\tau_0} \left(\frac{\tau}{\tau_0} - \frac{\tau^2}{\tau_0{}^2} \right) \left(\frac{1-\theta}{1-\theta_b} \right) = -E_2(\tau) - E_2(\tau_0 - \tau)$$

$$- \int_0^{\tau_0} \theta^4(t) E_1(|\tau - t|)\, dt + 2\theta^4(\tau) \tag{10-11}$$

while the dimensionless total heat flux at the wall, ψ, is given by equation (10-10) as

$$\psi = -4N\left(\frac{d\theta}{d\tau}\right)_{\tau=0} + 2 - 2E_3(\tau_0) - 2\int_0^{\tau_0} \theta^4(t)E_2(t)\,dt \qquad (10\text{-}12)$$

The boundary conditions for equation (10-11) are in turn

$$\theta(0) = \theta(\tau_0) = 1$$

The parameters that govern the present problem are seen to be τ_0, N, and θ_b. The parameter N plays the same role here as in the preceding chapter; that is, a measure of the relative importance of conduction to radiation heat transfer. It should be noted that while equation (10-11) can be formulated in terms of the single independent variable τ, the bulk temperature T_b is actually a function of x. Consequently, different axial locations along the duct will correspond to different values of the parameter

$$\theta_b = \frac{T_b}{T_w}$$

An approximate solution of equation (10-11) has been obtained by Viskanta (ref. 4) through expansion of $\theta^4(\tau)$ in a Taylor series. In presenting his results Viskanta replaced the parameter θ_b by $\theta_c = T_c/T_w$, where T_c is the centerline fluid temperature. Of course, the alternate parameter θ_c will also possess different values for different axial locations.

The effect of optical thickness on the temperature distribution is illustrated in Fig. 10-2 for the conditions $N = 0.01$ and $\theta_c = 0.5$ (heating of the fluid). Since the temperature profile is symmetric, only half the profile need be shown. For $\tau_0 = 0.1$ the temperature distribution differs from that for a nonparticipating fluid by less than one percent.

Figure 10-3 illustrates the radiation heat flux at the wall for $\theta_c = 0.5$, and it is seen that the wall radiation flux achieves a maximum at about $\tau_0 \simeq 2$. The vanishing of q_{Rw} for $\tau_0 = 0$ and $\tau_0 \to \infty$ is obvious, for these two limits correspond to a transparent and opaque fluid, respectively.

The conduction heat flux at the wall is shown in Fig. 10-4. Again this is for the condition $\theta_c = 0.5$. It may be observed that the greatest influence of the conduction-radiation interaction parameter N occurs for mid-values of optical thickness ($\tau_0 \simeq 1$), and this is a further result of radiative transfer being most pronounced for intermediate values of τ_0.

The total wall heat flux is shown in Fig. 10-5 for $\theta_c = 0.5$ in terms

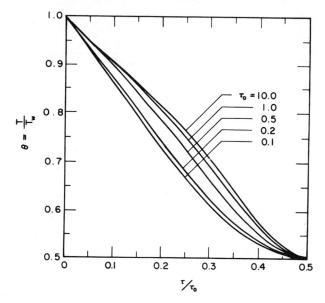

Fig. 10-2   Effect of optical thickness on temperature distribution for $N = 0.01$ and $\theta_c = 0.5$.   From Viskanta (ref. 4).

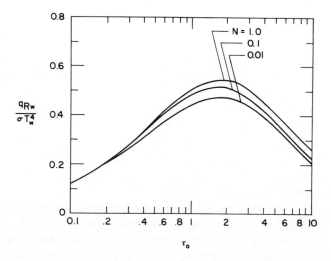

Fig. 10-3   Variation of radiation heat flux at the wall for $\theta_c = 0.5$. From Viskanta (ref. 4).

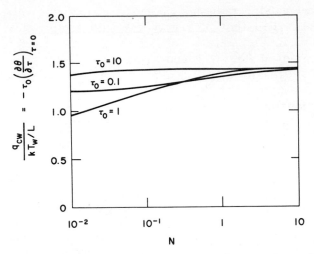

Fig. 10-4   Variation of conduction heat flux at the wall for $\theta_c = 0.5$.   From Viskanta (ref. 4).

of the conventional Nusselt number, which is defined for a parallel-plate channel as

$$\text{Nu} = \frac{2q_w L}{k(T_w - T_b)}$$

Upon recasting this in dimensionless form, we have

$$\text{Nu} = \frac{\psi \tau_0}{2N(1 - \theta_b)}$$

For $N \to \infty$ (nonradiating fluid) Viskanta obtained the value $\text{Nu} = 7.54$, which is in agreement with previous convective heat transfer results for nonradiating fluids.

Two-dimensional analyses (including q_{Rx}) of combined convection and radiation have been presented by Einstein for slug flow of a gray fluid through parallel-plate (ref. 6) and circular-tube (ref. 7) channels of finite length.   In both studies the integrodifferential energy equation was replaced by a system of algebraic equations by using the zonal method of Hottel and Cohen (ref. 8).

A nongray analysis for laminar flow in the entrance region of an isothermal circular tube having a black internal surface is given by deSoto and Edwards (ref. 9) for carbon dioxide.   Unlike the preceding analyses, the temperature profile within the gas was assumed to be independent of the radiation transfer process, and the temperature distribution is consequently described by the Graetz solution (ref. 10).

On the other hand, a radiating gas model was chosen which is much more realistic than that of a gray gas with temperature-invariant absorption coefficient.   Specifically, the exponential band model (ref. 11) was employed for CO_2,* and this accounts for both the wavelength and temperature dependence of the monochromatic absorption coefficient.

A portion of the results obtained by deSoto and Edwards is illustrated in Fig. 10-6 and corresponds to the following conditions:

$$\text{Wall temperature} = 2500 \, °\text{R}$$
$$\text{Gas inlet temperature} = 530 \, °\text{R}$$
$$\text{Gas pressure} = 1 \text{ atmosphere}$$
$$\text{Tube diameter} = 2 \text{ inches}$$
$$\text{Reynolds number} = 1000$$

Since the temperature distribution is assumed to be independent of

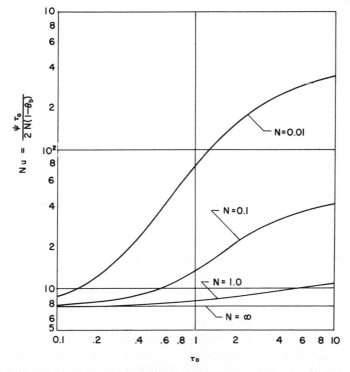

Fig. 10-5   Total wall heat flux for $\theta_c = 0.5$.   From Viskanta (ref. 4).

* See equation (1-35).

radiation, the conduction heat transfer is that given by the Graetz
solution for a nonradiating gas.

The results of two methods of calculating the radiation heat flux
are shown.   One uses the Graetz temperature distribution, whereas the
other is based on the less-realistic assumption that the gas is isothermal
at the volumetric mean temperature.   It may be seen that the isothermal assumption overpredicts the radiation heat flux to a considerable extent.   On the other hand, this assumption was shown in ref. 9
to be quite good for the reverse situation of a cold wall and a hot gas.

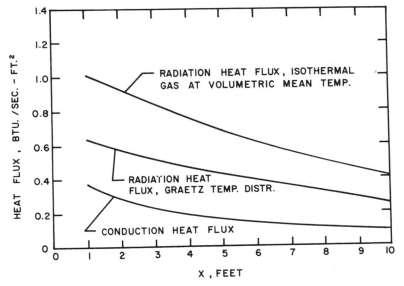

Fig. 10-6   Heat flux results for flow through a circular tube.
From deSoto and Edwards (ref. 9).

One should note from Fig. 10-6 that the radiation heat flux exceeds the
conduction flux by as much as a factor of three.

A combined experimental and analytical investigation concerning
turbulent flow of superheated steam in the thermal entrance region of an
annular duct has been presented by Nichols (ref. 12).   The experimental results were obtained for an annular duct having a radius ratio
of 0.2, with the inner surface maintained at 2000 °R and the outer
surface at approximately 750 °R.   The Reynolds number and pressure
were 20,000 and 3.22 atmospheres, respectively, and the temperature
profile was measured at a point 15 diameters downstream of the
entrance of the heated section.   The analytical portion of the investigation consisted of determining the first-order effect of radiation on the

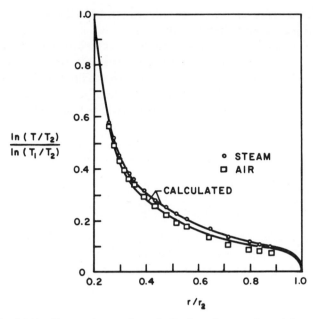

Fig. 10-7   Comparison of analytical and experimental results.
From Nichols (ref. 12).

temperature profile for a nonradiating gas, and a nongray solution was
obtained through use of the band approximation.

A comparison of Nichols' analytical and experimental results is
shown in Fig. 10-7 where T_1 and T_2 are the inner and outer surface
temperatures, respectively, and r_2 is the radius of the outer cylinder.
The air data represent a nonradiating gas.   It is seen that the effect of
gaseous radiation is to increase the dimensionless temperature within
the gas, which is in qualitative agreement with Viskanta's results shown
in Fig. 10-2.   The solid curves in Fig. 10-7 denote the analytical results
corresponding to the conditions for the nonradiating gas (air) and
radiating gas (steam) data.   Unfortunately, as a consequence of the
large turbulent energy transfer within the gas, the maximum effect due
to radiation is only 7 percent.   Nevertheless, the good agreement
between experimental and analytical results is quite evident.

10-3   Nonviscous, Nonconducting Flow over a Flat Plate

A very useful illustrative analysis concerning flow over external
surfaces involves the case of a radiating fluid that is assumed to be nonviscous as well as a nonconductor of heat.   Although this is certainly an

oversimplification to real flow problems, the model serves as a convenient stepping stone toward analyzing more realistic problems. The particular situation to be studied is illustrated in Fig. 10-8, which shows the flow of a fluid having the free-stream temperature T_∞ over an isothermal flat plate. As before, the plate surface is assumed to be black.

Since the fluid is assumed to be nonviscous, the uniform velocity field is maintained everywhere; that is, $u = u_\infty$ and $v = 0$. Upon deleting the viscous and thermal conduction terms from equation (10-4), and noting that $dp/dx = 0$, the applicable form of the energy equation becomes

$$u_\infty \frac{\partial T}{\partial x} = -\frac{1}{\rho c_p} \frac{\partial q_R}{\partial y} \tag{10-13}$$

Now, combining equations (7-20) and (7-28), setting $\omega_0 = 0$ and $\epsilon_w = 1$, and replacing τ_0 by ∞ since one is concerned with the flow of an infinite medium, the radiation flux is given by

$$q_R = 2\sigma T_w{}^4 E_3(\tau) + 2\sigma \int_0^\tau T^4(x, t) E_2(\tau - t)\, dt$$

$$- 2\sigma \int_\tau^\infty T^4(x, t) E_2(t - \tau)\, dt \tag{10-14}$$

and correspondingly

$$-\frac{\partial q_R}{\partial \tau} = 2\sigma T_w{}^4 E_2(\tau) + 2\sigma \int_0^\infty T^4(x, t) E_1(|\tau - t|)\, dt - 4\sigma T^4(x, \tau)$$

$$\tag{10-15}$$

An additional simplification will now be employed; it consists of assuming that temperature differences within the flow are sufficiently small such that T^4 may be expressed as a linear function of temperature. This is accomplished by expanding T^4 in a Taylor series about T_∞ and neglecting higher-order terms, thus

$$T^4 \simeq 4T_\infty{}^3 T - 3T_\infty{}^4 \tag{10-16}$$

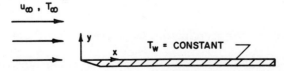

Fig. 10-8   Coordinate system for nonviscous, nonconducting flow over a flat plate.

Through use of this expression the energy equation becomes linear, and the temperature ratio T_w/T_∞ will not appear as a governing parameter. However, since temperature differences must be small in order for equation (10-16) to apply, the present analysis actually corresponds to the limiting case $T_w/T_\infty \to 1$. This linearized problem has been discussed by Goulard (ref. 13) and by Tien and Greif (ref. 14). A somewhat more detailed treatment will be given here.

Dimensionless quantities will now be defined as

$$\Theta = \frac{T - T_w}{T_\infty - T_w} \qquad \xi = \frac{2\sigma\kappa T_\infty^3 x}{\rho c_p u_\infty}$$

and after combining equations (10-13), (10-15), and (10-16), the dimensionless form of the energy equation becomes

$$\frac{\partial \Theta}{\partial \xi} = 4 \int_0^\infty \Theta(\xi, t) E_1(|\tau - t|)\, dt - 8\Theta(\xi, \tau) \qquad (10\text{-}17)$$

Since radiation transfer in the x direction is neglected, the boundary condition for this equation is

$$\Theta(0, \tau) = 1 \qquad (10\text{-}18)$$

Before proceeding, one may note that if ξ is redefined as

$$\xi = \frac{2\sigma\kappa T_\infty^3 t}{\rho c_p}$$

where t denotes time, then equations (10-17) and (10-18) represent an alternate physical problem. This consists of one-dimensional transient radiation transfer in a semi-infinite medium for the boundary condition of a step-change in surface temperature with time. A more complete discussion of the transient problem, including conduction as well as radiation heat transfer, is given by Lick (ref. 15).

With the dimensionless temperature $\Theta(\xi, \tau)$ described by equation (10-17), the radiation heat flux from the plate surface is given by equation (10-14) with $\tau = 0$. In linearized form this becomes

$$\frac{q_{Rw}}{\sigma(T_w^4 - T_\infty^4)} = 2 \int_0^\infty \Theta(\xi, t) E_2(t)\, dt \qquad (10\text{-}19)$$

where the relation

$$T_w^4 - T_\infty^4 \simeq 4 T_\infty^3 (T_w - T_\infty)$$

has been employed in accordance with equation (10-16). Even though a considerable number of simplifications have been employed, equation (10-17) does not appear to have a closed-form solution. Nevertheless, several approximate and/or limiting analyses are applicable, three of which will be illustrated and compared.

Series solution.   A series solution of equation (10-17) of the form

$$\Theta(\xi, \tau) = 1 + g_1(\tau)\xi + g_2(\tau)\xi^2 + \cdots \qquad (10\text{-}20)$$

may be assumed.   This satisfies equation (10-18), and upon substituting into equation (10-17) and collecting like powers of ξ, the first two functions $g_1(\tau)$ and $g_2(\tau)$ are found to be

$$g_1(\tau) = -4E_2(\tau) \qquad (10\text{-}21a)$$

$$g_2(\tau) = 16E_2(\tau) - 8 \int_0^\infty E_2(t)E_1(|\tau - t|)\, dt \qquad (10\text{-}21b)$$

Since conduction heat transfer has not been included, a temperature jump will occur at the boundary between the fluid and plate surface, and from equation (10-20) the jump temperature $T(\xi, 0)$ is expressed by

$$\frac{T(\xi, 0) - T_w}{T_\infty - T_w} = \Theta(\xi, 0) = 1 + g_1(0)\xi + g_2(0)\xi^2 + \cdots$$

With reference to equations (10-21), together with Appendix B,

$$g_1(0) = -4 \qquad g_2(0) = 12$$

and
$$\frac{T(\xi, 0) - T_w}{T_\infty - T_w} = \Theta(\xi, 0) = 1 - 4\xi + 12\xi^2 + \cdots \qquad (10\text{-}22)$$

A second quantity of interest is the radiation heat flux from the plate surface.   From equations (10-19) and (10-20)

$$\frac{q_{Rw}}{\sigma(T_w{}^4 - T_\infty{}^4)} = 1 + 2\xi \int_0^\infty g_1(t)E_2(t)\, dt$$

$$+ 2\xi^2 \int_0^\infty g_2(t)E_2(t)\, dt + \cdots \qquad (10\text{-}23)$$

The first integral may be evaluated with the aid of Appendix B, while the value of the second integral has been obtained numerically, giving

$$\int_0^\infty g_1(t)E_2(t)\, dt = -0.818$$

$$\int_0^\infty g_2(t)E_2(t)\, dt = 1.17$$

Consequently, the surface radiation flux is

$$\frac{q_{Rw}}{\sigma(T_w{}^4 - T_\infty{}^4)} = 1 - 1.64\xi + 2.34\xi^2 + \cdots \qquad (10\text{-}24)$$

Obviously both equations (10-22) and (10-24) are restricted to small values of ξ.

Radiation slip solution. Consider for the moment the applicability of the optically thick limit to the present problem. For a gray fluid equation (7-42) becomes

$$q_R = -\frac{4\sigma}{3}\frac{\partial T^4}{\partial \tau}$$

and making use of equation (10-16), we have for linearized radiation

$$q_R = -\frac{16\sigma T_\infty{}^3}{3}\frac{\partial T}{\partial \tau} \qquad (10\text{-}25)$$

The energy equation applicable to optically thick conditions is obtained through substitution of equation (10-25) into equation (10-13), and recasting the result in dimensionless form:

$$\frac{\partial \Theta}{\partial \xi} = \frac{8}{3}\frac{\partial^2 \Theta}{\partial \tau^2} \qquad (10\text{-}26)$$

We are now in a position to illustrate the physical interpretation of optically thick radiation with reference to the problem at hand. For example, regardless of whether or not optically thick conditions prevail, the convection-radiation energy transport will result in the penetration of a temperature field from the plate surface into the fluid. In other words, a "radiation boundary layer" will exist adjacent to the plate surface, and temperature gradients will be restricted to within this layer. The optically thick limit will therefore be applicable only if the optical thickness of this layer is large. Now, since equation (10-26) is a diffusion equation,* it is clear that under optically thick conditions the temperature field will penetrate an optical distance of the order $\xi^{1/2}$. This implies that optically thick radiation corresponds to large values of ξ. This finding is physically quite logical, for one would expect the penetration depth to increase in the flow direction, and ξ may be regarded as a dimensionless streamwise coordinate. An additional argument leading to the same conclusion is given by Goulard (ref. 13).

Recall now that the radiation slip method constitutes a means of extending the optically thick limit through the introduction of a temperature jump at the surface, and within the context of the present problem this represents an extension to smaller ξ values. Of course, the radiation slip method is still restricted to moderately large values of ξ.

---

* Since the two terms in equation (10-26) are equal, then $\tau = O(\xi^{1/2})$, with τ denoting the optical penetration depth of the temperature field.

Proceeding with the radiation slip solution, equation (10-26) constitutes the appropriate energy equation, while for a gray fluid and black surface the temperature jump at the surface is given by equation (7-48) as

$$T^4(\xi, 0) - T_w{}^4 = \frac{2}{3}\left(\frac{\partial T^4}{\partial \tau}\right)_{\tau=0}$$

or, in linearized form,

$$T(\xi, 0) - T_w = \frac{2}{3}\left(\frac{\partial T}{\partial \tau}\right)_{\tau=0}$$

The boundary conditions applicable to equation (10-26) readily follow:

$$\Theta = \frac{2}{3}\frac{d\Theta}{d\tau}, \qquad \tau = 0 \qquad\qquad (10\text{-}27a)$$

$$\Theta = 1, \qquad\qquad \tau \to \infty \qquad\qquad (10\text{-}27b)$$

$$\Theta = 1, \qquad\qquad \xi = 0 \qquad\qquad (10\text{-}27c)$$

and from ref. 16 the solution of equation (10-26) is

$$\Theta(\xi, \tau) = \exp\left(\frac{3\tau}{2} + 6\xi\right)\operatorname{erfc}\left(\frac{\tau}{4}\sqrt{\frac{3}{2\xi}} + \sqrt{6\xi}\right) + \operatorname{erf}\left(\frac{\tau}{4}\sqrt{\frac{3}{2\xi}}\right)$$

The jump temperature is readily evaluated as

$$\frac{T(\xi, 0) - T_w}{T_\infty - T_w} = \Theta(\xi, 0) = \exp(6\xi)\operatorname{erfc}(\sqrt{6\xi}) \qquad (10\text{-}28)$$

and upon applying equation (10-25) at $\tau = 0$, the radiation flux at the wall may be expressed as

$$\frac{q_{Rw}}{\sigma(T_w{}^4 - T_\infty{}^4)} = \frac{4}{3}\left(\frac{\partial\Theta}{\partial\tau}\right)_{\tau=0}$$

such that from equation (10-27a)

$$\frac{q_{Rw}}{\sigma(T_w{}^4 - T_\infty{}^4)} = 2\,\Theta(\xi, 0) \qquad\qquad (10\text{-}29)$$

Equations (10-28) and (10-29) augment the previous series solution, for the series results are restricted to small ξ, while the radiation slip solution is applicable to moderately large values of ξ.

Exponential kernel solution. A final method employs the exponential kernel substitution, thus yielding an approximate solution that is applicable for all values of ξ. With reference to equations (10-17) and (10-19), it is seen that the present case deals with the pair of exponential integrals $E_1(t)$ and $E_2(t)$, whereas previous applications of the exponential kernel approximation (see Sections 8-1 and 9-3) dealt with $E_2(t)$ and $E_3(t)$. The reason for this difference is that previously one was concerned with the equation for the radiation flux q_R, whereas equation (10-17) contains the derivative $\partial q_R/\partial \tau$.

One approach is that employed by Lick (ref. 15) in analyzing transient radiation transfer. This method consists of substituting the exponential approximations for $E_2(t)$ and $E_3(t)$, as given by equations (8-6) and (8-7), into equation (10-14) and then using this approximate flux equation for evaluating $\partial q_R/\partial \tau$. An alternate method will be illustrated here, and this deals directly with equations (10-17) and (10-19), together with approximations for $E_1(t)$ and $E_2(t)$. Following the same procedure used in arriving at equations (8-6) and (8-7), one obtains

$$E_1(t) \simeq 2e^{-2t} \tag{10-30a}$$

$$E_2(t) \simeq e^{-2t} \tag{10-30b}$$

and equation (10-17) becomes

$$\frac{\partial \Theta}{\partial \xi} = 8e^{-2\tau} \int_0^\tau \Theta(\xi, t)e^{2t}\, dt + 8e^{2\tau} \int_\tau^\infty \Theta(\xi, t)e^{-2t}\, dt - 8\Theta(\xi, \tau) \tag{10-31}$$

This approximate form of the energy equation may be recast as an ordinary differential equation by first taking the Laplace transform with respect to ξ. Letting $\overline{\Theta}(s, \tau)$ denote the Laplace transform of $\Theta(\xi, \tau)$; that is,

$$\overline{\Theta}(s, \tau) = \int_0^\infty \Theta(\xi, \tau)e^{-s\xi}\, d\xi$$

equation (10.31) becomes

$$(8 + s)\overline{\Theta}(s, \tau) - 1 = 8e^{-2\tau} \int_0^\tau \overline{\Theta}(s, t)e^{2t}\, dt + 8e^{2\tau} \int_\tau^\infty \overline{\Theta}(s, t)e^{-2t}\, dt \tag{10-32}$$

After differentiating twice, the integrals repeat themselves and may be eliminated. The resulting differential equation is

$$\frac{d^2\overline{\Theta}}{d\tau^2} - \frac{4s}{s + 8}\overline{\Theta} = -\frac{4}{s + 8}$$

for which the solution, satisfying the obvious condition $\overline{\Theta}(s, \infty) = 1/s$, is

$$\overline{\Theta} = \frac{1}{s} + C \exp\left(-2\tau \sqrt{\frac{s}{s + 8}}\right) \tag{10-33}$$

The remaining constant of integration, C, is evaluated by substituting equation (10-33) into (10-32), with the result

$$C = \frac{1}{\sqrt{s^2 + 8s}} - \frac{1}{s} \qquad (10\text{-}34)$$

Consider now the evaluation of the dimensionless jump temperature $\Theta(\xi, 0)$. Noting that

$$\Theta(\xi, 0) = \mathcal{L}^{-1}[\overline{\Theta}(s, 0)] = \mathcal{L}^{-1}\left(\frac{1}{\sqrt{s^2 + 8s}}\right)$$

the inverse transform yields

$$\frac{T(\xi, 0) - T_w}{T_\infty - T_w} = \Theta(\xi, 0) = \exp(-4\xi)I_0(4\xi) \qquad (10\text{-}35)$$

where I_n is the modified Bessel function of the first kind.

To determine the corresponding expression for the radiation flux from the plate surface, one notes that equations (10-19) and (10-30b) give

$$\frac{q_{Rw}}{\sigma(T_w{}^4 - T_\infty{}^4)} = 2 \int_0^\infty \Theta(\xi, t)e^{-2t}\, dt$$

The integral may be evaluated by setting $\tau = 0$ in equation (10-31), with the result

$$\frac{q_{Rw}}{\sigma(T_w{}^4 - T_\infty{}^4)} = \frac{1}{4}\frac{d\,\Theta(\xi, 0)}{d\xi} + 2\Theta(\xi, 0)$$

and from equation (10-35)

$$\frac{q_{Rw}}{\sigma(T_w{}^4 - T_\infty{}^4)} = \exp(-4\xi)[I_0(4\xi) + I_1(4\xi)] \qquad (10\text{-}36)$$

Equations (10-35) and (10-36) are thus the desired results as obtained by the exponential kernel approximation.

Results and discussion. Figure 10-9 shows a comparison of the jump temperature $T(\xi, 0)$ as obtained from the three separate analyses. The series solution, the radiation slip solution, and the exponential kernel solution are given by equations (10-22), (10-28), and (10-35), respectively. The exponential kernel solution represents a logical joining of the two limiting solutions. It may be noted that $T(\xi, 0) \to T_w$ as $\xi \to \infty$. This obviously follows from the fact that large ξ corresponds to optically thick radiation for which temperature continuity at the surface prevails.

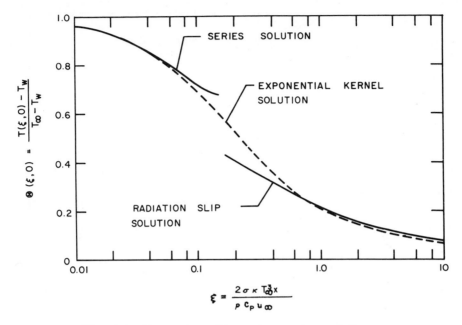

Fig. 10-9   Comparison of jump temperature solutions.

The radiation heat flux at the plate surface is illustrated in Fig. 10-10, where the results for the series solution, the radiation slip solution, and the exponential kernel solution have been evaluated from equations (10-24), (10-29), and (10-36), respectively.   Once again the exponential kernel solution represents a logical joining of the two limiting analyses.   However, the agreement is not as good as for the jump temperature illustrated in Fig. 10-9.

It is observed that the black-body relation

$$q_{Rw} = \sigma(T_w{}^4 - T_\infty{}^4)$$

is obtained for $\xi \to 0$; that is, near the leading edge of the plate.   This fact is easily explained on physical grounds.   Near the plate-leading edge the fluid temperature will be nearly uniform, for radiation exchange will not have had a sufficient opportunity to influence the free-stream fluid.   Furthermore, from equation (7-66) it is seen that the emittance of an infinite isothermal medium is unity, and thus the foregoing expression follows.   As the fluid progresses downstream in the flow direction, its temperature tends to approach toward T_w, such that radiation exchange between the fluid and the plate decreases with increasing ξ as illustrated in Fig. 10-10.

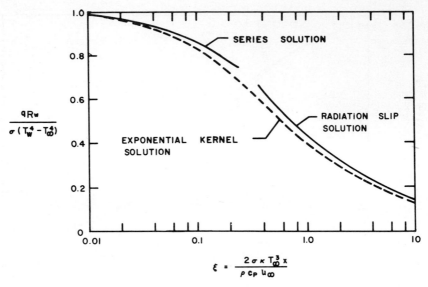

Fig. 10-10   Comparison of surface flux solutions.

10-4   Laminar Boundary Layer Flow

Consider now the extension of the problem treated in the preceding section to include effects of both viscosity and heat conduction. Specifically, laminar flow of a constant-property gray fluid over a black isothermal plate is considered.   Assuming the Eckert number to be small (Ec $\ll$ 1), the appropriate energy equation is obtained through combination of equations (10-5) and (10-15):

$$u\frac{\partial T}{\partial x} + v\frac{\partial T}{\partial y} = \alpha\frac{\partial^2 T}{\partial y^2} + \frac{2\sigma\kappa}{\rho c_p}\left[T_w{}^4 E_2(\tau) \right.$$

$$\left. + \int_0^\infty T^4(x, t)E_1(|\tau - t|)\,dt - 2T^4(x, \tau) \right] \quad (10\text{-}37)$$

The velocity components u and v are given by the well-known expressions (ref. 17)

$$u = u_\infty f' \qquad v = \frac{1}{2}\sqrt{\frac{u_\infty \nu}{x}}\,(\eta f' - f) \qquad (10\text{-}38)$$

where $f(\eta)$ is the dimensionless Blasius stream function, and

$$\eta = y\sqrt{\frac{u_\infty}{\nu x}}$$

It will now be convenient to rephrase equation (10-37) in terms of the dimensionless independent variables $\tau = \kappa y$ and

$$\xi = \frac{2\sigma\kappa T_\infty^3 x}{\rho c_p u_\infty}$$

with the result

$$f'\frac{\partial T}{\partial \xi} + \frac{1}{2}\sqrt{\frac{2 \Pr N}{\xi}}\,(\eta f' - f)\frac{\partial T}{\partial \tau} = 2N\frac{\partial^2 T}{\partial \tau^2} + \frac{1}{T_\infty^3}\left[T_w^4 E_2(\tau)\right.$$

$$\left. + \int_0^\infty T^4(\xi, t)E_1(|\tau - t|)\,dt - 2T^4(\xi, \tau)\right] \quad (10\text{-}39)$$

where

$$N = \frac{k\kappa}{4\sigma T_\infty^3}$$

is again a measure of the relative importance of conduction versus radiation, while Pr denotes the Prandtl number of the fluid ($\Pr = \nu/\alpha$). Although η appears separately in equation (10-39), it is expressible in terms of ξ and τ as

$$\eta = y\sqrt{\frac{u_\infty}{\nu x}} = \frac{\tau}{\sqrt{2 \Pr N\xi}} \quad (10\text{-}40)$$

Equation (10-39) constitutes a partial nonlinear, integrodifferential equation, and even with the simplifications that have heretofore been made, a complete solution to the problem represents an extremely formidable undertaking.   As an alternate to this, two limiting solutions that, although quantitatively restrictive, yield quite useful qualitative information will be discussed.

Optically thin boundary layer.   The first limiting solution corresponds to the physical restriction of an optically thin, thermal boundary layer.   However, rather than impose this restriction initially, it will be shown to be a consequence of a somewhat more general condition relating to the conduction-radiation parameter N.   For example, values of N (based on the Planck mean absorption coefficient) are illustrated in Fig. 10-11 for ammonia, carbon dioxide, and water vapor.   In all cases the magnitude of N is considerably less than unity, and it thus appears natural to seek a limiting solution applicable for $N \ll 1$.   It may be recalled that in Section 9-3 a boundary layer or singular perturbation solution due to Lick was presented for the problem of combined conduction and radiation, and that this solution was for $N \ll 1$.   The same general technique will be applied to the present problem.

Before actually proceeding with the singular perturbation solution, a brief physical discussion will be of benefit.   Basically it will be shown

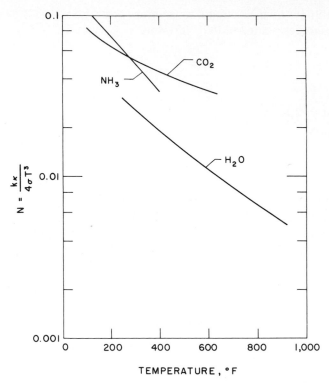

Fig. 10-11   Values of N for ammonia, carbon dioxide, and water vapor at one atmosphere.

that for $N \ll 1$ conduction heat transfer within the fluid is restricted to a boundary layer whose optical thickness is much less than unity. This boundary layer, however, represents only a portion of the entire temperature field, for radiation emitted by the plate will pass through this layer with only slight attenuation.   Consequently, it is necessary to consider also an adjacent outer region that is not optically thin but within which temperature gradients, and thus heat conduction, are small.

One may note a direct analogy between the present radiation problem and a velocity boundary layer, with the outer region corresponding to the potential flow outside of the velocity boundary layer. The procedures for evaluation of the boundary condition at the outer edge of either boundary layer are, in turn, analogous. With this introduction attention will now be directed toward the singular perturbation analysis.

Following Section 9-3, a good approximation to equation (10-39) for $N \ll 1$ is obtained by setting $N = 0$. Noting from equation (10-40) that

$$f'(\eta) \to f'(\infty) = 1 \qquad \text{for } N = 0, \, \tau > 0$$

equation (10-39) reduces to

$$\frac{\partial T_0}{\partial \xi} = \frac{1}{T_\infty{}^3} \left[T_w{}^4 E_2(\tau) + \int_0^\infty T_0{}^4(\xi, t) E_1(|\tau - t|) \, dt - 2T_0{}^4(\xi, \tau) \right]$$

$$(10\text{-}41)$$

where $T_0(\xi, \tau)$ represents the temperature distribution for $N = 0$. Physically equation (10-41) corresponds to nonviscous flow in the absence of heat conduction within the fluid, which is *specifically* the problem treated in the preceding section.

Since equation (10-41) will generally not yield continuity of temperature at the plate surface, it does not constitute a good approximation in the vicinity of the surface. This implies that an additional solution which includes conduction heat transfer is required near the plate; this region corresponds to the conventional definition of the thermal boundary layer.

Before continuing with the application of equation (10-39) to the boundary layer region, it is again emphasized that the *only* reason equation (10-41) is inapplicable near the surface is the fact that continuity of temperature is not satisfied. However, such is not the case if optically thick conditions prevail. Then temperature continuity at the surface occurs even in the absence of conduction. In other words, whenever $T_0(\xi, \tau)$ corresponds to optically thick radiation, $T_0(\xi, \tau)$ is by itself a valid approximation for $N \ll 1$.

By this reasoning, one evidently requires a boundary layer solution only when $T_0(\xi, \tau)$ represents a temperature field that is not optically thick; that is, for $\tau \leqslant O(1)$. Furthermore, as discussed in Section 10-3, optically thick conditions correspond to large ξ; consequently, one need now consider only $\xi \leqslant O(1)$. Without any great loss of generality, it will be assumed that $\text{Pr} = O(1)$.

Consider now the application of equation (10-39) to the boundary layer region. It is clear that with the change of variable

$$\bar{\tau} = \frac{\tau}{\sqrt{2 \, \text{Pr} \, N}}$$

the second derivative in equation (10-39) is of importance within a region of thickness $\bar{\tau} = O(1)$ and that this region corresponds to $\tau = O(\sqrt{2 \, \text{Pr} \, N}) \ll 1$. In fact, we may denote the optical thickness of the boundary layer by τ_δ with $\tau_\delta = O(\sqrt{2 \, \text{Pr} \, N})$, which illustrates that the boundary layer is optically thin. The boundary layer form of equation (10-39) thus reduces to*

$$f' \frac{\partial T}{\partial \xi} + \frac{1}{2}\, \xi^{-\frac{1}{2}}(\eta f' - f)\, \frac{\partial T}{\partial \bar{\tau}} = \frac{1}{\mathrm{Pr}}\, \frac{\partial^2 T}{\partial \bar{\tau}^2}$$

$$+ \frac{1}{T_\infty{}^3}\left[T_w{}^4 + \int_0^\infty T^4(\xi, t) E_1(t)\, dt - 2T^4(\xi, \bar{\tau}) \right] \quad (10\text{-}42)$$

An additional simplification to this equation may be made by noting that $T(\xi, \tau) = T_0(\xi, \tau)$ for $\tau > \tau_\delta$, with the result

$$\int_0^\infty T^4(\xi, t) E_1(t)\, dt = \int_0^\infty T_0{}^4(\xi, t) E_1(t)\, dt$$

$$- \int_0^{\tau_\delta} T_0{}^4(\xi, t) E_1(t)\, dt + \int_0^{\tau_\delta} T^4(\xi, t) E_1(t)\, dt$$

Since the second and third integrals on the right side of this equation are of the order $\sqrt{2\,\mathrm{Pr}\,N}$ they may be deleted, while the first integral is evaluated by setting $\tau = 0$ in equation (10-41), giving

$$\int_0^\infty T_0{}^4(\xi, t) E_1(t)\, dt = T_\infty{}^3 \frac{dT_0(\xi, 0)}{d\xi} + 2T_0{}^4(\xi, 0) - T_w{}^4$$

Equation (10-42) now becomes

$$f' \frac{\partial T}{\partial \xi} + \frac{1}{2}\, \xi^{-\frac{1}{2}}(\eta f' - f)\, \frac{\partial T}{\partial \bar{\tau}} = \frac{1}{\mathrm{Pr}}\, \frac{\partial^2 T}{\partial \bar{\tau}^2}$$

$$+ \frac{dT_0(\xi, 0)}{d\xi} + \frac{2}{T_\infty{}^3}\, T_0{}^4(\xi, 0) - \frac{2}{T_\infty{}^3}\, T^4(\xi, \bar{\tau}) \quad (10\text{-}43)$$

and this is the desired form of the boundary layer energy equation.

It remains to match properly the boundary layer or inner solution $T(\xi, \bar{\tau})$ with the free stream or outer solution $T_0(\xi, \tau)$. General principles of asymptotic matching are discussed by Van Dyke (ref. 18). For the present case it is sufficient to state that the inner solution for large values of $\bar{\tau}$ must match the outer solution written in terms of the inner variable; that is, $T_0(\xi, \bar{\tau}\sqrt{2\,\mathrm{Pr}\,N})$. Thus, with $N \ll 1$, we have

$$T(\xi, \infty) = T_0(\xi, 0)$$

---

* Note that for $\bar{\tau} = O(1)$

$$E_2(\tau) = E_2(\bar{\tau}\sqrt{2\,\mathrm{Pr}\,N}) = 1 + O(\sqrt{2\,\mathrm{Pr}\,N})$$

and

$$E_1(|\tau - t|) = E_1(|\bar{\tau}\sqrt{2\,\mathrm{Pr}\,N} - t|) = E_1(t) + O(\sqrt{2\,\mathrm{Pr}\,N})$$

and the solution of equation (10-43) must satisfy the boundary conditions

$$T = T_w \qquad\qquad \bar{\tau} = 0 \qquad\qquad \text{(10-44a)}$$

$$T \to T_0(\xi, 0) \qquad\qquad \bar{\tau} \to \infty \qquad\qquad \text{(10-44b)}$$

Before discussing solutions to equation (10-43), there is in addition to ξ and N a third quantity that is sometimes employed in treating boundary layer problems involving radiating gases, and this is defined as

$$\zeta = \kappa \sqrt{\frac{\alpha x}{u_\infty}} \qquad\qquad \text{(10-45)}$$

The variable ζ actually does not constitute a separate quantity, for ζ is related to ξ and N through the relation

$$\zeta = \sqrt{2N\xi}$$

It follows that the present problem corresponds to $\zeta \ll 1$, and this has a definite physical interpretation. Recall from boundary layer theory that the thickness of the thermal boundary layer is of $O(\sqrt{\alpha x/u_\infty})$. Thus, from equation (10-45), the optical thickness of the boundary layer is of $O(\zeta)$, which serves to illustrate further that the thermal boundary layer is optically thin for $N \ll 1$.

It may be observed that a considerable simplification has been gained in going from equation (10-39) to equation (10-43), since equation (10-43) is solely a differential equation. This equation does, however, require *a priori* information concerning $T_0(\xi, \tau)$. For linearized radiation this is described by the analyses given in Section 10-3, such that

$$\frac{T_0(\xi, \tau) - T_w}{T_\infty - T_w} = \Theta(\xi, \tau)$$

The restriction of linearized radiation will now be imposed on the present boundary layer problem. By combining equations (10-16) and (10-43), and defining

$$\theta = \frac{T(\xi, \bar{\tau}) - T_w}{T_\infty - T_w}$$

the linearized energy equation becomes

$$f' \frac{\partial\theta}{\partial\xi} + \frac{1}{2}\xi^{-\frac{1}{2}}(\eta f' - f)\frac{\partial\theta}{\partial\bar{\tau}} = \frac{1}{\text{Pr}}\frac{\partial^2\theta}{\partial\bar{\tau}^2}$$

$$+ \frac{d\Theta(\xi, 0)}{d\xi} + 8\Theta(\xi, 0) - 8\theta(\xi, \bar{\tau}) \qquad \text{(10-46)}$$

The boundary conditions follow from equations (10-44):

$$\theta = 0 \qquad \bar{\tau} = 0 \qquad \text{(10-47a)}$$

$$\theta \to \Theta(\xi, 0) \qquad \bar{\tau} \to \infty \qquad \text{(10-47b)}$$

Before discussing solutions of equation (10-46), consider the evaluation of the heat flux from the plate surface. This will consist of both radiation and conduction contributions, and it will be convenient to consider these two quantities separately.

From equation (10-14) with $\tau = 0$, the surface radiation flux is given by

$$q_{Rw} = \sigma T_w^4 - 2\sigma \int_0^\infty T^4(\xi, t) E_2(t)\, dt$$

or, again recalling that $T(\xi, \tau) = T_0(\xi, \tau)$ for $\tau > \tau_\delta$,

$$q_{Rw} = \sigma T_w^4 - 2\sigma \int_0^\infty T_0^4(\xi, t) E_2(t)\, dt + 2\sigma \int_0^{\tau_\delta} T_0^4(\xi, t) E_2(t)\, dt$$

$$- 2\sigma \int_0^{\tau_\delta} T^4(\xi, t) E_2(t)\, dt$$

Since the second two integrals are of $O(\sqrt{2\,\mathrm{Pr}\,N})$, they may be neglected, and putting the above result in dimensionless and linearized form

$$\frac{q_{Rw}}{\sigma(T_w^4 - T_\infty^4)} = 2 \int_0^\infty \Theta(\xi, t) E_2(t)\, dt$$

This is identical to equation (10-19), and the radiation flux is thus given by the solutions of the previous section for nonviscous, nonconducting flow. Such a result is physically evident from the fact that the boundary layer is optically thin.

Turning next to the conduction heat transfer from the plate surface, one has

$$q_{Cw} = -k\left(\frac{\partial T}{\partial y}\right)_{y=0} \qquad \text{(10-48)}$$

If use is made of the conventional definition of the Nusselt number, then

$$\mathrm{Nu} = \frac{q_{Cw} x}{k(T_w - T_\infty)}$$

Dimensionless forms of equation (10-48) are, in terms of the inner solution,

$$\frac{\mathrm{Nu}}{\sqrt{\mathrm{Re}}} = \left(\frac{\partial \theta}{\partial \eta}\right)_{\eta=0} = \xi^{\frac{1}{2}}\left(\frac{\partial \theta}{\partial \bar{\tau}}\right)_{\bar{\tau}=0} \qquad \text{(10-49)}$$

where $\mathrm{Re} = u_\infty x/\nu$.

These expressions, of course, require the solution of equation (10-46), and three methods of solution are presented in ref. 19. For the sake of brevity the reader is referred to ref. 19 for details. The three solution methods are summarized as follows:

(1)   A three-term power series of the form

$$\theta(\xi, \eta) = \theta_0(\eta) + \theta_1(\eta)\xi + \theta_2(\eta)\xi^2 + \cdots \qquad (10\text{-}50)$$

utilizing, for the boundary conditions, the comparable series solution for $\Theta(\xi, \tau)$.

(2)   An asymptotic solution expressed as

$$\theta(\xi, \bar{\tau}) \sim \phi_0(\bar{\tau})\xi^{-\frac{1}{2}} + \phi_1(\bar{\tau})\xi^{-1} + \phi_2(\bar{\tau})\xi^{-\frac{3}{2}} + \cdots \qquad (10\text{-}51)$$

and for which the radiation slip solution is employed for $\Theta(\xi, \tau)$.

(3)   An approximate solution applicable for all values of ξ which is based on the assumption of local similarity. In this case the exponential kernel solution for $\Theta(\xi, \tau)$ was employed.

Through use of these three methods of solution, composite results for $\mathrm{Nu}/\sqrt{\mathrm{Re}}$ as a function of ξ may easily be obtained, and this is illustrated in Fig. 10-12 for $\mathrm{Pr} = 1$. The limiting value $\mathrm{Nu}/\sqrt{\mathrm{Re}} = 0.332$ is obtained for $\xi = 0$, and this represents negligible radiation interaction. This is physically logical, since from the definition

$$\xi = \frac{2\sigma\kappa T_\infty^3 x}{\rho c_p u_\infty}$$

it follows that ξ is a measure of the relative role of radiation versus convection. At the other extreme one finds that $\mathrm{Nu}/\sqrt{\mathrm{Re}} \to 0.652$ as $\xi \to \infty$. This limit corresponds to negligible convection within the boundary layer (i.e., a balance between conduction and radiation). Thus the maximum effect of radiation interaction is to increase the conduction heat transfer by slightly less than a factor of two.

Recall now that the present linearized solution corresponds to $T_w/T_\infty \to 1$. Additional results have been presented in ref. 20 for non-linear radiation. For $\mathrm{Pr} = 1$ the conduction heat transfer can be expressed in terms of the Nusselt number as

$$\frac{\mathrm{Nu}}{\sqrt{\mathrm{Re}}} = 0.332 + \theta_1'(0)\xi + \cdots \qquad (10\text{-}52)$$

where $\theta_1'(0)$ is a function of T_w/T_∞. This corresponds to what would be obtained from equation (10-50) except that only two, rather than

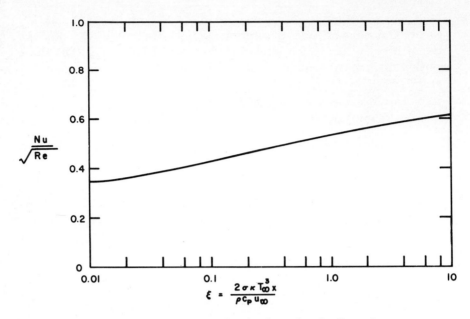

Fig. 10-12   Surface conduction heat flux for Pr = 1.

three, terms are considered.   Of course, $\theta_1'(0)$ for $T_w/T_\infty \to 1$ will be the same as for linearized radiation.

The effect of the parameter T_w/T_∞ on the first-order radiation-interaction coefficient $\theta_1'(0)$ is illustrated in Fig. 10-13.   It is seen that the maximum increase in conduction heat transfer due to radiation interaction occurs for $T_w/T_\infty \simeq 0.5$ to 1.   An especially interesting result is that for $T_w/T_\infty > 1.7$ the interaction effect is reversed, and the presence of radiation causes a *reduction* in conduction heat transfer relative to that for no interaction (at least for small ξ).   This indeed illustrates the importance of the nonlinear nature of interaction problems of this type.

Optically thick boundary layer.   Although it has previously been shown that the optically thin boundary layer is often a physically realistic model, it will, for the sake of completeness, be of interest to consider the opposite limit of an optically thick boundary layer.   Again assuming the Eckert number to be small, combination of equation (10-5) with the optically thick expression

$$q_R = -\frac{4\sigma}{3\kappa}\frac{\partial T^4}{\partial y}$$

yields $$u\frac{\partial T}{\partial x} + v\frac{\partial T}{\partial y} = \alpha\frac{\partial^2}{\partial y^2}\left(T + \frac{4\sigma}{3k\kappa}T^4\right) \qquad (10\text{-}53)$$

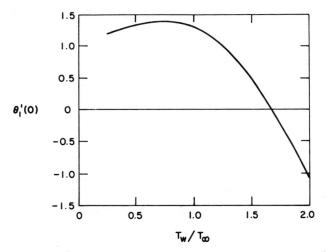

Fig. 10-13   Variation of $\theta_1'(0)$ for Pr = 1.   (From ref. 20.)

as the appropriate form of the energy equation.   The velocity components are given by equations (10-38), and employing the similarity transformation

$$T(\eta) = T(x, y)$$

equation (10-53) becomes

$$\frac{1}{\mathrm{Pr}} \frac{d^2}{d\eta^2}\left(T + \frac{T^4}{3T_\infty{}^3 N}\right) + \frac{1}{2}f\frac{dT}{d\eta} = 0 \qquad (10\text{-}54)$$

The boundary conditions readily follow as

$$T(0) = T_w \qquad T(\infty) = T_\infty \qquad (10\text{-}55)$$

Before proceeding, consideration must be given to the magnitude of N which characterizes the present problem.   Clearly, either from the definition of N or from equation (10-54), it follows that $N \gg 1$ corresponds to negligible radiation, while $N \ll 1$ denotes negligible conduction heat transfer.   Although the limit $N \ll 1$ should be associated with the optically thin boundary layer model, it is easily verified that for $N \ll 1$ equation (10.54) reduces to an equation representing optically thick radiation, negligible conduction, and nonviscous flow. Indeed, this corresponds precisely to the preceding outer solution for $\xi \gg 1$, and recall that for $\xi \gg 1$, $T_0(\xi, \tau)$ is by itself a valid solution for $N \ll 1$.   In other words, for $\xi \gg 1$ the singular perturbation problem is completely described by the outer solution $T_0(\xi, \tau)$, and this coincides with equation (10-54) when $N \ll 1$ in equation (10-54).   The present optically thick analysis thus constitutes an alternative method of solution only for $N \geqslant O(1)$.

With this restriction to $N \geqslant O(1)$, equation (10-54) illustrates that the depth of penetration of the temperature field is* $\eta = O(1/\sqrt{\mathrm{Pr}})$; that is, the boundary layer thickness is of $O(\sqrt{\alpha x/u_\infty})$. Inasmuch as the boundary layer is assumed to be optically thick, it is then required that $\kappa\sqrt{\alpha x/u_\infty} \gg 1$, and with reference to equation (10-45), this requirement corresponds to $\zeta \gg 1$.

Numerical solutions of equation (10-54) have been obtained by Viskanta and Grosh (ref. 21). An interesting study of the applicability of the optically thick limit to boundary layer flows has recently been made by Novotny, Kelleher, and Schimmel (ref. 25). In this investigation (which actually dealt with free convection) it was found that the optically thick model yields correct limiting values for the surface heat flux, whereas incorrect values result for the separate conduction and radiation contributions. In other words, correct large ζ results are obtained for the sum $q_{Cw} + q_{Rw}$, but not for q_{Cw} and q_{Rw} separately (see footnote on p. 209).

For purposes of comparing the present analysis with the preceding optically thin results, it is sufficient to obtain a solution of equation (10-54) for the condition of linearized radiation. Upon defining

$$\theta = \frac{T - T_w}{T_\infty - T_w}$$

and making use of equation (10-16), equation (10-54) reduces to

$$\frac{1}{\mathrm{Pr}}\left(1 + \frac{4}{3N}\right)\frac{d^2\theta}{d\eta^2} + \frac{1}{2}f\frac{d\theta}{d\eta} = 0 \qquad (10\text{-}56)$$

with the boundary conditions

$$\theta(0) = 0 \qquad \theta(\infty) = 1$$

The local heat transfer from the plate surface may now be expressed in terms of the surface derivative $\theta'(0)$. Letting q_w denote the surface heat flux (i.e., $q_w = q_{Cw} + q_{Rw}$), then the sum of the conduction and radiation heat transfer rates at the surface becomes

$$q_w = -k\left(\frac{\partial T}{\partial y}\right)_{y=0} - \frac{4\sigma}{3\kappa}\left(\frac{\partial T^4}{\partial y}\right)_{y=0}$$

* Note that the two terms in equation (10-54) must be of equal magnitude and that $f(\eta) = O(\eta)$.

or, upon employing equation (10-16)

$$q_w = -\left(k + \frac{16\sigma T_\infty^3}{3\kappa}\right)\left(\frac{\partial T}{\partial y}\right)_{y=0} \tag{10-57}$$

Correspondingly, this may be expressed in dimensionless form as

$$\frac{q_w}{k(T_w - T_\infty)}\sqrt{\frac{\nu x}{u_\infty}} = +\left(1 + \frac{4}{3N}\right)\theta'(0) \tag{10-58}$$

The quantity $\theta'(0)$ is readily evaluated by noting that equation (10-56), with

$$\frac{1}{\mathrm{Pr}}\left(1 + \frac{4}{3N}\right)$$

replaced by $1/\mathrm{Pr}$, is simply the equation describing the nonradiating thermal boundary layer. Hence $\theta'(0)$ is obtained directly from tabulated results such as given, for example, in ref. 22.

Discussion of Results. A comparison of the dimensionless heat flux results, as given by both the optically thin ($\zeta \ll 1$) and optically thick ($\zeta \gg 1$) boundary layer models, is illustrated in Fig. 10-14 for $\mathrm{Pr} = 1$. The optically thin curves were obtained by combining the results given in Figs. 10-10 and 10-12. Recall again that the radiation heat flux as given in Fig. 10-10 applies to the optically thin boundary layer model. The optically thick curves represent equation (10-58), which expresses the dimensionless heat flux in terms of N. This may be rephrased in terms of ζ and ξ, since

$$N = \frac{\zeta^2}{2\xi}$$

Note that the limit $\zeta = \infty$ corresponds to a completely opaque fluid, such that there is no radiation influence.

The manner in which the surface heat flux has been made dimensionless in Fig. 10-14 corresponds to that conventionally used when radiation transfer is absent; that is, the dimensionless heat flux employed in Fig. 10-14 is simply a Nusselt number. An alternative dimensionless representation of the surface heat flux, corresponding to that utilized when conduction is absent, is shown in Fig. 10-15. Note from this figure that for large ξ the two limiting solutions are converging, which is in agreement with previous arguments, since for $\xi \gg 1$ the problem approaches that of optically thick radiation, negligible conduction, and nonviscous flow. Furthermore, the curve for $\zeta = 0$ corresponds to Fig. 10-10, since $\zeta = 0$ implies $N = 0$.

Although the present section has been devoted to rather elementary

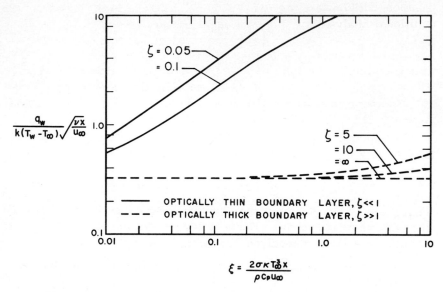

$$\frac{q_w}{k(T_w - T_\infty)}\sqrt{\frac{\nu x}{u_\infty}}$$

$\zeta = 0.05$
$= 0.1$

$\zeta = 5$
$= 10$
$= \infty$

—— OPTICALLY THIN BOUNDARY LAYER, $\zeta \ll 1$
--- OPTICALLY THICK BOUNDARY LAYER, $\zeta \gg 1$

$$\xi = \frac{2\sigma\kappa T_\infty^3 x}{\rho c_p u_\infty}$$

Fig. 10-14   Comparison of surface heat transfer results for Pr = 1.

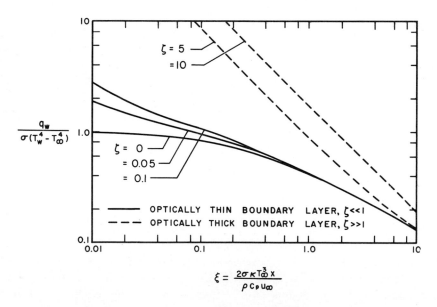

$$\frac{q_w}{\sigma(T_w^4 - T_\infty^4)}$$

$\zeta = 5$
$= 10$

$\zeta = 0$
$= 0.05$
$= 0.1$

—— OPTICALLY THIN BOUNDARY LAYER, $\zeta \ll 1$
--- OPTICALLY THICK BOUNDARY LAYER, $\zeta \gg 1$

$$\xi = \frac{2\sigma\kappa T_\infty^3 x}{\rho c_p u_\infty}$$

Fig. 10-15   Comparison of surface heat transfer results for Pr = 1.

solutions for the limiting cases $N \ll 1$ (or $\zeta \ll 1$) and $\zeta \gg 1$, several additional analyses are available in the literature.   A complete summary of these analyses is given by Viskanta (ref. 23), while a particularly interesting investigation is that of Oliver and McFadden (ref. 24). This involves an iterative solution of equation (10-39) through substitution of the temperature profile for no radiation interaction into the radiation term in equation (10-39) and thus obtaining a first approximation.   This process was repeated to the third approximation.*

In obtaining the iterative solution, no account was taken of temperature gradients occurring outside the thermal boundary layer.   For example, if the boundary layer were optically thin, this would amount to neglecting the outer solution $T_0(\xi, \tau)$.   Instead it was assumed that a representative radiation flux could be prescribed at the outer edge of the thermal boundary layer.

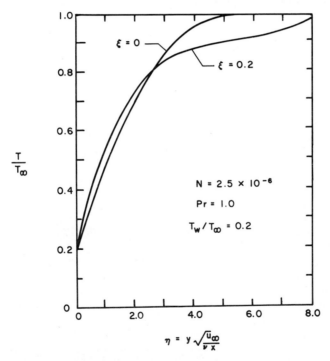

Fig. 10-16   Boundary layer temperature profiles.   From Oliver and McFadden (ref. 24).

* This actually constitutes a regular perturbation.   For example, letting $\hat{\xi} = N\xi$ in equation (10-39), the quantity $1/N$ appears only as a multiplier of the radiation term in the equation.   Consequently, in addition to the previously discussed singular perturbation, one may also perform a regular perturbation in powers of $1/N$.

Figure 10-16 illustrates a typical temperature profile for the case in which the radiation flux incident on the outer edge of the boundary layer is $\sigma T_\infty{}^4$. The profile for $\xi = 0$ represents no radiation interaction and is included for comparative purposes. One may note that radiation interaction results in an increase in the temperature gradient at the wall, with a corresponding increase in conduction heat transfer. This is in qualitative agreement with previous findings. The value of N given in Fig. 10-16 clearly corresponds to an optically thin boundary layer, and equation (10-52), together with Fig. 10-13, also illustrates an increase in conduction for $T_w/T_\infty = 0.2$.

REFERENCES

1. H. Schlichting, *Boundary Layer Theory*, McGraw-Hill, New York, 1960, pp. 299–302.

2. L. S. Wang, Differential methods for combined radiation and conduction. Ph.D. thesis in engineering, University of California, Berkeley, September 1965.

3. R. Viskanta, Heat transfer in a radiating fluid with slug flow in a parallel-plate channel. *Appl. Sci. Res.*, Sect. A, **13**, 291–311 (1964).

4. R. Viskanta, Interaction of heat transfer by conduction, convection and radiation in a radiating fluid. *J. Heat Transfer*, **C85**, 318–328 (1963).

5. R. A. Seban and T. T. Shimazaki, Heat transfer to a fluid flowing turbulently in a smooth pipe with walls at constant temperature. *Trans. ASME*, **73**, 803–809 (1951).

6. T. H. Einstein, Radiant heat transfer to absorbing gases enclosed between parallel plates with flow and conduction. *NASA TR R-154* (1963).

7. T. H. Einstein, Radiant heat transfer to absorbing gases enclosed in a circular pipe with conduction, gas flow and internal heat generation. *NASA TR R-156* (1963).

8. H. C. Hottel and E. S. Cohen, Radiant heat exchange in a gas-filled enclosure; Allowance for nonuniformity of gas temperature. *Am. Inst. Chem. Engr. J.*, **4**, 3–14 (1958).

9. S. deSoto and D. K. Edwards, Radiative emission and absorption in non-isothermal nongray gases in tubes. Proceedings of the 1965 Heat Transfer and Fluid Mechanics Institute, Stanford University Press, Palo Alto, Calif., 1965, pp. 358–372.

10. M. Jakob, *Heat Transfer*, vol. I, Wiley, New York, 1949.

11. D. K. Edwards and W. A. Menard, Comparison of models for correlation of total band absorption. *Appl. Optics*, **3**, 621 (1964).

12. L. D. Nichols, Temperature profile in the entrance region of an annular passage considering the effects of turbulent convection and radiation. *Intern. J. Heat Mass Transfer*, **8**, 589–608 (1965).

13. R. Goulard, The transition from black body to Rosseland formulations in optically thick flows. *Intern. J. Heat Mass Transfer*, **6**, 927–930 (1963).

14. C. L. Tien and R. Greif, On the transition from black body to Rosseland formulations in optically thick flows. *Intern. J. Heat Mass Transfer*, **7**, 1145–1146 (1964).

15. W. Lick, Transient energy transfer by conduction and radiation. *Intern. J. Heat Mass Transfer*, **8**, 119–128 (1965).

16. H. S. Carslaw and J. C. Jaeger, *Conduction of Heat in Solids*, Oxford University Press, London, 1948, pp. 51–53.

17. H. Schlichting, *Boundary Layer Theory*, McGraw-Hill, New York, 1960.

18. M. Van Dyke, *Perturbation Methods in Fluid Mechanics*, Academic Press, New York, 1964.

19. R. D. Cess, The interaction of thermal radiation in boundary layer heat transfer. Proceedings, Third International Heat Transfer Conference, Chicago, August 1966, pp. 154–163.

20. R. D. Cess, Radiation effects upon boundary-layer flow of an absorbing gas. *J. Heat Transfer*, **C86**, 469–475 (1964).

21. R. Viskanta and R. J. Grosh, Boundary layer in thermal radiation absorbing and emitting media. *Intern. J. Heat Mass Transfer*, **5**, 795 (1962).

22. S. S. Kutateladze, *Fundamentals of Heat Transfer*, Academic Press, New York, 1963.

23. R. Viskanta, Radiation transfer and interaction of convection with radiation heat transfer. In *Advances in Heat Transfer*, vol. 3, Academic Press, New York, 1966.

24. C. C. Oliver and P. W. McFadden, The interaction of radiation and convection in the laminar boundary layer. *J. Heat Transfer*, **C88**, 205–213 (1966).

25. J. L. Novotny, M. D. Kelleher, and P. Schimmel, Radiation Interaction in Free-Convection Stagnation Flow, presented at 5th U.S. National Congress of Applied Mechanics, University of Minnesota, June 1966.

PROBLEMS

10-1.   Consider the generalized exponential kernel approximation

$$E_1(t) \simeq ae^{-bt}, \qquad E_2(t) \simeq (a/b)e^{-bt}$$

Employing the above in equation (10-17), show that in order to satisfy the physical requirement that $\Theta(\xi, \infty) = 1$, it is necessary that $a = b$. Note that this condition is satisfied by equations (10-30).

10-2.   Consider again the exponential kernel approximation given above.   Show that in order to satisfy the condition

$$\frac{q_{Rw}}{\sigma(T_w^4 - T_\infty^4)} = 1 \text{ for } \xi \to 0$$

it is required that $b^2 = 2a$. This, combined with the requirement of the preceding problem, yields equations (10-30).

10-3.   Solve the problem discussed in Section 10-3 by assuming optically thick radiation.   Show that, in the limit of large ξ, the radiation slip solution, equation (10-29), agrees with the optically thick solution. Hint: For large x, $\exp(x^2)\, \text{erfc}\,(x) = \pi^{-1/2}x^{-1}$.

10-4.   Compare equations (10-29) and (10-36) in the limit of large ξ.

10-5.   Show that for $N \to 0$, equations (10-56) and (10-26) are identical.

CHAPTER 11

INFRARED RADIATIVE ENERGY
TRANSFER IN GASES

The object of this chapter is to incorporate spectroscopic gas properties into the basic equations of radiative transfer, so as to present a realistic means of analyzing radiative transfer within real (nongray) gases. For this purpose specific restriction will be made to infrared gaseous radiation, which results from molecular transitions involving both vibrational and rotational energies, with emphasis on the application of molecular band models to radiative transfer analyses. It will be necessary to heavily augment the discussion of Section 1-3 concerning infrared vibration-rotation bands, clearly illustrating the physical nature of these absorption-emission bands, as well as describing the spectroscopic parameters that are employed to analytically model such bands.

This description of vibration-rotation bands is presented in Section 11-1, together with the formulation of band absorptance models. In Section 11-2 the basic equations of radiative transfer, as derived in Section 7-3, are applied to infrared gaseous radiation through incorporation of the previously discussed band absorptance, while Section 11-3 deals with several illustrative applications. The survey article by Cess and Tiwari (ref. 1) has been utilized extensively throughout this chapter.

11-1 Band Absorptance Models

The purpose of this section is to formulate and discuss spectroscopic models describing the total band absorptance for infrared radiating gases. As will be seen in Section 11-2, the band absorptance plays an essential role in formulating the equations for the radiative energy flux. First, however, it will be necessary to review and augment the discussion of Section 1-3 concerning the basic structure of vibration-rotation bands.

Band absorption. Infrared absorption and emission of thermal radiation is a consequence of coupled vibrational and rotational energy transitions. Quite obviously a diatomic molecule is the simplest molecule that will undergo such transitions. However, symmetric diatomic molecules, such as H_2, O_2 and N_2, have no permanent dipole moment and thus are transparent to infrared radiation. An exception to this concerns pressure-induced bands. For example, infrared transmission by hydrogen is important within the atmospheres of the major planets. For unsymmetric diatomic molecules, such as CO, the infrared spectrum will consist of a fundamental vibration-rotation band occurring at the fundamental vibrational frequency of the molecule; that is, the band is produced by energy transitions between adjacent vibrational levels. Vibrational transitions spanning three vibrational levels produce the first overtone band, located at twice the fundamental frequency of the molecule, and subsequent overtone bands occur at higher multiples of the fundamental frequency. In general, the overtone bands are much weaker than the fundamental band.

The picture is much the same for polyatomic molecules, except that these have more vibrational degrees of freedom. Carbon dioxide, for example, is a linear triatomic molecule and thus possesses four vibrational degrees of freedom. The two bending frequencies, however, are identical, while one of the stretching modes is symmetric and thus has no permanent dipole moment. Consequently, carbon dioxide has two fundamental bands. In addition to fundamental and overtone bands, the infrared spectrum of polyatomic molecules also includes combination and difference bands, which occur at linear combinations or differences of the fundamental frequencies. Again choosing carbon dioxide as an example, the important infrared bands are the 15μ and 4.3μ fundamental bands and the 2.7μ and 1.9μ combination bands (see Table 1-3 and Fig. 1-12).

While the location of a vibration-rotation band is described by the associated vibrational transition, the band structure is governed by simultaneous rotational transitions that accompany the vibrational transition. As a consequence of the unequal spacing of rotational energy levels, the coupled vibration-rotation transitions occur at discrete frequencies centered about the vibrational frequency. The resulting band structure in turn consists of an array of discrete rotational lines.

Before proceeding, it should be mentioned that while a vibrational transition is always coupled with a rotational transition, rotational transitions do occur by themselves. Since the rotational transition energies are generally small, the resulting spectrum is normally in the microwave region and has no influence on infrared radiation. There are exceptions, such as water vapor, which possesses a pure rotation band in the far infrared.

In order to describe the absorption characteristics of a vibration-rotation band, it is first necessary to consider the variation of the spectral absorption coefficient for a single line. For infrared radiation, the most important line-broadening mechanism is pressure broadening, for which the variation of the spectral absorption coefficient with wave number is given by the Lorentz line profile as

$$\frac{\kappa_{\omega j}}{P} = \frac{S_j}{\pi} \frac{\gamma_j}{\gamma_j^2 + (\omega - \omega_j)^2} \tag{11-1}$$

Here κ_ω denotes the volumetric absorption coefficient and ω is wave number ($\omega = 1/\lambda$). The rotational quantum number is denoted by j, such that the subscript j refers to a specific line within the band. Thus the wave-number location of the line is ω_j, and γ_j and S_j refer to the half-width and intensity of the line, respectively. For the time being no distinction will be made between total and partial pressures. The line intensity is defined as

$$S_j = \int_{-\infty}^{\infty} \frac{\kappa_{\omega j}}{P} \, d(\omega - \omega_j) \tag{11-2}$$

and this is consistent with equation (11-1). The line intensity may be described in terms of the molecular number density and Einstein coefficients (or transition probabilities); for a perfect gas it follows that S_j is a function solely of temperature.

From kinetic theory the line half-width may be shown to vary with temperature and pressure according to

$$\gamma_j \sim \frac{P}{\sqrt{T}} \tag{11-3}$$

More detailed quantum mechanical calculations again show the linear dependency on pressure, but indicate that the inverse square-root variation with temperature often is not correct. Again considering CO_2 as an example, Yamamoto, Tanaka, and Aoki (ref. 2) have shown that the temperature dependency of the line half-width may be described by $\gamma_j \sim T^{-n_j}$, and that n_j approaches 0.75 for small j, decreasing with increasing j to approximately 0.3, and then increases with a further increase in j to the kinetic theory value of 0.5.

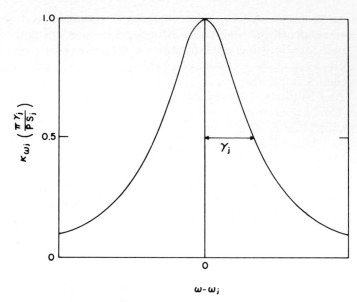

Fig. 11-1   Lorentz line profile.   From Cess and Tiwari (ref. 1).

The Lorentz line profile, as described by equation (11-1), is illustrated in Fig. 11-1. There are two points worth noting. The first is that increasing pressure broadens the line and, with respect to a complete band consisting of many lines, leads to a smearing out of the discrete line structure for sufficiently high pressures. Figures 1-9 and 1-10 illustrate this high-pressure limit for carbon monoxide bands. The second point is that the maximum absorption coefficient, occurring at $\omega = \omega_j$, is invariant with pressure, since

$$(\kappa_{\omega j})_{\omega = \omega_j} = \frac{PS_j}{\pi\gamma_j} \tag{11-4}$$

from equation (11-1), while $\gamma_j \sim P$.

It remains to describe the variation of line intensity with rotational quantum number; for present purposes the simple model of a harmonic oscillator and rigid rotor will be assumed. Following Penner (ref. 3), and assuming a large number of lines (large j), the variation of S_j with j is given by

$$S_j = \frac{ShcBj}{kT} \exp\left(-\frac{hcBj^2}{kT}\right) \tag{11-5}$$

where h is Planck's constant, k is Boltzmann's constant, and B is the rotational constant of the molecule. Furthermore, S denotes the intensity of the total band (or integrated band absorption) such that

$$S = \int_{-\infty}^{\infty} \frac{\kappa_\omega}{P}\, d(\omega - \omega_0) \qquad (11\text{-}6)$$

where ω_0 is the wave number at the band center. It should be realized, of course, that the integration limits in equation (11-6) imply integration over the entire band, as opposed to equation (11-2), where the limits indicate integration over a single line.

A further consequence of the rigid rotor is that the lines are equally spaced, with the spacing $d = 2B$. Consequently the line locations may be expressed in terms of wave number by

$$\omega - \omega_0 = \pm 2jB \qquad (11\text{-}7)$$

Combination of equations (11-6) and (11-7) allows a continuous representation of S_j with wave number, as illustrated in Fig. 11-2. For an actual band the two branches (P and R branches) are not symmetric, while vibrational modes involving bending exhibit a third central branch (Q branch). Nevertheless, the present simplified model serves the purpose for which it is intended; that is, to illustrate the basic features of the total band absorptance.

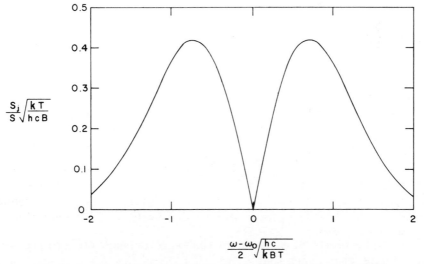

Fig. 11-2. Variation of S_j for a harmonic oscillator and rigid rotor. From Cess and Tiwari (ref. 1).

With regard to the variation of κ_ω over the entire band, this will consist of the superposition of the contributions of the individual lines, such that

$$\kappa_\omega = \sum_j \kappa_{\omega j} = \frac{P}{\pi} \sum_j \frac{S_j \gamma_j}{\gamma_j{}^2 + (\omega - \omega_j)^2} \qquad (11\text{-}8)$$

Furthermore

$$S = 2 \sum_{j=0}^{\infty} S_j$$

where the factor of two is included to account for both branches. Again assuming a large number of lines, the summation may be replaced by an integration, and employing equation (11-5)

$$S = 2 \int_0^\infty S_j \, dj = S$$

which illustrates that the separate applications of the assumption of a large number of lines are mutually consistent. Since S_j is a function solely of temperature, then the above expression illustrates that the band intensity is also a function only of temperature.

Band absorptance.   The spectral band absorptance is defined as

$$\alpha_\omega = 1 - \exp(-\kappa_\omega y) \qquad (11\text{-}9)$$

The physical interpretation of α_ω is that it is the fraction of energy which is absorbed when a beam of radiant energy passes through an isothermal slab of thickness y.   The total band absorptance is, in turn,

$$A = \int_{-\infty}^{\infty} \alpha_\omega \, d(\omega - \omega_0) \qquad (11\text{-}10)$$

where integration over the single band is again implied.   The physical interpretation of the total band absorptance is not as simple as that of its spectral counterpart α_ω.   For present purposes it will suffice to state that the total band absorptance will be needed later to generate the kernel function in the radiative flux equations.

A convenient form of equation (11-10) is

$$A = \int_{-\infty}^{\infty} \left[1 - \exp\left(-\frac{\kappa_\omega}{P} P y \right) \right] d(\omega - \omega_0) \qquad (11\text{-}11)$$

From equation (11-8) it is evident that κ_ω / P depends on both pressure and temperature, such that

$$A = A(Py, P, T) \qquad (11\text{-}12)$$

It is important to note the dual role that pressure plays. Its appearance in the pressure path length, Py, is simply due to the fact that absorption is dependent on the number of molecules that are present along a line of sight. The second dependency on pressure comes about because the line structure of the band is a function of pressure. For sufficiently high pressures the line structure is smeared out, so that in this limit pressure enters solely through the pressure path length Py. This will later be illustrated in a quantitative manner.

In the two following subsections simple band models will be employed to illustrate certain basic features of the total band absorptance. There is, however, one important limiting form of A that is completely independent of the band model, and this applies when $\kappa_\omega y \ll 1$; that is, for the conventional optically thin limit. On expanding the exponential in equation (11-11),

$$A = Py \int_{-\infty}^{\infty} \frac{\kappa_\omega}{P} d(\omega - \omega_0) = PyS \qquad (11\text{-}13)$$

This is referred to as the linear limit; the important feature is that in this limit the total band absorptance is independent of rotational line structure.

A second limiting form for the total band absorptance is that of strong, nonoverlapping lines. Although the actual limiting result for A depends on the band model employed, the conditions for achieving this limit may be discussed in general terms. The limit requires that two separate conditions be satisfied. The first is the requirement of strong lines, for which complete absorption occurs in the vicinity of the line centers. From equation (11-9), this is equivalent to requiring that $\kappa_{\omega j} y \gg 1$ for $\omega = \omega_j$, and upon combining this with equation (11-4), the strong line condition becomes

$$\frac{S_j Py}{\pi \gamma_j} \gg 1 \qquad (11\text{-}14)$$

The second condition pertains to nonoverlapping lines, and the motivation for this limit is to be able to employ the expression

$$A = \sum_j A_j \qquad (11\text{-}15)$$

where A_j is the absorptance of a single line

$$A_j = \int_{-\infty}^{\infty} [1 - \exp(-\kappa_{\omega j} y)] d(\omega - \omega_j) \qquad (11\text{-}16)$$

where the integration is performed over the individual lines. Equation (11-15) is, of course, applicable only if the integrands in equation (11-16) do not overlap, since equation (11-15) is simply a summation of individual line absorptances. What is required, then, is that the integrand in equation (11-16) approach zero for $\omega - \omega_j = 0(d)$, where d is the line spacing. With reference to equation (11-1), the nonoverlapping limit will be satisfied provided

$$\frac{S_j Py}{\pi} \frac{\gamma_j}{\gamma_j{}^2 + d^2} \ll 1 \tag{11-17}$$

At this point it should be noted that if we were to allow $\gamma_j \geq 0(d)$, equation (11-17) would yield

$$\frac{S_j Py}{\pi \gamma_j} \ll 1$$

which is a direct contradiction to the strong-line condition of equation (11-14). Hence, to avoid this contradiction it is necessary to require that $\gamma_j \ll d$. The conditions that must be satisfied in order to achieve nonoverlapping lines are thus

$$\frac{\gamma_j}{d} \ll 1, \qquad \frac{S_j Py\gamma_j}{\pi d^2} \ll 1 \tag{11-18}$$

where the second condition follows from equation (11-17). The above conditions, together with equation (11-14), thus describe the strong nonoverlapping line limit. The application of these three conditions in deriving this limit will be illustrated in the following subsection.

Elsasser model. The simplest model that accounts for line structure is the Elsasser band model, for which equally spaced lines of equal intensity and equal half-width are assumed. A portion of such a band is illustrated in Fig. 11-3, where the broken curves represent the absorption coefficients of the individual lines, while the solid curve is the absorption coefficient as given by equation (11-8), and this may be rephrased as

$$\frac{\kappa_\omega}{P} = \frac{S_j\gamma}{\pi} \sum_{j=0}^{\infty} \frac{1}{\gamma^2 + (\omega - \omega_0 \pm jd)^2} \tag{11-19}$$

The subscript j has been dropped from γ_j in accord with the previous assumption, but it is retained in S_j, even though S_j is independent of j, to denote that this is a line intensity. Carrying the summation to infinity does not preclude restriction to a finite bandwidth (finite

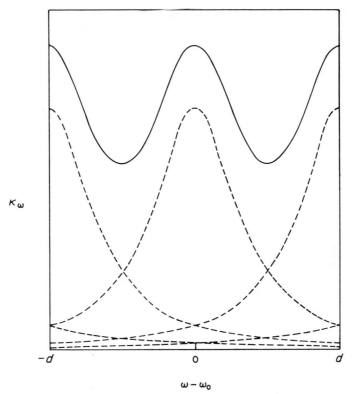

κ_ω

$-d$ O d

$\omega - \omega_0$

Fig. 11-3.   Elsasser band model.   From Cess and Tiwari (ref. 1).

number of lines), but is merely consistent with the earlier assumption of a large number of lines.   The above series may be expressed in closed form (ref. 4) as

$$\frac{\kappa_\omega}{P} = \frac{S_j}{d}\left[\frac{\sinh\,(\pi\beta/2)}{\cosh\,(\pi\beta/2)\,-\,\cos\,(\pi z/2)}\right] \tag{11-20}$$

where

$$\beta = \frac{4\gamma}{d}, \qquad z = \frac{4(\omega\,-\,\omega_0)}{d}$$

The quantity β is a particularly significant parameter, since it represents the role of line structure.   Recalling that $\gamma \sim P$, the limit of large pressure corresponds to $\beta \to \infty$; this is the limit for which line structure is smeared out, and equation (11-20) reduces to

$$\frac{\kappa_\omega}{P} = \frac{S_j}{d} \tag{11-21}$$

The ratio S_j/d has an alternate interpretation. If an average absorption coefficient is defined over a line spacing as

$$\frac{\bar{\kappa}_\omega}{P} = \frac{1}{d} \int_{-d/2}^{d/2} \frac{\kappa_\omega}{P} \, d(\omega - \omega_j)$$

it follows from equation (11-20) that $\bar{\kappa}_\omega/P = S_j/d$.

The wave number width of the total band will be denoted by A_0 and, letting n be the number of lines in the band, $A_0 = nd$. Furthermore

$$S = \sum_j S_j = nS_j$$

and the total absorptance of the Elsasser band follows from equations (11-11) and (11-20) as

$$A = \frac{A_0}{2} \int_0^\infty \left\{ 1 - \exp\left[-\frac{u \sinh (\pi\beta/2)}{\cosh (\pi\beta/2) - \cos (\pi z/2)} \right] \right\} dz \quad (11\text{-}22)$$

where u is a dimensionless pressure path length defined by

$$u = \frac{SPy}{A_0}$$

Equation (11-22) has a form that is characteristic of all band models, namely that the total absorptance may be expressed as

$$A = A_0 \bar{A}(u, \beta) \tag{11-23}$$

where $\bar{A}(u, \beta)$ is a dimensionless function. Recall from previous discussion that pressure enters into the band absorptance in two ways, through both the pressure path length and the line structure effect. This dual role is clearly illustrated by equation (11-23), since u is a dimensionless pressure path length while β is a line structure parameter.

Consider now limiting forms of the total band absorptance. The linear limit, applicable for $u \ll 1$, readily follows from equation (11-22) as

$$\bar{A} = u; \quad u \ll 1 \tag{11-24}$$

and this is consistent with equation (11-13). Note once again that line structure plays no role in the limit of small path lengths. In the large path length limit, $u \gg 1$, equation (11-22) yields

$$\bar{A} = 1; \quad u \gg 1 \tag{11-25}$$

Physically, of course, this represents total absorption within the finite-width band. It should be emphasized, however, that more realistic band models yield considerably different results for $u \gg 1$, as will be illustrated in the next subsection.

A third limit corresponds to strong nonoverlapping lines, and following Penner (ref. 3) or Goody (ref. 4), equation (11-22) may be reduced to

$$\bar{A} = \mathrm{erf}\left(\frac{1}{2}\frac{\sqrt{\pi u \beta}}{2}\right) \tag{11-26}$$

subject to certain constraints. As discussed by Penner (ref. 3), these consist of $\beta \ll 1$ and $u/\beta \gg 1$. With reference to equations (11-14) and (11-18), the remaining requirement for the strong nonoverlapping line limit is $\beta u \ll 1$, for which equation (11-26) reduces to

$$\bar{A} = \sqrt{\beta u}; \qquad \beta \ll 1, \quad u/\beta \gg 1, \quad \beta u \ll 1 \tag{11-27}$$

This is also referred to as the square-root limit.

A final limiting form of equation (11-22) is that for which line structure is smeared out, and letting $\beta \to \infty$ then

$$\bar{A} = 1 - e^{-u}; \qquad \beta \gg 1 \tag{11-28}$$

which, as should be expected, is simply Beer's law.

The Elsasser band absorptance is illustrated in Fig. 11-4, and the various limiting forms are clearly evident. For $u \ll 1$, the linear limit

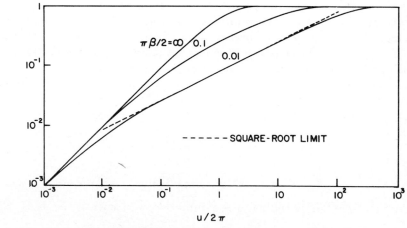

Fig. 11-4. Total band absorptance for Elsasser model. From Cess and Tiwari (ref. 1).

is obtained, with the band absorptance being independent of line structure. The three constraints on the square-root limit are also apparent. This limit requires $\beta \ll 1$, but there is a departure from the square-root limit for either large or small u. The departure for small u denotes a violation of the requirement that $u/\beta \gg 1$, such that the strong-line condition is no longer satisfied. For large u the condition $\beta u \ll 1$ is not fulfilled, such that equation (11-15) is no longer applicable. The present large path length limit, $\bar{A} = 1$, simply denotes total absorption within the band, and this is a consequence of the Elsasser bands having a prescribed finite width. For a more realistic band model, as described in the following subsection, the total band absorptance will asymptotically approach a function of u rather than unity.

Rigid-rotor, harmonic-oscillator model. Assuming the molecular model of a rigid rotor and harmonic oscillator, the distribution of line intensities is given by equation (11-5), while the line spacing corresponds to equation (11-7). Lorentz line shapes will again be assumed. In contrast to the Elsasser model, there is no defined bandwidth, since the line intensities approach zero in the band wings. Thus the bandwidth parameter A_0 will not correspond to a simple specified width, but instead will arise as an effective width resulting from the line intensity distribution of equation (11-5). No attempt at a complete formulation of the total band absorptance will be made; however the limiting expressions will be presented. Since the linear limit is completely general, equation (11-24) is applicable to the present band model.

Considering the large path length limit, it will be convenient to initially assume a high pressure such that line structure is smeared out, which corresponds to the limit $\beta \to \infty$. Thus, in accord with equation (11-21), and upon combining equations (11-5), (11-7), and (11-9), the spectral band absorptance is described by

$$\alpha_\omega = 1 - \exp\left(-u\xi e^{-\xi^2}\right) \qquad (11\text{-}29)$$

where again $u = SPy/A_0$, while

$$\xi = \frac{\omega - \omega_0}{A_0}$$

$$A_0 = \sqrt{4kBT/hc} \qquad (11\text{-}30)$$

Furthermore, since the band is symmetric, and with $\bar{A} = A/A_0$, then

$$\bar{A} = 2 \int_0^\infty \alpha_\omega(\xi)\, d\xi \qquad (11\text{-}31)$$

The combination of equations (11-29) and (11-31) thus describes the total band absorptance in the limit as $\beta \to \infty$; numerical results are given by Penner (ref. 3). Concerning an asymptotic expression for large u, a physically useful procedure is that employed by Edwards and Menard (ref. 5). Upon defining

$$\xi_1 = \sqrt{\ln u}$$

equation (11-31) may be written as

$$\bar{A} = 2\Gamma_1 + 2\Gamma_2 \tag{11-32}$$

where

$$\Gamma_1 = \int_0^{\xi_1} [1 - \exp(-u\xi e^{-\xi^2})] \, d\xi$$

$$\Gamma_2 = \int_{\xi_1}^{\infty} [1 - \exp(-u\xi e^{-\xi^2})] \, d\xi$$

It may readily be shown that for $u \gg 1$

$$\Gamma_1 \to \sqrt{\ln u} \tag{11-33a}$$

$$\Gamma_2 < \tfrac{1}{2} \tag{11-33b}$$

Thus the large path length limit follows as

$$\bar{A} = 2\sqrt{\ln u}; \qquad u \gg 1 \tag{11-34}$$

While equation (11-34) has been derived subject to the condition that $\beta \to \infty$, it is easily shown that this restriction may be removed. With reference to Fig. 11-5, Γ_1 denotes the area of the saturated portion of the band, and the inclusion of line structure will not alter equation (11-33a) as a proper asymptotic limit. Thus only Γ_2 will be affected, but again with reference to Fig. 11-5, if the region $\xi > \xi_1$ is considered to consist of a series of Elsasser bands, then the inclusion of line structure will result in a decrease in Γ_2, and equation (11-33b) is again valid. Equation (11-34) therefore constitutes the asymptotic limit for the total band absorptance regardless of the value of the line structure parameter β.

The third limit is the square-root limit, and recall that this limit corresponds to strong nonoverlapping lines. For strong lines, the Lorentz line profile, equation (11-1), may be rephrased as (Goody, ref. 4)

$$\frac{\kappa_{\omega j}}{P} = \frac{S_j \gamma_j}{(\omega - \omega_j)^2}$$

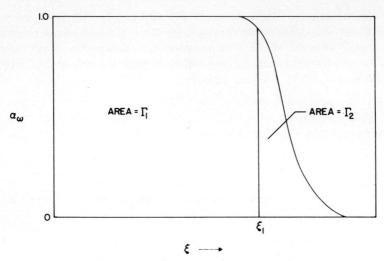

Fig. 11-5. Spectral band absorptance for large path lengths.
From Cess and Tiwari (ref. 1).

and upon substituting this into equation (11-16)

$$A_j = 2\sqrt{S_j\gamma Py}$$

where, as with the Elsasser model, it is assumed that γ_j is independent
of the rotational quantum number j; that is, $\gamma_j = \gamma$. In turn, from
equation (11-15)

$$A = 2\sum_{j=0}^{\infty} A_j = 4\sqrt{\gamma Py}\sum_{j=0}^{\infty}\sqrt{S_j}$$

Again employing the assumption of a large number of rotational lines,

$$A = 4\sqrt{\gamma Py}\int_0^{\infty}\sqrt{S_j}\,dj$$

and upon combining this with equation (11-5), noting from equation
(11-7) that $d = 2B$, it follows that

$$A = 2^{3/4}\Gamma\left(\frac{3}{4}\right)\sqrt{SPyA_0\left(\frac{4\gamma}{d}\right)} \qquad (11\text{-}35)$$

Thus, upon again letting $\beta = 4\gamma/d$ denote the line-structure parameter,
the square-root limit follows from equation (11-35) as

$$A = 2.06\sqrt{\beta u}; \qquad \beta \ll 1, \quad u/\beta \gg 1, \quad \beta u \ll 1 \qquad (11\text{-}36)$$

The primary utility of the present molecular model of a rigid rotor and harmonic oscillator has been to illustrate limiting solutions of the total band absorptance for a semirealistic molecular model. One important conclusion is that line structure appears only in the square-root limit.

Band absorptance correlations. While the preceding subsection dealt with limiting forms of the total band absorptance, it is necessary to have an expression for $\bar{A}(u, \beta)$ which is applicable for all values of u and β. Numerous such expressions are available, and we consider here only a few of these, which are based on constructing an expression for $\bar{A}(u, \beta)$ that satisfies certain limiting conditions. For present purposes the same limits as employed by Edwards and Menard (ref. 5) will be used, and these are

$$\bar{A} = u; \qquad u \ll 1 \tag{11-37a}$$

$$\bar{A} = 2\sqrt{\beta u}; \qquad \beta \ll 1, \quad u/\beta \gg 1, \quad \beta u \ll 1 \tag{11-37b}$$

$$\bar{A} = \ln u; \qquad u \gg 1 \tag{11-37c}$$

While Edwards and Menard interpreted the above in terms of a reordered exponential distribution of line intensities (see, for example, equation (1-35)), a slightly different explanation will be given here. The first limit, equation (11-37a), is simply the general linear limit, while equation (11-37b) is essentially equation (11-36) for the rigid rotor and harmonic oscillator. The third limit is of a different form than equation (11-34) for the rigid rotor and harmonic oscillator. For moderately large values of u the two expressions are in reasonable agreement, but for increasingly large u they begin to diverge. The rationale for using equation (11-37c) is twofold. First, Edwards and Menard (ref. 5) have shown that a logarithmic limit is attained for nonrigid rather than rigid rotation, and second, existing empirical correlations are of the same form as equation (11-37c).

The first band absorptance correlation to satisfy all three limits is that proposed by Edwards and Menard (ref. 5), and this consists of an analytic interpolation of the form

$$\beta \leq 1: \bar{A} = u; \qquad \bar{A} < \beta$$
$$\bar{A} = 2\sqrt{u\beta} - \beta; \qquad \beta < \bar{A} < (2 - \beta)$$
$$\bar{A} = \ln(\beta u) + (2 - \beta); \qquad \bar{A} > (2 - \beta)$$

$$\beta > 1: \bar{A} = u; \qquad \bar{A} < 1$$
$$\bar{A} = \ln u + 1; \qquad \bar{A} > 1$$

By comparing the above correlation with experimental data over a large range of pressures and temperatures, Edwards and Balakrishnan (ref. 6)* have empirically determined the necessary correlation quantities $S(T)$, $A_0(T)$, and $\beta(T, P_e)$, where P_e is the effective broadening pressure, for a number of bands of CO, CO_2, H_2O, NO, SO_2, and CH_4.

A continuous band absorptance correlation has been proposed by Tien and Lowder (ref. 7), and this is of the form

$$\bar{A} = \ln\left[uf(\beta)\left(\frac{u + 2}{u + 2f(\beta)}\right) + 1\right] \qquad (11\text{-}38)$$

where

$$f(\beta) = 2.94[1 - \exp(-2.60\beta)]$$

The choice of equation (11-38) was based on the specification of five limiting conditions, and the form of $f(\beta)$ was chosen to give agreement with the correlation of Edwards and Menard. The square-root limit, equation (11-37b), was not, however, one of the specified conditions, and equation (11-38) does not satisfy this requirement.

Recently, numerous functional forms have been suggested for $\bar{A}(u, \beta)$, and a comprehensive survey of all of these is not attempted here. A fairly simple continuous correlation, which does satisfy all of the constraints imposed by equation (11-37), is given by Cess and Ramanathan (ref. 8) as

$$\bar{A} = 2\ln\left[1 + \frac{u}{2 + \sqrt{u(1 + 1/\beta)}}\right] \qquad (11\text{-}39)$$

This formulation has proved useful in a number of atmospheric radiation problems.

11-2   Basic Equations

In this section the basic equations will be developed which describe the radiative flux and its divergence within an infrared absorbing-emitting gas. In formulating the radiative flux equation, the kernel function will be expressed in terms of the total band absorptance, an approach that is analogous to the formulations of Goody (ref. 4), Gille and Goody (ref. 9), and Wang (refs. 10 and 11), in which the kernel function is expressed in terms of a modified gas emissivity. The

---

* The correlation quantities α_0 and ω_0 of ref. 6 correspond to the present nomenclature through $S = \alpha_0/RT$ and $A_0 = \omega_0$, where R is the gas constant.

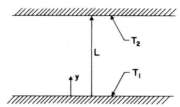

Fig. 11-6.   Physical model and coordinate system.   From Cess
and Tiwari (ref. 1).

physical model and coordinate system is that illustrated in Fig. 11-6.
This consists of a gas bounded by two plates whose surfaces are assumed
to be black.

The applicable expression for the monochromatic radiative flux
$q_{R\omega}$, expressed in terms of wave number $\omega(\omega = 1/\lambda)$ rather than wavelength λ, follows from equations (7-16) and (7-28) as

$$q_{R\omega} = 2e_{1\omega}E_3(\kappa_\omega y) - 2e_{2\omega}E_3[\kappa_\omega(L - y)]$$

$$+ 2\left\{ \int_0^y e_\omega(z)\kappa_\omega E_2[\kappa_\omega(y - z)\ dz] \right.$$

$$\left. - \int_y^L e_\omega(z)\kappa_\omega E_2[\kappa_\omega(z - y)]\ dz \right\} \qquad (11\text{-}40)$$

with $\gamma_\omega = 0$ since we are considering a nonscattering gas, while the
surface radiosities have been replaced by black-body emissive powers
since, for illustrative purposes, we are considering only black bounding
surfaces.   The monochromatic absorption coefficient, κ_ω, will be taken
to be independent of temperature within equation (11-40), and this
effectively constitutes a linearization for small temperature differences.

As discussed in Chapter 8, the exponential kernel substitution, as
given by equations (8-6) and (8-7), constitutes a useful approximation
for gray gases, and it will further be employed in the present formulation.   Thus

$$E_2(x) \simeq \tfrac{3}{4} \exp\left(-\frac{3x}{2}\right) \qquad (11\text{-}41\text{a})$$

$$E_3(x) \simeq \tfrac{1}{2} \exp\left(-\frac{3x}{2}\right) \qquad (11\text{-}41\text{b})$$

The total radiative flux is further given by

$$q_R = \int_0^\infty q_{R\omega}\ d\omega \qquad (11\text{-}42)$$

For illustrative purposes we initially consider a gas having a single vibration-rotation band centered at the wave number ω_0. Upon combining equations (11-40), (11-41), and (11-42), the total radiative flux may be expressed as

$$q_R = e_1 - e_2 + \tfrac{3}{2} \int_0^y [e_{\omega_0}(z) - e_{1\omega_0}] \int_{\Delta\omega} \kappa_\omega \exp\left[-\tfrac{3}{2}\kappa_\omega(y - z)\right] d\omega dz$$

$$- \tfrac{3}{2} \int_y^L [e_{\omega_0}(z) - e_{2\omega_0}] \int_{\Delta\omega} \kappa_\omega \exp\left[-\tfrac{3}{2}\kappa_\omega(z - y)\right] d\omega dz \quad (11\text{-}43)$$

with $e = \sigma T^4$, while $\Delta\omega$ indicates integration over the single band, and equation (11-43) makes use of the fact that $e_\omega(T)$ is a slowly varying function of wave number over the single band.

The motivation for employing the exponential kernel substitution in the present formulation is that it allows the kernel function within equation (11-43) to be expressed in terms of the total band absorptance since, from equation (11-11),

$$\frac{dA}{dy} = A'(y) = \int_{\Delta\omega} \kappa_\omega \exp\left(-\kappa_\omega y\right) d\omega$$

and this is the form of the kernel function within equation (11-43). Thus, letting

$$\xi = \frac{y}{L}, \qquad u_0 = \frac{SPL}{A_0}$$

and employing the dimensionless band absorptance, $\bar{A}(u, \beta)$, as defined by equation (11-23), the final form of the radiative flux equation becomes

$$q_R(\xi) = e_1 - e_2 + \tfrac{3}{2}A_0u_0 \left\{ \int_0^\xi [e_{\omega_0}(\xi') - e_{1\omega_0}]\bar{A}'[\tfrac{3}{2}u_0(\xi - \xi')]\, d\xi' \right.$$

$$\left. - \int_\xi^1 [e_{\omega_0}(\xi') - e_{2\omega_0}]\bar{A}'[\tfrac{3}{2}u_0(\xi' - \xi)\, d\xi'] \right\} \quad (11\text{-}44)$$

where $\bar{A}'(u)$ denotes the derivative of $A(u)$ with respect to u. Note that rotational line structure is included within equation (11-44) through the band absorptance.

Equation (11-44) possesses two convenient limiting forms. One is the conventional optically thin limit, while the other, the large path length limit, corresponds to $u_0 \gg 1$ and differs considerably from the optically thick or Rosseland limit. These two limiting forms of equation (11-44) are discussed in the following subsections.

Optically thin limit. Recall from Section 7-5 that in the optically thin limit one is generally concerned with the divergence of the radiative flux, such that on differentiating equation (11-44) and employing the linear limit $\bar{A}'(u) = 1$, the appropriate expression for $u_0 \ll 1$ becomes

$$\frac{dq_r}{d\xi} = (3/2)A_0 u_0 [2e_{\omega 0}(\xi) - e_{1\omega 0} - e_{2\omega 0}] \qquad (11\text{-}45)$$

An alternate approach to the optically thin limit is given by Cess, Mighdoll, and Tiwari (ref. 12), which does not make use of the exponential kernel substitution as given by equation (11-41), and it is shown that the factor 3/2 appearing in equation (11-45) is replaced by 2 in the exact formulation. Further comments on the applicability of the exponential kernel substitution to infrared radiative transfer are given by Grief and Habib (ref. 13). It should be observed that equation (11-45) is independent of rotational line structure, and this is consistent with the previous discussion on the invariance of $\bar{A}(u, \beta)$ with the line structure parameter β in the linear limit.

Large path length limit. Even though the optically thick (Rosseland) limit does not apply to vibration-rotation bands, since optically nonthick radiation will always occur within the wings of the bands, a large path length limit does exist and is achieved for $u_0 \gg 1$. Employing the method of steepest descent, it may be shown that the asymptotic form of the integrals appearing in equation (11-44) corresponds to the use of the logarithmic limit for the band absorptance, equation (11-37c). For illustrative purposes it will again be convenient to deal with the divergence of the radiative flux. Thus, upon differentiating equation (11-14), performing a subsequent integration by parts, and utilizing the asymptotic expression $\bar{A}(u) = \ln u$, one obtains

$$\frac{dq_R}{d\xi} = A_0 \int_0^1 \frac{de_{\omega 0}}{d\xi'} \frac{d\xi'}{\xi - \xi'} \qquad (11\text{-}46)$$

In arriving at equation (11-46), continuity of temperature has been assumed between the gas and the bounding surfaces. This is physically realistic, since $u_0 \gg 1$ implies that the central portion of the band is optically thick, which would ensure temperature continuity. A more quantitative treatment of this point will be given in Section 11-3.

A simplification associated with equation (11-46) is that, as for the optically thin limit, the radiative transfer process is independent of line structure since the line structure parameter does not appear within the equation. The reason for this is that the band absorptance becomes invariant with line structure for large u_0, and it is this asymptotic result for the band absorptance that yields equation (11-46). Note also that equation (11-46) is independent of both pressure and band intensity, and this will be discussed in more detail in the following section.

11-3 Radiative Transfer Illustration

This section presents a simple illustrative application of conservation of energy within an infrared radiating gas, with primary emphasis on the basic features of the radiative transfer process. For this purpose we choose a case for which radiative transfer is the sole mechanism of energy transfer within the gas, the local temperature within the gas being a consequence of a uniform heat source per unit volume Q within the gas. Thus energy is added to the gas by this heat source, and the energy is in turn transferred through the gas to the bounding surfaces by radiative transfer. The two bounding surfaces are assumed to be at the same temperature, $T_2 = T_1$.

The energy equation for this situation is simply

$$\frac{dq_R}{dy} = Q$$

and since the problem is symmetric,

$$q_R = \frac{QL}{2}(2\xi - 1) \tag{11-47}$$

where again $\xi = y/L$. For a single-band spectrum, the radiative flux is described by equation (11-44), but this may be extended to multiple-band spectra by simply summing equation (11-44) over the individual bands. Furthermore, since small temperature differences have been assumed in arriving at equation (11-44), one may additionally employ the linearization.

$$e_{\omega i} - e_{1\omega i} \simeq \left(\frac{de_{\omega i}}{dT}\right)_{T_1}(T - T_1) \tag{11-48}$$

where the subscript i refers to the ith band, such that ω_i is the wave number location of the band. The subsequent extension of equation (11-44), recalling that $T_2 = T_1$, thus yields

$$q_R = \tfrac{3}{2}\sum_{i=1}^{n}\left[A_{0i}u_{0i}\left(\frac{de_{\omega i}}{dT}\right)_{T_1}\right]\left\{\int_0^\xi [T(\xi') - T_1]\bar{A}'[\tfrac{3}{2}u_{0i}(\xi - \xi')]\,d\xi'\right.$$
$$\left. - \int_\xi^1 [T(\xi') - T_1]\bar{A}'(\tfrac{3}{2}u_{0i}(\xi' - \xi))\,d\xi'\right\} \tag{11-49}$$

where n represents the number of vibration-rotation bands in the spectrum.

Upon combining equations (11-47) and (11-49), conservation of energy is described by the integral equation

$$\xi - \tfrac{1}{2} = \tfrac{3}{2} \sum_{i=1}^{n} \left(\frac{H_i u_{0i}}{H}\right) \left\{ \int_0^{\xi} \phi(\xi') \bar{A}' \left[\frac{3u_{0i}}{2} (\xi - \xi')\right] d\xi' \right.$$
$$\left. - \int_{\xi}^{1} \phi(\xi') \bar{A}' \left[\frac{3u_{0i}}{2} (\xi' - \xi)\right] d\xi' \right\} \tag{11-50}$$

where

$$H_i = A_{0i} \left(\frac{de_{\omega i}}{dT}\right)_{T_1} \tag{11-51a}$$

$$H = \sum_{i=1}^{n} H_i \tag{11-51b}$$

$$\phi = \frac{T - T_1}{QL/H} \tag{11-51c}$$

Employing the band absorptance correlation of Tien and Lowder (ref. 7), together with the empirical correlations for $S_i(T)$, $A_{0i}(T)$, and $\beta_i(T, P_e)$ given by Edwards et al. (ref. 14), Cess and Tiwari (ref. 15) have numerically solved equation (11-50) for application to CO, CO_2, H_2O, and CH_4. Before discussing these results, however, it will be convenient to first investigate the optically thin and large path length solutions.

Optically thin solution. Following Section 11-2, the optically thin solution of equation (11-50) is achieved by letting $\bar{A}'(u) = 1$, and it easily follows by differentiating equation (11-50) that

$$\phi = \frac{H}{3 \sum_{i=1}^{n} H_i u_{0i}}$$

or

$$T - T_1 = \frac{Q}{3P \sum_{i=1}^{n} S_i(T_1)(de_{\omega i}/dT)_{T_1}} \tag{11-52}$$

The fact that equation (11-52) predicts that the gas temperature will be independent of location is consistent with the prior discussion (Section 8-1) concerning the gray gas.

As discussed in Section 11-2, equation (11-52) is independent of the line structure parameter β_i. One may further note that the optically thin limit is also independent of the bandwidth parameter A_{0i}. An indication of the relative ability of gases to transfer radiative energy is clearly given by equation (11-52), since a lower gas temperature implies a greater capability to transmit energy. Thus the appropriate gas property that serves to measure the ability of a gas to radiate energy in the optically thin limit is the quantity

$$K = \sum_{i=1}^{n} S_i(T)(de_{\omega i}/dT) \tag{11-53}$$

Carbon dioxide, for example, has a larger value of K than does water vapor.

Large path length solution. As discussed in Section 11-2, the large path length limit is achieved when $u_{0i} \gg 1$ for each band of importance, and this limit corresponds to employing $\bar{A}'(u) = 1/u$ within equation (11-50), with the result that

$$\xi - \tfrac{1}{2} = \int_{0}^{1} \phi(\xi') \frac{d\xi'}{\xi - \xi'} \tag{11-54}$$

Aside from the obvious simplification in form in going from equation (11-50) to (11-54), there are other more striking consequences associated with equation (11-54). For example, of the three correlation quantities A_{0i}, β_i, and S_i, only A_{0i} appears within equation (11-54), through the definition of $\phi(\xi)$. The dependence on this single correlation quantity for the large path length limit has also been illustrated by Edwards et al. (ref. 14) in dealing with laminar flow between parallel plates. The absence of the line structure parameter β_i has been discussed in Section 11-2, while the invariance of the band absorptance S_i is physically logical, since the central portion of the band is saturated in the large path length limit, and consequently the radiative transfer process should not depend on the total band area.

A further simplification associated with equation (11-54) is that the temperature profile within the gas is independent of pressure. This is not the case with regard to the general formulation, equation (11-50), for which pressure appears both in the dimensionless path length u_{0i} and in the line structure parameter β_i. This invariance of temperature profile with pressure can also be found from the results of Edwards et al. (ref. 14), and experimental confirmation has been presented by Schimmel, Novotny, and Olsofka (ref. 16).

Equation (11-54) constitutes a singular integral equation with a Cauchy-type kernel, for which the solution is given by Mikhlin (ref. 17) as

$$\phi(\xi) = \frac{1}{\pi} [\xi(1 - \xi)]^{1/2} + C[\xi(1 - \xi)]^{-1/2}$$

where C is an arbitrary constant that arises since the solution of equation (11-54) is not unique. However, to satisfy the physical requirement of finite temperature everywhere within the gas, $C = 0$, and

$$\phi(\xi) = \frac{1}{\pi} [\xi(1 - \xi)]^{1/2} \qquad (11\text{-}55)$$

Note that this temperature profile yields the result that the gas temperature at a surface is equal to the surface temperature; this absence of a temperature slip is, of course, characteristic of optically thick radiation. As discussed in Section 11-2, this is a consequence of the fact that optically thick radiation is occurring within certain spectral regions. Optically nonthick radiation exists, however, in other spectral regions (band wings), with the result that equation (11-55) differs substantially from the temperature profile that would be predicted using a Rosseland-type (or diffusion) equation.

Upon recasting equation (11-55) as

$$T - T_1 = \frac{QL}{\pi H} [\xi(1 - \xi)]^{1/2}$$

it is apparent that the gas property which measures the ability of a gas to transfer radiative energy in the large path length limit is

$$H = \sum_{i=1}^{n} A_{0i} \left(\frac{de\omega_i}{dt} \right) \qquad (11\text{-}56)$$

as opposed to the optically thin transport property K, defined by equation (11-53).

Results. For the sake of brevity, numerical solutions to equation (11-50) will be presented solely in terms of the center-line temperature; that is, $T_c = T(\xi = \frac{1}{2})$. In the case of a single-band gas, the summation sign is removed in equation (11-50), and results may be expressed in terms of the single pair of parameters u_0 and β. This is illustrated by the solid curves in Fig. 11-7, and the results apply to any situation in which radiative transfer within the gas is the result of a single band. For small u_0 the results approach the optically thin limit as described by equation (11-52). The maximum influence of the line structure parameter β exists for intermediate values of u_0, while

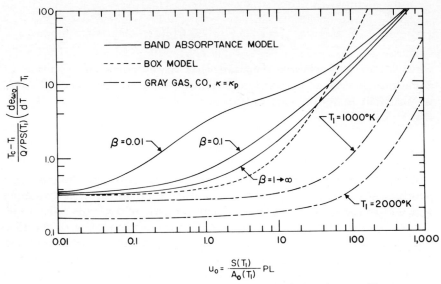

Fig. 11-7.   Comparison of results for a single-band gas.   From
Cess and Tiwari (ref. 1).

in the large path length limit the solution again becomes independent
of β.

Also illustrated in Fig. 11-7 is a solution employing the box model,
for which a constant absorption coefficient $\bar{\kappa}$ is assumed within a
finite bandwidth $\Delta\omega$.   The relation between this width and the bandwidth parameter A_0 was taken to be $\Delta\omega = (214/38)A_0$, which is
appropriate for CO (ref. 12), and the value of the mean coefficient is in
turn $\bar{\kappa} = SP/\Delta\omega$.   Clearly, such a model does not account for line
structure.   Since the box model preserves the band intensity, it reduces
to the correct optically thin limit, but a significant departure between
the two solutions takes place for increasing u_0.   This is easily explained
on physical grounds.   In the central portion of the band the box model
underpredicts the value of the spectral absorption coefficient, and it
thus will yield optically thin results for greater values of u_0 than will
the solution employing the band absorptance.   At large values of u_0
the box model overpredicts the centerline temperature because of the
neglect of the band wings.   For large path lengths the wing regions
contribute primarily to radiative transfer.   Since the box model
neglects the wings, it underestimates the ability of the gas to transfer
radiant energy for large u_0 values, and consequently it overpredicts
the centerline temperature.

The inapplicability of the optically thick (Rosseland) limit should again be emphasized. From the box model, it readily follows that $(T_c - T_1) \sim L^2$ for large u_0 (ref. 12), and this corresponds to optically thick radiation occurring throughout the finite-width band. From the large path length solution of equation (11-55), however, $(T_c - T_1) \sim L$, such that the occurrence of nonthick radiation within the band wings significantly influences the nature of the radiative transfer process for large u_0.

Also shown in Fig. 11-7 are gray gas results (ref. 12), where, for lack of a more rational choice, the mean absorption coefficient has been chosen as the Planck mean coefficient, which is defined by equation (1-29). The gray gas solution is that given in Problem (8-3); specific comparisons are made for CO. It is quite obvious that the gray solution constitutes a rather large departure from reality.

With respect to multiple-band spectra, dimensionless centerline temperatures for CO_2, H_2O, and CH_4, as obtained from equation (11-50), are illustrated in Figs. 11-8 through 11-11. Since the abscissa variable is the pressure path length, the separate influence of pressure on the centerline temperature is due solely to the alteration of the line structure of the bands due to pressure broadening. As the pressure is increased, the discrete line structure is eliminated and, as illustrated in Figs. 11-8 through 11-11, pressure ceases to be a separate parameter in the high-pressure limit. This, of course, is analgous to the large β limit of Fig. 11-7.

In the large path length limit, the dimensionless centerline temperature follows from equation (11-55) as

$$\frac{T_c - T_1}{QL/H} = \frac{1}{2\pi} = 0.159 \qquad (11\text{-}56)$$

Figures 11-8 through 11-11 consequently serve to illustrate the conditions under which the large path length limit constitutes a useful means of describing the radiative transfer process. Although these figures correspond to a specific physical problem, the limits of applicability of the large path length limit should be qualitatively indicative of other physical situations. Additional numerical results are given by Cess and Tiwari (ref. 15).

A comparison of the relative ability of various gases to transmit radiative energy may be obtained by considering the dimensional quantity $(T_c - T_1)/QL$. This is shown in Fig. 11-12 for a temperature of 500°K and a pressure of 1 atm. Recall that a lower centerline temperature implies a greater ability of the gas to transmit radiative

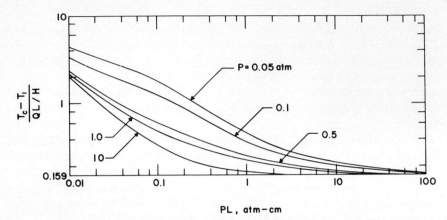

Fig. 11-8.   Centerline temperature results for CO_2 with $T_1 =$ 300°K.   From Cess and Tiwari (ref. 1).

energy, and that in the optically thin limit the radiative transfer capability of a given gas is dependent upon the magnitude of K given by equation (11-53).   For the four gases considered, CO_2 has the largest value of K, followed respectively by H_2O, CH_4, and CO.   This is consistent with the results shown in Fig. 11-12 for small path lengths; that is, CO_2 has the lowest centerline temperature, and so on.   As the

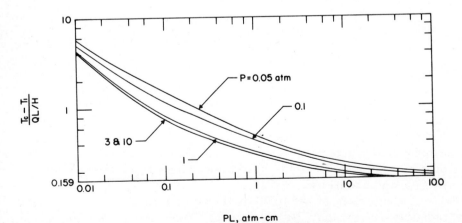

Fig. 11-9   Centerline temperature results for CO_2 with $T_1 =$ 1000°K.   From Cess and Tiwari (ref. 1).

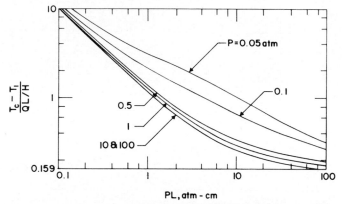

Fig. 11-10   Centerline temperature results for H_2O with $T_1 =$ 1000°K.   From Cess and Tiwari (ref. 1).

path length is increased, however, CO_2 undergoes a transition from being the most capable to being nearly the least capable transmitter of radiative energy, since CO_2 has a small relative value for H, as defined by equation (11-56), indicating that it is a poor radiator for large path lengths.

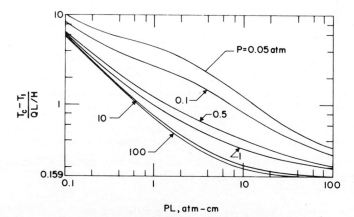

Fig. 11-11   Centerline temperature results for CH_4 with $T_1 =$ 1000°K.   From Cess and Tiwari (ref. 1).

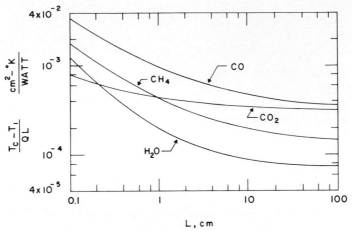

Fig. 11-12   Comparative results for $P = 1$ atm and $T_1 = 500°K$.
From Cess and Tiwari (ref. 1).

REFERENCES

1. R. D. Cess and S. N. Tiwari, Infrared radiative energy transfer in gases. In *Advances in Heat Transfer*, vol. 8, Academic Press, New York, 1972.

2. G. Yamamoto, M. Tanaka, and T. Aoki, Estimation of rotational line widths of carbon dioxide bands. *J. Quant. Spectr. Radiative Transfer*, **9**, 371 (1969).

3. S. S. Penner, *Quantitative Molecular Spectroscopy and Gas Emissivities*, Addison-Wesley, Reading, Mass., 1959.

4. R. M. Goody, *Atmospheric Radiation*, Oxford University Press, London, 1964.

5. D. K. Edwards and W. A. Menard, Comparison of models for correlation of total band absorption. *Appl. Optics*, **3**, 621 (1964).

6. D. K. Edwards and A. Balakrishnan, Thermal radiation by combustion gases. *Intern. J. Heat Mass Transfer*, **16**, 25 (1973).

7. C. L. Tien and J. E. Lowder, A correlation for total band absorptance of radiating gases. *Intern. J. Heat Mass Transfer*, **9**, 698 (1966).

8. R. D. Cess and V. Ramanathan, Radiative transfer in the atmosphere of Mars and that of Venus above the cloud deck. *J. Quant. Spectr. Radiative Transfer*, **12**, 933 (1972).

9. J. Gille and R. Goody, Convection in a radiating gas. *J. Fluid Mech.*, **22**, 47 (1964).

10. L. S. Wang, The role of emissivities in radiative transport calculations. *J. Quant. Spectr. Radiative Transfer*, **8**, 1233 (1968).

11. L. S. Wang, An integral equation of radiative equilibrium in infrared radiating gases. *J. Quant. Spectr. Radiative Transfer*, **8**, 851 (1968).

12. R. D. Cess, P. Mighdoll, and S. H. Tiwari, Infrared radiative heat transfer in nongray gases. *Intern. J. Heat Mass Transfer*, **10**, 1521 (1967).

13. R. Grief and I. S. Habib, Infrared radiative transport: Exact and approximate results. *J. Heat Transfer*, **91**, 282 (1969).

14. D. K. Edwards, L. K. Glassen, W. C. Hauser, and J. S. Tuchscher, Radiation heat transfer in nonisothermal nongray gases. *J. Heat Transfer*, **89**, 219 (1967).

15. R. D. Cess and S. N. Tiwari, The large path length limit for infrared gaseous radiation. *Appl. Sci. Res.*, **19**, 439 (1968).

16. W. P. Schimmel, J. L. Novotny, and F. A. Olsofka, Interferometric study of radiation-conduction interaction. Proceedings of the Fourth International Heat Transfer Conference, Paris-Versailles, 1970.

17. S. G. Mikhlin, *Integral Equations*, Pergamon Press, Oxford, 1967.

PROBLEMS

11-1. Prove that $\kappa_\omega/P = S_j/d$ follows from equation (11-20), where $\bar{\kappa}_\omega$ is the absorption coefficient averaged over a line spacing.

11-2. Show that equation (11-28) follows from equation (11-22) for $\beta \rightarrow \infty$.

11-3. Verify that equation (11-32) reduces to the proper linear limit.

11-4. Show that the expression

$$\bar{A} = 2 \ln \left[1 + \frac{u}{\sqrt{4 + u(1 + 1/\beta)}} \right]$$

satisfies the conditions imposed by equation (11-37).

11-5. Show that equation (11-46) is independent of the coefficients a and b within the exponential kernel substitution $E_2(x) \simeq ae^{-bx}$.

11-6. Derive the exact counterpart to equation (11-45); that is, do not make use of the exponential kernel substitution.

APPENDIXES

APPENDIX A

ANGLE FACTOR CATALOGUE

The geometrical configurations for which information is to be presented are displayed schematically in Figs. A-1 and A-2 along with dimensional nomenclature. The first of the figures relates to interchange between pairs of finite surfaces, while the second relates to interchange between an infinitesimal element and a finite surface. Following these figures, algebraic representations for the angle factors are given. Finally, in Figs. A-3 through A-10, numerical values of the angle factors have been plotted for those configurations for which such information is available.

In preparing this appendix the authors have drawn heavily on material contained in refs. 4 and 6 of Chapter 4 as well as on other sources.

Configuration 1: $X = a/c$, $Y = b/c$

$$F_{A_1 - A_2}\left(\frac{\pi X Y}{2}\right) = \ln \left[\frac{(1 + X^2)(1 + Y^2)}{1 + X^2 + Y^2}\right]^{\frac{1}{2}}$$

$$+ Y\sqrt{1 + X^2} \tan^{-1}\left(\frac{Y}{\sqrt{1 + X^2}}\right)$$

$$+ X\sqrt{1 + Y^2} \tan^{-1}\left(\frac{X}{\sqrt{1 + Y^2}}\right)$$

$$- Y \tan^{-1} Y - X \tan^{-1} X$$

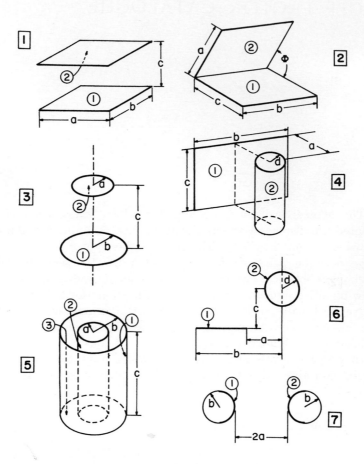

Fig. A-1   Schematic representations of configurations 1 through 7.

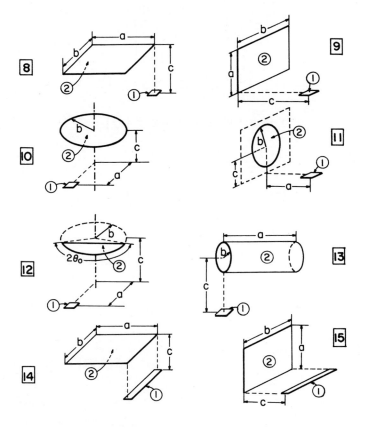

Fig. A-2   Schematic representations of configurations 8 through 15.

Configuration 2: $\quad X = a/b, \; Y = c/b, \; Z = X^2 + Y^2 - 2XY \cos \Phi$

$$
\begin{aligned}
F_{A_1 - A_2}(\pi Y) = &-\frac{\sin 2\Phi}{4}\left[XY \sin \Phi + \left(\frac{\pi}{2} - \Phi\right)(X^2 + Y^2) \right. \\
&+ Y^2 \tan^{-1}\left(\frac{X - Y \cos \Phi}{Y \sin \Phi}\right) \\
&+ \left. X^2 \tan^{-1}\left(\frac{Y - X \cos \Phi}{X \sin \Phi}\right)\right] \\
&+ \frac{\sin^2 \Phi}{4}\left\{ \left(\frac{2}{\sin^2 \Phi} - 1\right) \ln \left[\frac{(1 + X^2)(1 + Y^2)}{1 + Z}\right] \right. \\
&+ Y^2 \ln \left[\frac{Y^2(1 + Z)}{(1 + Y^2)Z}\right] + X^2 \ln \left[\frac{X^2(1 + X^2)^{\cos 2\Phi}}{Z(1 + Z)^{\cos 2\Phi}}\right]\right\} \\
&+ Y \tan^{-1}\left(\frac{1}{Y}\right) + X \tan^{-1}\left(\frac{1}{X}\right) - \sqrt{Z}\tan^{-1}\left(\frac{1}{\sqrt{Z}}\right) \\
&+ \frac{\sin \Phi \sin 2\Phi}{2} X\sqrt{1 + X^2 \sin^2 \Phi} \\
&\times \left[\tan^{-1}\left(\frac{X \cos \Phi}{\sqrt{1 + X^2 \sin^2 \Phi}}\right)\right. \\
&+ \left. \tan^{-1}\left(\frac{Y - X \cos \Phi}{\sqrt{1 + X^2 \sin^2 \Phi}}\right)\right] \\
&+ \cos \Phi \int_0^Y \sqrt{1 + \xi^2 \sin^2 \Phi}\left[\tan^{-1}\left(\frac{X - \xi \cos \Phi}{\sqrt{1 + \xi^2 \sin^2 \Phi}}\right)\right. \\
&+ \left. \tan^{-1}\left(\frac{\xi \cos \Phi}{\sqrt{1 + \xi^2 \sin^2 \Phi}}\right)\right] d\xi
\end{aligned}
$$

Configuration 3: $\quad X = a/c, \; Y = c/b, \; Z = 1 + (1 + X^2)Y^2$

$$F_{A_1 - A_2} = \tfrac{1}{2}(Z - \sqrt{Z^2 - 4X^2 Y^2})$$

Configuration 4: $\quad X = a/d, \; Y = b/d, \; Z = c/d$

$$A = Z^2 + X^2 + \xi^2 - 1, \quad B = Z^2 - X^2 - \xi^2 + 1$$

$$F_{A_1 - A_2} = \frac{2}{Y}\int_0^{Y/2} f(\xi)\, d\xi$$

$$
\begin{aligned}
f(\xi) = &\frac{X}{X^2 + \xi^2} - \frac{X}{\pi(X^2 + \xi^2)} \\
&\times \left\{ \cos^{-1}\frac{B}{A} - \frac{1}{2Z}\left[\sqrt{A^2 + 4Z^2}\cos^{-1}\left(\frac{B}{A\sqrt{X^2 + \xi^2}}\right)\right.\right. \\
&+ \left.\left. B \sin^{-1}\left(\frac{1}{\sqrt{X^2 + \xi^2}}\right) - \frac{\pi A}{2}\right]\right\}
\end{aligned}
$$

Configuration 5: $X = b/a, \ Y = c/a$

$$A = Y^2 + X^2 - 1, \ B = Y^2 - X^2 + 1$$

$$F_{A_1 - A_2} = \frac{1}{X} - \frac{1}{\pi X} \left\{ \cos^{-1} \frac{B}{A} - \frac{1}{2Y} \left[\sqrt{(A + 2)^2 - (2X)^2} \ \cos^{-1} \frac{B}{XA} \right. \right.$$

$$\left. \left. + B \sin^{-1} \frac{1}{X} - \frac{\pi A}{2} \right] \right\}$$

$$F_{A_1 - A_1} = 1 - \frac{1}{X} + \frac{2}{\pi X} \tan^{-1} \left(\frac{2\sqrt{X^2 - 1}}{Y} \right)$$

$$- \frac{Y}{2\pi X} \left\{ \frac{\sqrt{4X^2 + Y^2}}{Y} \sin^{-1} \left[\frac{4(X^2 - 1) + \frac{Y^2}{X^2}(X^2 - 2)}{Y^2 + 4(X^2 - 1)} \right] \right.$$

$$\left. - \sin^{-1} \left(\frac{X^2 - 2}{X^2} \right) + \frac{\pi}{2} \left(\frac{\sqrt{4X^2 + Y^2}}{Y} - 1 \right) \right\}$$

$$F_{A_1 - A_3} = \tfrac{1}{2}(1 - F_{A_1 - A_2} - F_{A_1 - A_1})$$

Configuration 6: $X = c/d, \ Y = a/d, \ Z = b/d$

$$F_{A_1 - A_2} = \frac{1}{Z - Y} \left(\tan^{-1} \frac{Z}{X} - \tan^{-1} \frac{Y}{X} \right)$$

Configuration 7: $X = 1 + (a/b)$

$$F_{A_1 - A_2} = \frac{2}{\pi} \left(\sqrt{X^2 - 1} - X + \frac{\pi}{2} - \cos^{-1} \frac{1}{X} \right)$$

Configuration 8: $X = a/c, \ Y = b/c$

$$F_{dA_1 - A_2} = \frac{1}{2\pi} \left[\frac{X}{\sqrt{1 + X^2}} \tan^{-1} \left(\frac{Y}{\sqrt{1 + X^2}} \right) \right.$$

$$\left. + \frac{Y}{\sqrt{1 + Y^2}} \tan^{-1} \left(\frac{X}{\sqrt{1 + Y^2}} \right) \right]$$

Configuration 9: $X = a/b, \ Y = c/b$

$$A = 1/\sqrt{X^2 + Y^2}$$

$$F_{dA_1 - A_2} = \frac{1}{2\pi} \left(\tan^{-1} \frac{1}{Y} - A Y \tan^{-1} A \right)$$

Configuration 10: $X = c/a,\ Y = b/c,\ Z = 1 + (1 + Y^2)X^2$

$$F_{dA_1 - A_2} = \frac{1}{2}\left(1 - \frac{Z - 2Y^2X^2}{\sqrt{Z^2 - 4Y^2X^2}}\right)$$

Configuration 11: $X = a/c,\ Y = b/c$

$$F_{dA_1 - A_2} = \frac{X}{2}\left[\frac{1 + Y^2 + X^2}{\sqrt{(1 + Y^2 + X^2)^2 - 4Y^2}} - 1\right]$$

Configuration 12: $X = a/b,\ Y = c/b,\ Z = [(1 + X^2 + Y^2)^2 - 4X^2]^{\frac{1}{2}}$

$$F_{dA_1 - A_2} = \frac{\theta_0}{2\pi} + \frac{(1 - X^2 + Y^2)}{\pi Z}\tan^{-1}\left[\frac{Z\tan(\theta_0/2)}{1 + X^2 + Y^2 - 2X}\right]$$

$$+ \frac{X - \cos\theta_0}{\pi\sqrt{(X - \cos\theta_0)^2 + Y^2}}\tan^{-1}\left[\frac{\sin\theta_0}{\sqrt{(X - \cos\theta_0)^2 + Y^2}}\right]$$

Configuration 13: $X = a/b,\ Y = c/b$

$$A = (1 + Y)^2 + X^2,\ B = (1 - Y)^2 + X^2$$

$$F_{dA_1 - A_2} = \frac{1}{\pi Y}\tan^{-1}\left(\frac{X}{\sqrt{Y^2 - 1}}\right)$$

$$+ \frac{X}{\pi}\left\{\frac{(A - 2Y)}{Y\sqrt{AB}}\tan^{-1}\left[\sqrt{\frac{A(Y - 1)}{B(Y + 1)}}\right] - \frac{1}{Y}\tan^{-1}\left[\sqrt{\frac{(Y - 1)}{(Y + 1)}}\right]\right\}$$

Configuration 14: $X = b/c,\ Y = a/c$

$$F_{dA_1 - A_2} = \frac{1}{\pi X}\left[\sqrt{1 + X^2}\tan^{-1}\left(\frac{Y}{\sqrt{1 + X^2}}\right) - \tan^{-1}Y\right.$$

$$\left. + \frac{XY}{\sqrt{1 + Y^2}}\tan^{-1}\left(\frac{X}{\sqrt{1 + Y^2}}\right)\right]$$

Configuration 15: $X = a/b,\ Y = c/b,\ Z = X^2 + Y^2$

$$F_{dA_1 - A_2} = \frac{1}{\pi}\left\{\tan^{-1}\left(\frac{1}{Y}\right) + \frac{Y}{2}\ln\left[\frac{Y^2(Z + 1)}{(Y^2 + 1)Z}\right] - \frac{Y}{\sqrt{Z}}\tan^{-1}\left(\frac{1}{\sqrt{Z}}\right)\right\}$$

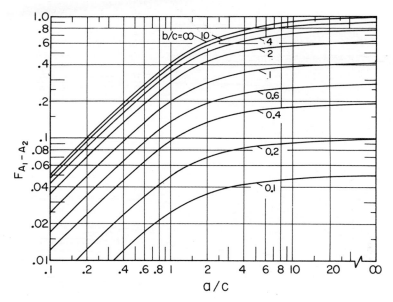

Fig. A-3   Angle factors for configuration 1.

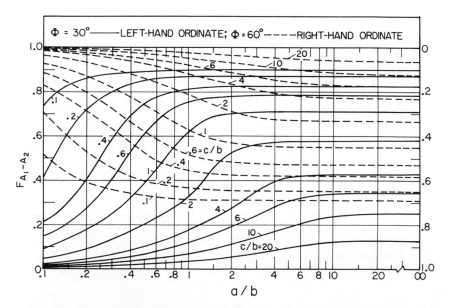

Fig. A-4a   Angle factors for configuration 2, $\Phi = 30$ and $60°$.

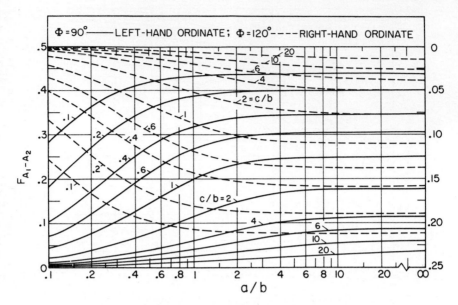

Fig. A-4b   Angle factors for configuration 2, $\Phi = 90$ and $120°$.

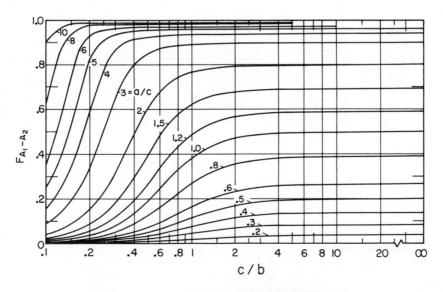

Fig. A-5   Angle factors for configuration 3.

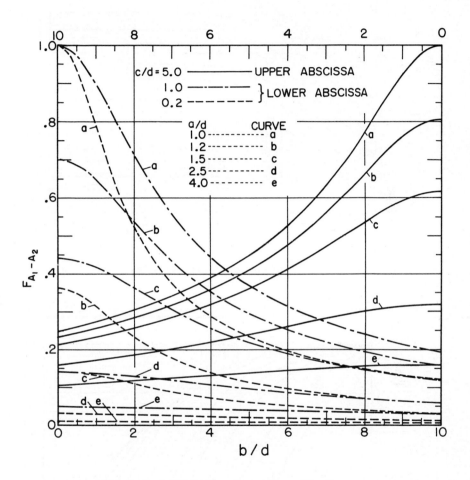

Fig. A-6   Angle factors for configuration 4.

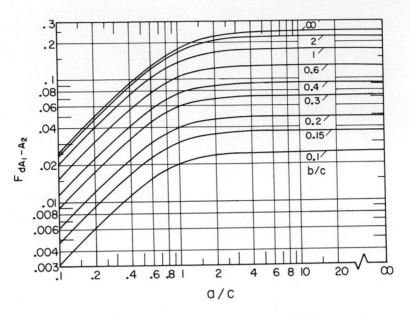

Fig. A-7   Angle factors for configuration 8.

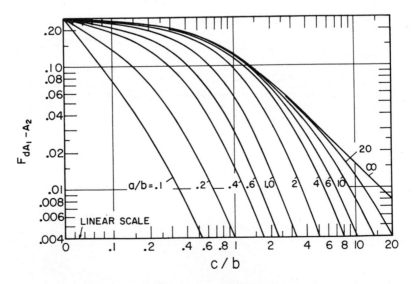

Fig. A-8   Angle factors for configuration 9.

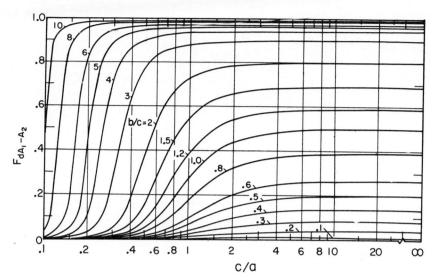

Fig. A-9   Angle factors for configuration 10.

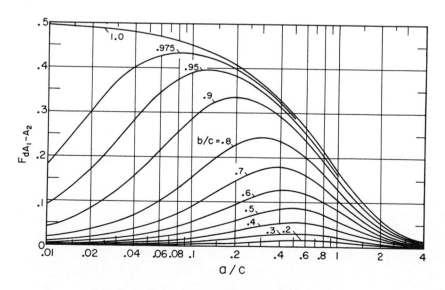

Fig. A-10   Angle factors for configuration 11.

APPENDIX B

THE EXPONENTIAL INTEGRALS

The nth exponential integral is defined for positive real arguments by

$$E_n(t) = \int_0^1 \mu^{n-2} e^{-t/\mu} \, d\mu \tag{B-1}$$

and consideration will be given only to positive integral values of n. For a detailed discussion of the exponential integrals, the reader is referred to Kourganoff (ref. 1) and Chandrasekhar (ref. 2).

Rephrasing equation (B-1) as

$$E_n(t) = \int_0^1 \mu^n \mu^{-2} e^{-t/\mu} \, d\mu = \frac{1}{t} \int_0^1 \mu^n \, d(e^{-t/\mu})$$

and integrating by parts, one obtains the recurrence relation

$$nE_{n+1}(t) = e^{-t} - tE_n(t) \tag{B-2}$$

The initial values are found from equation (B-1) to be

$$E_n(0) = \int_0^1 \mu^{n-2} \, d\mu = \frac{1}{n-1} \tag{B-3}$$

while differentiation of equation (B-1) yields

$$E_n{}'(t) = -E_{n-1}(t), \quad n > 1 \tag{B-4}$$

$$E_1{}'(t) = -\frac{1}{t} e^{-t} \tag{B-5}$$

Conversely,

$$\int E_n(t) \, dt = -E_{n+1}(t) \tag{B-6}$$

Series expansions of $E_1(t)$, $E_2(t)$, and $E_3(t)$ are of the form

$$E_1(t) = -\gamma - \ln t + t - \frac{t^2}{2 \cdot 2!} + \frac{t^3}{3 \cdot 3!} + \cdots \qquad \text{(B-7)}$$

$$E_2(t) = 1 + (\gamma - 1 + \ln t)t - \frac{t^2}{1 \cdot 2!} + \frac{t^3}{2 \cdot 3!} + \cdots \qquad \text{(B-8)}$$

$$E_3(t) = \frac{1}{2} - t + \frac{1}{2}\left(-\gamma + \frac{3}{2} - \ln t\right)t^2 + \frac{t^3}{1 \cdot 3!} + \cdots \qquad \text{(B-9)}$$

Table B-1   Values of $E_n(t)$*

t	$E_1(t)$	$E_2(t)$	$E_3(t)$	$E_4(t)$
0	∞	1.0000	0.5000	0.3333
0.01	4.0380	0.9497	0.4903	0.3284
0.02	3.3547	0.9131	0.4810	0.3235
0.03	2.9591	0.8817	0.4720	0.3188
0.04	2.6813	0.8535	0.4633	0.3141
0.05	2.4679	0.8278	0.4549	0.3095
0.06	2.2953	0.8040	0.4468	0.3050
0.07	2.1508	0.7818	0.4388	0.3006
0.08	2.0269	0.7610	0.4311	0.2962
0.09	1.9187	0.7412	0.4236	0.2919
0.10	1.8229	0.7225	0.4163	0.2877
0.20	1.2227	0.5742	0.3519	0.2494
0.30	0.9057	0.4691	0.3000	0.2169
0.40	0.7024	0.3894	0.2573	0.1891
0.50	0.5598	0.3266	0.2216	0.1652
0.60	0.4544	0.2762	0.1916	0.1446
0.70	0.3738	0.2349	0.1661	0.1268
0.80	0.3106	0.2009	0.1443	0.1113
0.90	0.2601	0.1724	0.1257	0.0978
1.00	0.2194	0.1485	0.1097	0.0861
1.25	0.1464	0.1035	0.0786	0.0628
1.50	0.1000	0.0731	0.0567	0.0460
1.75	0.0695	0.0522	0.0412	0.0339
2.00	0.0489	0.0375	0.0301	0.0250
2.25	0.0348	0.0272	0.0221	0.0185
2.50	0.0249	0.0198	0.0163	0.0138
2.75	0.0180	0.0145	0.0120	0.0103
3.00	0.0130	0.0106	0.0089	0.0077
3.25	0.0095	0.0078	0.0066	0.0057
3.50	0.0070	0.0058	0.0049	0.0043

* This table is an abbreviation of the more complete tabulation given by Kourganoff (ref. 1).

where $\gamma = 0.5772$ is Euler's constant. From equation (B-7) it is seen that $E_1(t)$ has a logarithmic singularity at the origin. For large values of t, there is the asymptotic expansion

$$E_n(t) \sim \frac{e^{-t}}{t}\left[1 - \frac{n}{t} + \frac{n(n+1)}{t^2} + \cdots\right] \qquad \text{(B-10)}$$

Values of the first four exponential integrals are given in Table B-1, while Table B-2 gives a brief listing (ref. 1) for the integral

$$J_{mn} = \int_0^\infty E_m(t)E_n(t)\, dt$$

which occasionally arises in problems dealing with radiative transfer in semi-infinite media.

Table B-2   Values of the Integral J_{mn}

$J_{11} = 1.3863$	$J_{22} = 0.2046$
$J_{12} = 0.5000$	$J_{23} = 0.1250$
$J_{13} = 0.2954$	$J_{33} = 0.0773$

REFERENCES

1. V. Kourganoff, *Basic Methods in Transfer Problems*, Dover Publications, New York, 1963.

2. S. Chandrasekhar, *Radiative Transfer*, Dover Publications, New York, 1960.

INDEXES

AUTHOR INDEX

SUBJECT INDEX